计算机网络实用技术人才培养丛书

锐 捷 职 业 认 证 系 列

网络设备互连

Interconnecting Networking Devices (IND)

实验指南

◎ 高 峡 钟啸剑 李永俊 编著

科学出版社
www.sciencep.com

内容简介

本书是《网络设备互连学习指南》一书的配套教材，内容为《网络设备互连学习指南》课程中所阐述的理论知识的配套实验内容。

本书共 34 个实验，内容包括交换机和路由器的基本操作实验，各项网络技术和协议的配套实验，局域网的综合实验：交换综合实验和路由的园区网综合实验，以及备份、恢复交换机和路由器的操作系统与配置文件的扩展实验。

本书可作为网络设计师、网络工程师、系统集成工程师以及相关技术人员在实际构建园区网络中的技术参考用书。由于本书的专业性、实用性、易读性，现已被选为锐捷网络有限公司 RCNA（锐捷认证网络工程师）认证指定教材。

本书配套电子课件、各章的习题答案可访问 http://www.labclub.com.cn、http://university.ruijie.com.cn 获得。

需要本书或技术支持的读者，请与北京清河 6 号信箱（邮编：100085）发行部联系，电话：010-62978181（总机）转发行部，82702675（邮购），传真：010-82702698，E-mail：tbd@bhp.com.cn。

图书在版编目（CIP）数据

网络设备互连实验指南/高峡，钟啸剑，李永俊编著.—北京：科学出版社，2009.4

（计算机网络实用技术人才培养丛书）

ISBN 978-7-03-024166-5

Ⅰ.网… Ⅱ.①高… ②钟… ③李… Ⅲ.计算机网络—实验—指南 Ⅳ.TP393-62

中国版本图书馆 CIP 数据核字（2009）第 026924 号

责任编辑：邓 伟 / 责任校对：马 君

责任印刷：双 青 / 封面设计：青青果园

科学出版社 出版

北京东黄城根北街 16 号

邮政编码：100717

http://www.sciencep.com

北京市四季青双青印刷厂印刷

科学出版社发行 各地新华书店经销

*

2009 年 4 月第 一 版 开本：787×1092 1/16

2009 年 4 月第一次印刷 印张：19.75

印数：1—4 000 字数：454 975

定价：32.00 元

《网络设备互连实验指南》编委会

Interconnecting Networking Devices (IND)

“计算机网络实用技术人才培养丛书”

编审委员会

前　言

随着信息化的高速发展，人们已经把更多的生活、娱乐和学习等事务转移到了网络这个平台上。小到一个家庭，大到一个企业，甚至是一所高校，为了提高工作效率，进行更多的信息交流，就需要构建一个园区网，从而实现内部的高效沟通。如果希望能进一步和互联网中的其他地区甚至其他国家的人、组织机构进行信息交流，则需要将内部的园区网接入到互联网中。

本书主要内容包括在构建中小型网络时所用到的各种网络技术的配置方式，所有的技术点都在实验中进行展现。实验中包括对当前网络状态的需求分析，以及为了满足这种需求所需要的技术以及配置方法，使读者能够更深入地理解各种技术在网络中的作用，以及如何使用这些技术对网络进行优化等。

本书是由计算机网络资深技术专家高峡、张选波、方洋，和具有丰富教学经验的钟啸剑、李永俊老师基于多年的网络工程经验、教学经验以及对网络技术的深刻理解联合编写而成。

本书目标

本书以实际的网络环境为依托，针对每一个技术点都有相应的实验进行支撑，通过工程案例采用其由浅入深的方法来讲解，从如何进行基本操作到如何实施园区网综合项目都进行了详细的讲解。在实验讲解中，注重如何引发读者思考，自行得出结论，使读者不但能够按照实验讲解中的步骤完成实验，更加深了对园区网技术的应用理解。

本书读者

本书的读者对象可以是本科类院校、高职类院校的学生、教师，也可以为准备参加RCNA考试的专业人士，以及希望学习更多园区网构建相关知识的技术人员。

阅读方法

本书将所有内容分为 3 个部分，每一部分所包含的实验都具有各自的功能，不但针对各种技术或协议提供了多个实验案例，更有综合实验和扩展实验供读者练习，以加深理解扩展知识。读者可以选择逐页的阅读方式，也可以灵活地、有选择地针对某些实验进行阅读。

本书配合《网络设备互连学习指南》一书使用，可达到理论和实践的融合，帮助读者更深入的了解技术内容、提高实践动手能力。

本书资源

为了保证课程在学校内的有效实施，及课程教学资源的长期提供，本课程建设了专门的课程实施教学俱乐部的网络资源共享基地，用来支持课程实施过程中的项目资源更新。读者可以访问 http://www.labclub.com.cn、http://university.ruijie.com.cn，免费获得配套电子课件、各章节的复习题答案以及更多的教学资源。

本书结构

本书共 34 个实验，前 26 个实验为各种交换、路由协议和相关技术提供实验案例，后面包含了 2 个综合实验和 6 个扩展实验。本书的具体结构如下。

实验 1～实验 2 是关于交换机的基本操作的实验案例，包括交换机的基本配置和在交换机上配置 Telnet 等实验，理论知识可以参阅《网络设备互连学习指南》一书的“第 2 章 交换技术”。

实验 3～实验 5 是关于 VLAN 技术的实验案例，包括单臂路由、使用 SVI 实现 VLAN 间路由、跨交换机实现 VLAN 间路由等实验，理论知识可以参阅《网络设备互连学习指南》一书的“第 3 章 虚拟局域网（VLAN）”。

实验 6～实验 7 是关于交换网络的冗余链路的实验案例，包括配置 RSTP、配置端口聚合等。理论知识可以参阅《网络设备互连学习指南》一书的“第 4 章 局域网中的冗余链路”。

实验 8～实验 10 是关于路由器的基本操作的实验案例，包括路由器的基本配置、在路由器上配置 Telnet、静态路由配置等。理论知识可以参阅《网络设备互连学习指南》一书的“第 6 章 路由技术”。

实验 11～实验 12 是关于路由信息协议（RIP）的实验案例，包括 RIP 路由协议基本配置、RIPv2 配置等。理论知识可以参阅《网络设备互连学习指南》一书的“第 7 章 RIP 路由协议”。

实验 13～实验 14 是关于开放式最短路径优先协议（OSPF）的实验案例，包括 OSPF 基本配置、OSPF 单区域配置等。理论知识可以参阅《网络设备互连学习指南》一书的“第 8 章 OSOF 路由协议”。

实验 15～实验 17 是关于点到点协议 PPP 的实验案例，包括广域网协议的封装、PPP PAP 认证的配置、PPP CHAP 认证的配置等。理论知识可以参阅《网络设备互连学习指南》一书的“第 9 章 点对点协议（PPP）”。

实验 18～实验 21 是关于网络安全控制的实验案例，包括交换机的端口安全、配置标准 IP ACL、配置扩展 IP ACL 等。理论知识可以参阅《网络设备互连学习指南》一书的“第 10 章 园区网安全”。

实验 22～实验 23 是关于网络地址转换 NAT 的实验案例：包括配置动态 NAPT 实现局域网访问互联网、利用 NAT 实现外网主机访问内网服务器等。理论知识可以参阅《网络设备互连学习指南》一书的“第 11 章 网络地址转换（NAT）”。

实验 24～实验 26 是无线局域网 WLAN 的实验案例，包括建立开放式的无线接入网络、配置单频多模的无线网络、搭建采用 WEP 加密方式的无线网络等。理论知识可以参阅《网络设备互连学习指南》一书的“第 12 章 无线局域网（WLAN）”。

综合实验是针对各项技术内容的综合练习，包括“综合实验 1 交换篇”和“综合实验 2 混合篇”。

扩展实验是管理交换机和路由器必须的扩展内容，包括利用 TFTP 升级现有交换机操作系统和升级现有路由器操作系统、利用 ROM 方式重写交换机操作系统和重写路由器操作系统、利用 TFTP 备份还原交换机配置文件和备份还原路由器配置文件 6 个实验。

“附录 A 术语表”介绍了书中出现的全部术语和常见计算机技术术语。

“附录 B 锐捷职业认证体系”介绍了通用技术认证和专项技术认证，及锐捷职业认证和途径。

命令语法规范

本书中使用的命令语法规范与产品命令参考手册中的命令语法相同。

- 竖线“|”：表示分隔符，用于分开可选择的选项。
- 星号“*”：表示可以同时选择多个选项。
- 方括号“[ ]”：表示可选项。
- 大括号“{ }”：表示必选项。
- **粗体字**表示按照显示的文字输入的命令和关键字。在配置的示例和输出中，粗体字表示需要用户手工输入的命令（例如 **show** 命令）。
- *斜体字*表示需要用户输入的具体值。

本书使用的图标

本书中使用的图标示例如下所示。

目　　录

实验 1　交换机的基本配置

【实验名称】

交换机的基本配置。

【实验目的】

掌握交换机命令行各种操作模式的区别，能够使用各种帮助信息，以及用命令进行基本的配置。

【背景描述】

某公司新来一位网管，公司要求其熟悉网络产品，并采用全系列锐捷网络产品。首先要登录交换机，了解并掌握交换机的命令行操作技巧，以及使用一些基本命令进行配置。

【需求分析】

需要在交换机上熟悉各种不同的配置模式，以及如何在配置模式间切换，使用命令进行基本的配置，并熟悉命令行界面的操作技巧。

【实验拓扑】

实验的拓扑图，如图 1-1 所示。

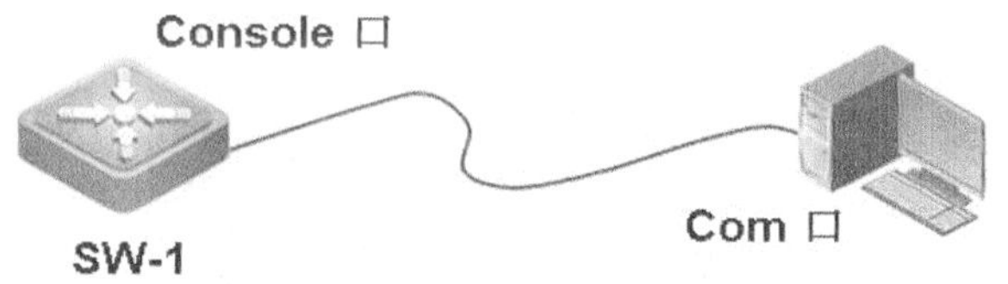

图 1-1

【实验设备】

三层交换机 1 台。

【预备知识】

交换机的命令行界面和基本操作。

【实验原理】

交换机的管理方式基本分为带内管理和带外管理两种。通过交换机的 Console 口管理交换机就属于带外管理，不占用交换机的网络接口，其特点是需要使用配置线缆，近距离配置。第一次配置交换机时必须利用 Console 端口进行配置。

交换机的命令行操作模式，主要包括：用户模式、特权模式、全局配置模式、端口模式等几种。

- ❑ 用户模式：进入交换机后得到的第一个操作模式，该模式下可以简单查看交换机的软、硬件版本信息，并进行简单的测试。用户模式提示符为 switch>。
- ❑ 特权模式：由用户模式进入的下一级模式，该模式下可以对交换机的配置文件进行管理，查看交换机的配置信息，进行网络的测试和调试等。特权模式提示符为 switch#。

- 全局配置模式：属于特权模式的下一级模式，该模式下可以配置交换机的全局性参数（如主机名、登录信息等）。在该模式中可以进入下一级的配置模式，对交换机具体的功能进行配置。全局模式提示符为 switch(config)#。
- 端口模式：属于全局配置模式的下一级模式，该模式下可以对交换机的端口进行参数配置。端口模式提示符为 switch(config-if)#。

交换机的基本操作命令包括如下。

- Exit 命令是退回到上一级操作模式。
- End 命令是指用户从特权模式以下级别直接返回到特权模式。
- 交换机命令行支持获取帮助信息、命令的简写、命令的自动补齐、快捷键功能。配置交换机的设备名称和配置交换机的描述信息必须在全局配置模式下执行。
- Hostname 配置交换机的设备名称。
- 当用户登录交换机时，需要向用户提示一些必要的信息。网管可以通过设置标题来达到这个目的。可以创建每日通知和登录两种类型的标题。
  - Banner motd：配置交换机每日提示信息。
  - Banner logi：配置交换机登录提示信息，位于每日提示信息之后。
- 查看交换机的系统和配置信息命令要在特权模式下执行。
  - Show version：查看交换机的版本信息，可以查看到交换机的硬件版本信息和软件版本信息，用于进行交换机操作系统升级时的依据。
  - Show mac-address-table：查看交换机当前的 MAC 地址表信息。
  - Show running-config：查看交换机当前生效的配置信息。

【实验步骤】

步骤 1　交换机各个操作模式直接的切换。

```
Swtich>enable
! 使用 enable 命令从用户模式进入特权模式
Swtich#configure terminal
Enter configuration commands, one per line.  End with CNTL/Z.
! 使用 configure terminal 命令从特权模式进入全局配置模式
Swtich(config)#interface fastEthernet 0/1
! 使用 interface 命令进入接口配置模式
Swtich(config-if)#
Swtich(config-if)#exit
! 使用 exit 命令退回上一级操作模式
Swtich(config)#interface fastEthernet 0/2
Swtich(config-if)#end
Swtich#
! 使用 end 命令直接退回特权模式
```

步骤 2　交换机命令行界面基本功能。

```
Switch> ?
! 显示当前模式下所有可执行的命令
  disable              Turn off privileged commands
```

```
  enable                 Turn on privileged commands
  exit                    Exit from the EXEC
  help                   Description of the interactive help system
  ping                   Send echo messages
  rcommand               Run command on remote switch
  show                   Show running system information
  telnet                 Open a telnet connection
traceroute             Trace route to destination
Swtich>en <tab>
Swtich>enable
! 使用tab键补齐命令
Swtich#con?
configure  connect
! 使用？显示当前模式下所有以“con”开头的命令
Swtich#conf t
Enter configuration commands, one per line.  End with CNTL/Z.
Swtich(config)#
! 使用命令的简写
Swtich(config)#interface ?
! 显示interface命令后可执行的参数
  Aggregateport     Aggregate port interface
  Dialer           Dialer interface
  FastEthernet      Fast IEEE 802.3
  GigabitEthernet   Gbyte Ethernet interface
  Loopback         Loopback interface
  Multilink         Multilink-group  interface
  Null             Null interface
  Tunnel           Tunnel interface
  Virtual-ppp       Virtual PPP interface
  Virtual-template  Virtual Template interface
  Vlan             Vlan interface
  range            Interface range command
Switch(config)#interface
Swtich(config)#interface fastEthernet 0/1
Switch(config-if)# ^Z
Switch#
! 使用快捷键“Ctrl+Z”可以直接退回到特权模式
Switch#ping 1.1.1.1
sending 5, 100-byte ICMP Echos to 1.1.1.1,
timeout is 2000 milliseconds.
.^C
Switch#
```

```
! 在交换机特权模式下执行 ping 1.1.1.1 命令，发现不能 ping 通目标地址，交换机默认情况下需要发送 5 个数据包，如不想等到 5 个数据包均不能 ping 通目标地址的反馈出现，可在数据包未发出 5 个之前按快捷键 Ctrl+C 终止当前操作。
```

步骤 3　配置交换机的名称和每日提示信息。

```
Switch(config)#hostname SW-1
! 使用 hostname 命令更改交换机的名称
SW-1(config)#banner motd $
! 使用 banner 命令设置交换机的每日提示信息，参数 motd 指定以哪个字符为信息的结束符
Enter TEXT message.  End with the character '$'.
Welcome to SW-1, if you are admin, you can config it.
If you are not admin, please EXIT!
$
SW-1(config)#
SW-1(config)#exit
SW-1#Nov 25 22:04:01 %SYS-5-CONFIG I: Configured from console by console
SW-1#exit
SW-1 CON0 is now available
Press RETURN to get started
Welcome to SW-1, if you are admin, you can config it.
If you are not admin, please EXIT!
SW-1>
```

步骤 4　配置接口状态。

锐捷全系列交换机 Fastethernet 接口默认情况下是 10Mbps/100Mbps 自适应端口，双工模式也是自适应（端口速率、双工模式可配置）。默认情况下，所有交换机端口均开启。

如果网络中有一些型号比较旧的主机，在使用 10Mbps 半双工的网卡，此时为了能够实现主机之间的正常访问，应当在交换机上进行相应的配置，把连接这些主机的交换机端口速率设为 10Mbps，传输模式设为半双工。

```
SW-1(config)#interface fastEthernet 0/1
! 进入端口 F0/1 的配置模式
SW-1(config-if)#speed 10
! 配置端口速率为 10Mbps
SW-1(config-if)#duplex half
! 配置端口的双工模式为半双工
SW-1(config-if)#no shutdown
! 开启端口，使端口转发数据。交换机端口默认已经开启。
SW-1(config-if)#description "This is a Accessport."
! 配置端口的描述信息，可作为提示。
SW-1(config-if)#end
SW-1#Nov 25 22:06:37 %SYS-5-CONFIG I: Configured from console by console
SW-1#
SW-1#show interface fastEthernet 0/1
Index(dec):1 (hex):1
```

```
FastEthernet 0/1 is UP   , line protocol is UP
Hardware is marvell FastEthernet
Description: "This is a Accessport."
Interface address is: no ip address
  MTU 1500 bytes, BW 10000 Kbit
  Encapsulation protocol is Bridge, loopback not set
  Keepalive interval is 10 sec , set
  Carrier delay is 2 sec
  RXload is 1 ,Txload is 1
  Queueing strategy: WFQ
  Switchport attributes:
    interface's description:""This is a Accessport.""
    medium-type is copper
    lastchange time:329 Day:22 Hour: 5 Minute: 2 Second
    Priority is 0
    admin duplex mode is Force Half Duplex, oper duplex is Half
    admin speed is 10Mbps, oper speed is 10 Mbps
    flow control admin status is OFF,flow control oper status is OFF
    broadcast Storm Control is OFF,multicast Storm Control is OFF,unicast Storm
Control is OFF
  5 minutes input rate 0 bits/sec, 0 packets/sec
  5 minutes output rate 0 bits/sec, 0 packets/sec
    0 packets input, 0 bytes, 0 no buffer, 0 dropped
    Received 0 broadcasts, 0 runts, 0 giants
    0 input errors, 0 CRC, 0 frame, 0 overrun, 0 abort
    0 packets output, 0 bytes, 0 underruns , 0 dropped
    0 output errors, 0 collisions, 0 interface resets
SW-1#
```

如果需要将交换机端口的配置恢复默认值，可以使用 default 命令。

```
SW-1(config)#interface fastEthernet 0/1
SW-1(config-if)#default bandwidth
! 恢复端口默认的带宽设置
SW-1(config-if)#default description
! 取消端口的描述信息
SW-1(config-if)#default duplex
! 恢复端口默认的双工设置
SW-1(config-if)#end
SW-1#Nov 25 22:11:13 %SYS-5-CONFIG I: Configured from console by console

SW-1#
SW-1#show interface fastEthernet 0/1
Index(dec):1 (hex):1
```

```
FastEthernet 0/1 is UP  , line protocol is UP
Hardware is marvell FastEthernet
Interface address is: no ip address
  MTU 1500 bytes, BW 100000 Kbit
  Encapsulation protocol is Bridge, loopback not set
  Keepalive interval is 10 sec , set
  Carrier delay is 2 sec
  RXload is 1 ,Txload is 1
  Queueing strategy: WFQ
  Switchport attributes:
    interface's description:""
    medium-type is copper
    lastchange time:329 Day:22 Hour:11 Minute:13 Second
    Priority is 0
    admin duplex mode is AUTO, oper duplex is Full
    admin speed is AUTO, oper speed is 100Mbps
    flow control admin status is OFF,flow control oper status is ON
    broadcast Storm Control is OFF,multicast Storm Control is OFF,unicast Storm
Control is OFF
  5 minutes input rate 0 bits/sec, 0 packets/sec
  5 minutes output rate 0 bits/sec, 0 packets/sec
    0 packets input, 0 bytes, 0 no buffer, 0 dropped
    Received 0 broadcasts, 0 runts, 0 giants
    0 input errors, 0 CRC, 0 frame, 0 overrun, 0 abort
    0 packets output, 0 bytes, 0 underruns , 0 dropped
    0 output errors, 0 collisions, 0 interface resets
SW-1#
```

步骤 5　查看交换机的系统和配置信息。

```
SW-1#show version
！查看交换机的系统信息
System description      : Ruijie Dual Stack Multi-Layer Switch(S3760-24) By Ruijie
Network
！交换机的描述信息（型号等）
System start time       : 2008-11-25 21:58:44
System hardware version : 1.0
！设备的硬件版本信息
System software version : RGNOS 10.2.00(2), Release(27932)
！操作系统版本信息
System boot version     : 10.2.27014
System CTRL version     : 10.2.24136
System serial number    : 0000000000000
SW-1#
```

```
SW-1#show running-config
! 查看交换机的配置信息
Building configuration...
Current configuration : 1279 bytes
!
version RGNOS 10.2.00(2), Release(27932)(Thu Dec 13 10:31:41 CST 2007 -ngcf32)
hostname SW-1
!
vlan 1
!
no service password-encryption
!
interface FastEthernet 0/1
!
interface FastEthernet 0/2
!
interface FastEthernet 0/3
!
interface FastEthernet 0/4
!
interface FastEthernet 0/5
!
interface FastEthernet 0/6
!
interface FastEthernet 0/7
!
interface FastEthernet 0/8
!
interface FastEthernet 0/9
!
interface FastEthernet 0/10
!
interface FastEthernet 0/11
!
interface FastEthernet 0/12
!
interface FastEthernet 0/13
!
interface FastEthernet 0/14
!
interface FastEthernet 0/15
!
```

```
interface FastEthernet 0/16
!
interface FastEthernet 0/17
!
interface FastEthernet 0/18
!
interface FastEthernet 0/19
!
interface FastEthernet 0/20
!
interface FastEthernet 0/21
!
interface FastEthernet 0/22
!
interface FastEthernet 0/23
!
interface FastEthernet 0/24
!
interface GigabitEthernet 0/25
!
interface GigabitEthernet 0/26
!
interface GigabitEthernet 0/27
!
interface GigabitEthernet 0/28
!
!
line con 0
line vty 0 4
 login
!
!
banner motd ^C
Welcome to SW-1, if you are admin, you can config it.
If you are not admin, please EXIT!
^C
!
end
```

步骤 6　保存配置。

以下 3 条命令都可以保存配置。

```
SW-1#copy running-config startup-config
SW-1#write memory
```

```
SW-1#write
```

【注意事项】

（1）命令行操作进行自动补齐或命令简写时，要求所简写的字母能够区别该命令。如switch#conf 可以代表 configure，但 switch#co 无法代表 configure，因为 co 开头的命令有两个 copy 和 configure，设备无法区别。

（2）注意区别每个操作模式下可执行的命令种类。交换机不可以跨模式执行命令。

（3）配置设备名称的有效字符是 22 个字节。

（4）配置每日提示信息时，注意终止符不能在描述文本中出现。如果输入结束的终止符后再输入字符，则这些字符将被系统丢弃。

（5）交换机端口在默认情况下是开启的，AdminStatus 是 UP 状态，如果该端口没有实际连接其他设备，OperStatus 是 down 状态。

（6）show running-config 查看的是当前生效的配置信息，该信息存储在 RAM（随机存储器里），当交换机掉电，重新启动时会重新生成新的配置信息。

实验 2　在交换机上配置 Telnet

【实验名称】

在交换机上配置 Telnet。

【实验目的】

学习如何在交换机上启用 Telent，实现通过 Telnet 远程访问交换机。

【背景描述】

企业园区网覆盖范围较大时，交换机会分别放置在不同的地点，如果每次配置交换机都到交换机所在地点现场配置，管理员的工作量会很大。这时可以在交换机上进行 Telnet 配置，以后再需要配置交换机时，管理员可以远程以 Telnet 方式登录配置。

【需求分析】

需要掌握如何配置交换机的密码，以及如何配置 Telnet，掌握以 Telnet 的方式远程访问交换机的方法。

【实验拓扑】

实验的拓扑图，如图 2-1 所示。

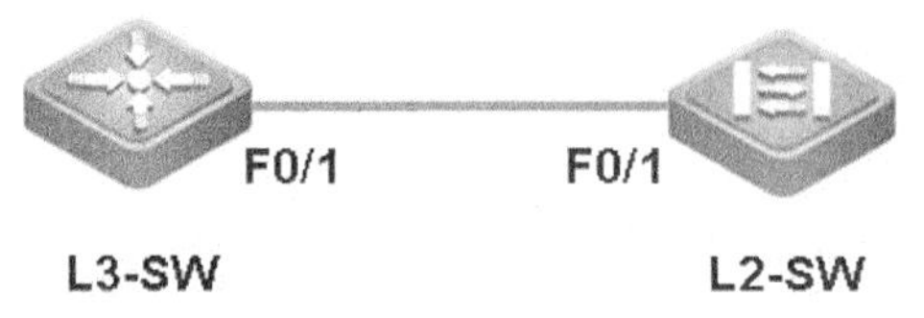

图 2-1

【实验设备】

二层交换机 1 台。

三层交换机 1 台。

【预备知识】

交换机的基本配置方法。

【实验原理】

在两台交换机上配置 VLAN 1 的 IP 地址，用双绞线将两台交换机的 F0/1 端口连接起来，分别配置 Telnet，然后就可以实现在每台交换机上以 Telnet 的方式登录另一台交换机。

【实验步骤】

步骤 1　在两台交换机上配置主机名、管理 IP 地址。

```
S3760(config)#hostname L3-SW
! 配置 3 层交换机的主机名
```

```
L3-SW(config)#interface vlan 1
！配置 3 层交换机的管理 IP 地址
L3-SW(config-if)#ip address 192.168.1.1 255.255.255.0
L3-SW(config-if)#no shutdown
L3-SW(config-if)#end
S2126G (config)#hostname L2-SW
！配置 2 层交换机的主机名
L2-SW(config)#interface vlan 1
！配置 2 层交换机的管理 IP 地址
L2-SW(config-if)#ip address 192.168.1.2 255.255.255.0
L2-SW(config-if)#no shutdown
L2-SW(config-if)#end
```

步骤 2　在三层交换机上配置 Telnet。

```
L3-SW(config)#enable password 0 star
！配置 enable 的密码
L3-SW(config)#line vty 0 4
！进入线程配置模式
L3-SW(config-line)#password 0 star
！配置 Telnet 的密码
L3-SW(config-line)#login
！启用 Telnet 的用户名密码验证
L3-SW(config-line)#exit
```

步骤 3　在二层交换机上配置 Telnet。

```
L2-SW(config)#enable secret level 1 0 star
！配置 Telnet 的密码，即 level 1 的密码
L2-SW(config)#enable secret 0 star
！配置 enable 密码，默认是 level 15 的密码
```

步骤 4　使用 Telnet 远程登录。

可以在 3 层交换机 L3-SW 上，以 Telnet 的方式登录 2 层交换机 L2-SW，进行验证。

```
L3-SW#telnet 192.168.1.2
Trying 192.168.1.2, 23...
User Access Verification
Password:
！提示输入 Telnet 密码，输入设置的密码 star
L2-SW>enable
Password:
！提示输入 enable 密码，输入设置的密码 star
L2-SW#
！现在已经进入了二层交换机 L2-SW，可以正常的进行配置
L2-SW#
L2-SW#exit
！使用 exit 命令退出 Telnet 登录
```

```
L3-SW#
```

可以在 2 层交换机 L2-SW 上，以 Telnet 的方式登录 3 层交换机 L3-SW，进行验证。

```
L2-SW#telnet 192.168.1.1
Trying 192.168.1.1 ... Open
User Access Verification
Password:
! 提示输入 Telnet 密码，输入设置的密码 star
L3-SW>en
Password:
! 提示输入 enable 密码，输入设置的密码 star
L3-SW#
L3-SW#exit
[Connection to 192.168.1.1 closed by foreign host]
! 使用 exit 命令退出 Telnet 登录
L2-SW#
```

【注意事项】

如果没有配置 enable 的密码，就不能登录到交换机进行配置。可以进入用户模式，但无法进入特权模式。二层交换机此时的提示信息为“% No password set”，三层交换机此时的提示信息为“Password required, but none set”。

【参考配置】

```
L3-SW#show running-config
Building configuration...
Current configuration : 1359 bytes
!
version RGNOS 10.2.00(2), Release(27932)(Thu Dec 13 10:31:41 CST 2007 -ngcf32)
hostname L3-SW
!
vlan 1
!
no service password-encryption
!
enable password star
!
!
interface FastEthernet 0/1
!
interface FastEthernet 0/2
!
interface FastEthernet 0/3
!
interface FastEthernet 0/4
```

```
!
interface FastEthernet 0/5
!
interface FastEthernet 0/6
!
interface FastEthernet 0/7
!
interface FastEthernet 0/8
!
interface FastEthernet 0/9
!
interface FastEthernet 0/10
!
interface FastEthernet 0/11
!
interface FastEthernet 0/12
!
interface FastEthernet 0/13
!
interface FastEthernet 0/14
!
interface FastEthernet 0/15
!
interface FastEthernet 0/16
!
interface FastEthernet 0/17
!
interface FastEthernet 0/18
!
interface FastEthernet 0/19
!
interface FastEthernet 0/20
!
interface FastEthernet 0/21
!
interface FastEthernet 0/22
!
interface FastEthernet 0/23
!
interface FastEthernet 0/24
!
interface GigabitEthernet 0/25
```

```
!
interface GigabitEthernet 0/26
!
interface GigabitEthernet 0/27
!
interface GigabitEthernet 0/28
!
interface VLAN 1
 ip address 192.168.1.1 255.255.255.0
!
!
line con 0
line vty 0 4
 login
 password star
!
!
end
```

L2-SW#show running-config

```
System software version : 1.66(8) Build Dec 22 2006 Rel
Building configuration...
Current configuration : 380 bytes
!
version 1.0
!
hostname L2-SW
vlan 1
!
!
enable secret level 1 5 $2H.Y*T73C,tZ[V/4D+S(\W&QG1X)sv'
enable secret level 15 5 $2,1u ;C3&-8U0<D4'.tj9=GQ+/7R:>H
!
interface fastEthernet 0/1
!
interface vlan 1
 no shutdown
 ip address 192.168.1.2 255.255.255.0
!
!
end
```

实验 3　跨交换机实现 VLAN

【实验名称】

跨交换机实现 VLAN。

【实验目的】

掌握如何在交换机上划分基于端口的 VLAN、给 VLAN 内添加端口，理解跨交换机之间 VLAN 的特点。

【背景描述】

假设某企业有两个主要部门：销售部和技术部，其中销售部门的个人计算机系统连接在不同的交换机上，两个部门之间需要相互进行通信，但为了数据安全起见，销售部和技术部需要进行相互隔离，要在交换机上做适当配置来实现这一目标。

【需求分析】

通过划分 Port VLAN 来实现交换机的端口隔离，然后使在同一个 VLAN 里的计算机系统能跨交换机进行相互通信，而在不同 VLAN 里的计算机系统不能进行相互通信。

【实验拓扑】

实验的拓扑图，如图 3-1 所示。

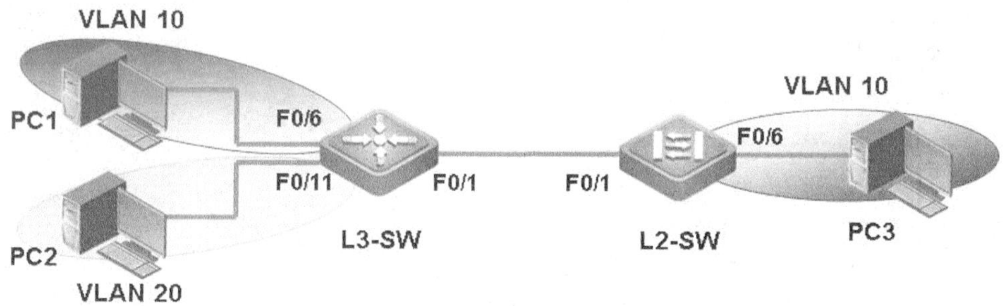

图 3-1

【实验设备】

三层交换机 1 台。

二层交换机 1 台。

【预备知识】

交换机的基本配置方法、VLAN 的工作原理和配置方法、Trunk 的工作原理和配置方法。

【实验原理】

VLAN（Virtual Local Area Network，虚拟局域网）是指在一个物理网段内，进行逻辑划分，并划分成若干个虚拟局域网。VLAN 最大的特性是不受物理位置的限制，可以进行灵活的划分。

VLAN 具备了一个物理网段所具备的特性。同一 VLAN 内的主机可以相互直接访问，不同 VLAN 是的主机之间互相访问，必须经路由设备进行转发。广播数据包只可以在同一 VLAN 内进行传播，不能传输到其他 VLAN 中。

Port Vlan 是实现 VLAN 的方式之一，它利用交换机的端口进行 VLAN 的划分，一个端口只能属于一个 VLAN。

Tag Vlan 是基于交换机端口的另外一种类型，主要用于实现跨交换机同一 VLAN 内主机之间的直接访问，同时对于不同 VLAN 的主机进行隔离。Tag Vlan 遵循了 IEEE802.1q 协议的标准。在利用配置了 Tag vlan 的接口进行数据传输时，需要在数据帧内添加 4 个字节的 802.1q 标签信息，用于标识该数据帧属于哪个 VLAN，以便在端交换机接收到数据帧后进行准确的过滤。

【实验步骤】

步骤 1　配置两台交换机的主机名。

```
Switch#configure terminal
Enter configuration commands, one per line.  End with CNTL/Z.
Switch(config)#hostname L2-SW
L2-SW(config)#
S3750#configure terminal
Enter configuration commands, one per line.  End with CNTL/Z.
S3750(config)#hostname L3-SW
L3-SW(config)#
```

步骤 2　在三层交换机上划分 VLAN 添加端口。

```
L3-SW(config)#vlan 10
L3-SW(config-vlan)#name xiaoshou
！划分销售部的 VLAN 10
L3-SW(config-vlan)#vlan 20
L3-SW(config-vlan)#name jishu
！划分技术部的 VLAN 20
L3-SW(config-vlan)#exit
L3-SW(config)#
L3-SW(config)#interface range fastEthernet 0/6-10
！将端口 Fa0/6 至 Fa0/10 划分到 VLAN 10
L3-SW(config-if-range)#switchport mode access
L3-SW(config-if-range)#switchport access vlan 10
L3-SW(config-if-range)#exit
L3-SW(config)#interface range fastEthernet 0/11-15
！将端口 Fa0/11 至 Fa0/15 划分到 VLAN 20
L3-SW(config-if-range)#switchport mode access
L3-SW(config-if-range)#switchport access vlan 20
L3-SW(config-if-range)#exit
L3-SW(config)#
```

步骤 3　在二层交换机上划分 VLAN 添加端口。

```
L2-SW(config)#vlan 10
```

```
L2-SW(config-vlan)#name xiaoshou
! 划分销售部的 VLAN 10
L2-SW(config-vlan)#vlan 20
L2-SW(config-vlan)#name jishu
! 划分技术部的 VLAN 20
L2-SW(config-vlan)#exit
L2-SW(config)#
L2-SW(config)#interface range fastEthernet 0/6-10
! 将端口 Fa0/6 至 Fa0/10 划分到 VLAN 10
L2-SW(config-if-range)#switchport mode access
L2-SW(config-if-range)#switchport access vlan 10
L2-SW(config-if-range)#exit
L2-SW(config)#
```

步骤 4　设置交换机之间的链路为 Trunk。

```
L3-SW(config)#interface fastEthernet 0/1
L3-SW(config-if)#switchport mode trunk
L3-SW(config-if)#exit
L3-SW(config)#
L2-SW(config)#interface fastEthernet 0/1
L2-SW(config-if)#switchport mode trunk
L2-SW(config-if)#exit
L2-SW(config)#
```

步骤 5　查看 VLAN 和 Trunk 的配置。

```
L2-SW#show vlan
VLAN Name                             Status    Ports
---- -------------------------------- --------- ------------------------------
1    default                          active    Fa0/1 ,Fa0/2 ,Fa0/3
                                                Fa0/4 ,Fa0/5 ,Fa0/11
                                                Fa0/12,Fa0/13,Fa0/14
                                                Fa0/15,Fa0/16,Fa0/17
                                                Fa0/18,Fa0/19,Fa0/20
                                                Fa0/21,Fa0/22,Fa0/23
                                                Fa0/24
10   xiaoshou                         active    Fa0/1 ,Fa0/6 ,Fa0/7
                                                Fa0/8 ,Fa0/9 ,Fa0/10
20   jishu                            active    Fa0/1
L2-SW#
L2-SW#show interfaces fastEthernet 0/1 switchport
Interface  Switchport Mode      Access  Native   Protected VLAN lists
---------- ---------- --------- ------- -------- --------- ----------------
Fa0/1      Enabled    Trunk     1       1        Disabled  All
L3-SW#show vlan
```

```
VLAN Name                             Status    Ports
---- -------------------------------- --------- -----------------------------------
   1 VLAN0001                         STATIC    Fa0/1, Fa0/2, Fa0/3, Fa0/4
                                                Fa0/5, Fa0/16, Fa0/17, Fa0/18
                                                Fa0/19, Fa0/20, Fa0/21, Fa0/22
                                                Fa0/23, Fa0/24, Gi0/25, Gi0/26
                                                Gi0/27, Gi0/28
  10 xiaoshou                         STATIC    Fa0/1, Fa0/6, Fa0/7, Fa0/8
                                                Fa0/9, Fa0/10
  20 jishu                            STATIC    Fa0/1, Fa0/11, Fa0/12, Fa0/13
                                                Fa0/14, Fa0/15
L3-SW#
L3-SW#show interfaces fastEthernet 0/1 switchport
Interface                Switchport Mode      Access Native Protected VLAN lists
------------------------ ---------- --------- ------ ------ --------- ----------
FastEthernet 0/1         enabled    TRUNK     1      1      Disabled  ALL
```

步骤 6　验证配置。

PC3 和 PC1 都属于 VLAN 10，它们的 IP 地址都在 C 类网络 192.168.10.0/24 内，PC2 属于 VLAN 20，它的 IP 地址在 C 类网络 192.168.20.0/24 内，从 PC3 可以 ping 通 PC1，如图 3-2 所示，而从 PC3 不能 ping 通 PC2，如图 3-3 所示。

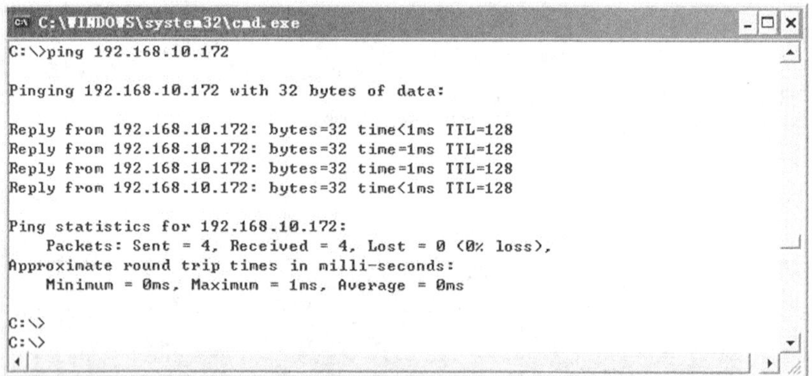

```
C:\WINDOWS\system32\cmd.exe
C:\>ping 192.168.10.172

Pinging 192.168.10.172 with 32 bytes of data:

Reply from 192.168.10.172: bytes=32 time<1ms TTL=128
Reply from 192.168.10.172: bytes=32 time=1ms TTL=128
Reply from 192.168.10.172: bytes=32 time=1ms TTL=128
Reply from 192.168.10.172: bytes=32 time<1ms TTL=128

Ping statistics for 192.168.10.172:
    Packets: Sent = 4, Received = 4, Lost = 0 (0% loss),
Approximate round trip times in milli-seconds:
    Minimum = 0ms, Maximum = 1ms, Average = 0ms

C:\>
C:\>
```

图 3-2

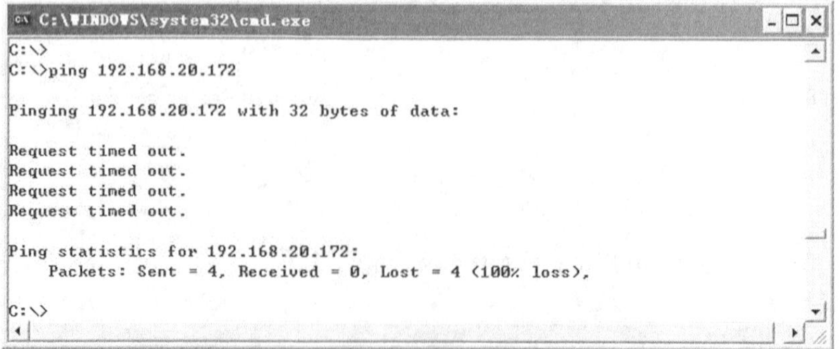

```
C:\WINDOWS\system32\cmd.exe
C:\>
C:\>ping 192.168.20.172

Pinging 192.168.20.172 with 32 bytes of data:

Request timed out.
Request timed out.
Request timed out.
Request timed out.

Ping statistics for 192.168.20.172:
    Packets: Sent = 4, Received = 0, Lost = 4 (100% loss),

C:\>
```

图 3-3

此时，如果把 PC1 的连线转移到属于 VLAN 20 的端口上，PC3 与 PC1 将不再能 ping 通，如图 3-4 所示。

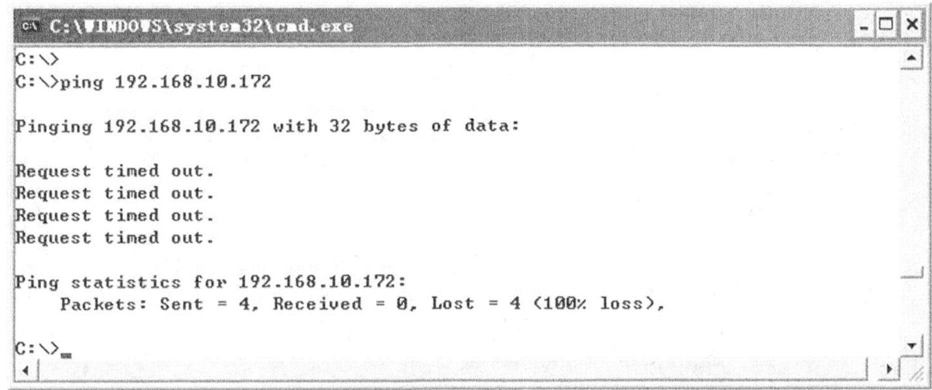

```
C:\WINDOWS\system32\cmd.exe
C:\>
C:\>ping 192.168.10.172

Pinging 192.168.10.172 with 32 bytes of data:

Request timed out.
Request timed out.
Request timed out.
Request timed out.

Ping statistics for 192.168.10.172:
    Packets: Sent = 4, Received = 0, Lost = 4 (100% loss),

C:\>_
```

图 3-4

【注意事项】

（1）交换机所有的端口在默认情况下都属于 ACCESS 端口，可直接将端口加入某个 VLAN。利用 switchport mode access/trunk 命令可以更改端口的 VLAN 模式。

（2）VLAN1 属于系统的默认 VLAN，不可以被删除。

（3）删除某个 VLAN，可使用 no 命令。例如：switch(config)#no vlan 10。

（4）删除当前某个 VLAN 时，注意先将属于该 VLAN 的端口加入另一个 VLAN 中，再删除 VLAN。

（5）两台交换机之间相连的端口应该设置为 tag vlan 模式。

（6）Trunk 接口在默认情况下支持所有 VLAN 的传输。

【参考配置】

```
L3-SW#show running-config
Building configuration...
Current configuration : 1481 bytes
!
version RGNOS 10.1.00(4), Release(18443)(Tue Jul 17 19:51:54 CST 2007
-ubu6server)
hostname L3-SW
!
vlan 1
!
vlan 10
 name xiaoshou
!
vlan 20
 name jishu
!
!
```

```
interface FastEthernet 0/1
 switchport mode trunk
!
interface FastEthernet 0/2
!
interface FastEthernet 0/3
!
interface FastEthernet 0/4
!
interface FastEthernet 0/5
!
interface FastEthernet 0/6
 switchport access vlan 10
!
interface FastEthernet 0/7
 switchport access vlan 10
!
interface FastEthernet 0/8
 switchport access vlan 10
!
interface FastEthernet 0/9
 switchport access vlan 10
!
interface FastEthernet 0/10
 switchport access vlan 10
!
interface FastEthernet 0/11
 switchport access vlan 20
!
interface FastEthernet 0/12
 switchport access vlan 20
!
interface FastEthernet 0/13
 switchport access vlan 20
!
interface FastEthernet 0/14
 switchport access vlan 20
!
interface FastEthernet 0/15
 switchport access vlan 20
!
interface FastEthernet 0/16
```

```
!
interface FastEthernet 0/17
!
interface FastEthernet 0/18
!
interface FastEthernet 0/19
!
interface FastEthernet 0/20
!
interface FastEthernet 0/21
!
interface FastEthernet 0/22
!
interface FastEthernet 0/23
!
interface FastEthernet 0/24
!
interface GigabitEthernet 0/25
!
interface GigabitEthernet 0/26
!
interface GigabitEthernet 0/27
!
interface GigabitEthernet 0/28
!
!
line con 0
line vty 0 4
 login
!
!
end
L2-SW#show running-config
System software version : 1.68 Build Apr 25 2007 Release
Building configuration...
Current configuration : 457 bytes
!
version 1.0
!
hostname L2-SW
vlan 1
!
```

```
vlan 10
 name xiaoshou
!
vlan 20
 name jishu
!
interface fastEthernet 0/1
 switchport mode trunk
!
interface fastEthernet 0/6
 switchport access vlan 10
!
interface fastEthernet 0/7
 switchport access vlan 10
!
interface fastEthernet 0/8
 switchport access vlan 10
!
interface fastEthernet 0/9
 switchport access vlan 10
!
interface fastEthernet 0/10
 switchport access vlan 10
!
end
```

实验 4　利用单臂路由实现 VLAN 间路由

【实验名称】

利用单臂路由实现 VLAN 间路由。

【实验目的】

掌握如何在路由器端口上划分子接口、封装 Dot1q（IEEE 802.1q）协议，实现 VLAN 间的路由。

【背景描述】

假设某企业有两个主要部门：销售部和技术部，员工都连接在 1 台二层交换机上，网络内有 1 台路由器用于连接 Internet。现在发现网络内的广播流量太多，需要对广播进行限制但不能影响 2 个部门进行相互通信，要在路由器上做适当配置来实现这一目标。

【需求分析】

需要在交换机上配置 VLAN，然后在路由器连接交换机的端口上划分子接口，给相应的 VLAN 设置 IP 地址，以实现 VLAN 间的路由。

【实验拓扑】

实验的拓扑图，如图 4-1 所示。

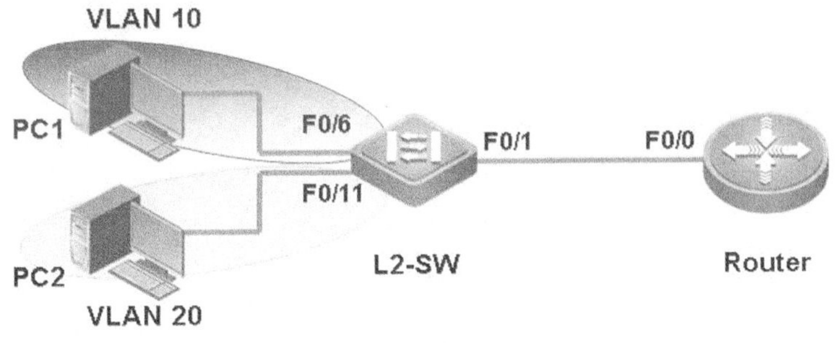

图 4-1

【实验设备】

路由器 1 台。

二层交换机 1 台。

【预备知识】

交换机的基本配置方法、VLAN 的工作原理和配置方法、Trunk 的工作原理和配置方法、单臂路由的工作原理和配置方法。

【实验原理】

在交换网络中，通过 VLAN 对一个物理网络进行了逻辑划分，不同的 VLAN 之间是无法直接访问的，必须通过三层的路由设备进行连接。一般利用路由器或三层交换机来实现不同 VLAN 之间的互相访问。

将路由器和交换机相连，使用 IEEE 802.1q 来启动路由器上的子接口成为干道模式，就可以利用路由器来实现 VLAN 之间的通信。

路由器可以从某一个 VLAN 接收数据包，并将这个数据包转发到另外的一个 VLAN，要实施 VLAN 间的路由，必须在一个路由器的物理接口上启用子接口，也就是将以太网物理接口划分为多个逻辑的、可编址的接口，并配置成干道模式，每个 VLAN 对应一个这种接口，这样路由器就能够知道如何到达这些互连的 VLAN。

【实验步骤】

步骤 1　配置交换机的主机名、划分 VLAN 和添加端口、设置 Trunk。

```
Switch#configure terminal
Switch(config)#hostname L2-SW
L2-SW(config)#vlan 10
L2-SW(config-vlan)#name xiaoshou
L2-SW(config-vlan)#vlan 20
L2-SW(config-vlan)#name jishu
L2-SW(config-vlan)#exit
L2-SW(config)#interface range fastEthernet 0/6-10
L2-SW(config-if-range)#switchport mode access
L2-SW(config-if-range)#switchport access VLAN 10
L2-SW(config-if-range)#exit
L2-SW(config)#interface range fastEthernet 0/11-15
L2-SW(config-if-range)#switchport mode access
L2-SW(config-if-range)#switchport access vlan 20
L2-SW(config-if-range)#exit
L2-SW(config)#interface fastEthernet 0/1
L2-SW(config-if)#switchport mode trunk
L2-SW(config-if)#end
```

步骤 2　在路由器上设置名称、划分子接口、配置 IP 地址。

```
RSR20#configure terminal
RSR20(config)#hostname Router
Router(config)#interface fastEthernet 0/0
Router(config-if)#no ip address
！去掉路由器主接口上的 IP 地址
Router(config-if)#no shutdown
Router(config-if)#exit
Router(config)#interface fastEthernet 0/0.10
！进入子接口 Fa0/0.10
Router(config-subif)#encapsulation dot1Q 10
！指定子接口 Fa0/0.10 对应 VLAN 10，并配置干道模式
Router(config-subif)#ip address 192.168.10.1 255.255.255.0
```

```
! 配置子接口 Fa0/0.10 的 IP 地址
Router(config-subif)#exit
Router(config)#interface fastEthernet 0/0.20
! 进入子接口 Fa0/0.20
Router(config-subif)#encapsulation dot1Q 20
! 指定子接口 Fa0/0.20 对应 VLAN 20，并配置干道模式
Router(config-subif)#ip address 192.168.20.1 255.255.255.0
! 配置子接口 Fa0/0.20 的 IP 地址
Router(config-subif)#end
```

步骤 3　查看交换机的 VLAN 和 Trunk 配置。

```
L2-SW#show vlan
VLAN Name                             Status    Ports
----------------------------------------------------------------------------
1    default                          active    Fa0/1 ,Fa0/2 ,Fa0/3
                                                Fa0/4 ,Fa0/5 ,Fa0/16
                                                Fa0/17,Fa0/18,Fa0/19
                                                Fa0/20,Fa0/21,Fa0/22
                                                Fa0/23,Fa0/24
10   xiaoshou                         active    Fa0/1 ,Fa0/6 ,Fa0/7
                                                Fa0/8 ,Fa0/9 ,Fa0/10
20   jishu                            active    Fa0/1 ,Fa0/11,Fa0/12
                                                Fa0/13,Fa0/14,Fa0/15
L2-SW#
L2-SW#show interfaces fastEthernet 0/1 switchport
Interface  Switchport Mode     Access  Native   Protected VLAN lists
---------- ---------- ------ ------- -------- --------- --------------------
Fa0/1      Enabled    Trunk    1       1        Disabled  All
```

步骤 4　查看路由器的路由表。

```
Router#show ip route
Codes:  C - connected, S - static,  R - RIP B - BGP
        O - OSPF, IA - OSPF inter area
        N1 - OSPF NSSA external type 1, N2 - OSPF NSSA external type 2
        E1 - OSPF external type 1, E2 - OSPF external type 2
        i - IS-IS, L1 - IS-IS level-1, L2 - IS-IS level-2, ia - IS-IS inter area
        * - candidate default
Gateway of last resort is no set
C    192.168.10.0/24 is directly connected, FastEthernet 0/0.10
C    192.168.10.1/32 is local host.
C    192.168.20.0/24 is directly connected, FastEthernet 0/0.20
C    192.168.20.1/32 is local host.
```

步骤 5　测试网络连通性。

给 PC1 和 PC2 分别配置 192.168.10.0/24 和 192.168.20.0/24 网段内的 IP 地址，并分别以

192.168.10.1 和 192.168.20.1 作为网关，例如 PC2 的 IP 地址配置为，如图 4-2 所示。

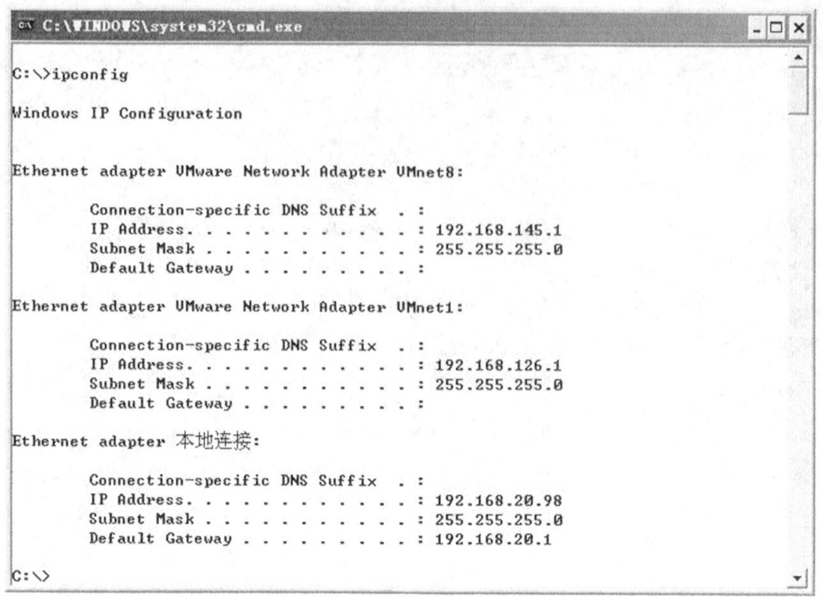

```
C:\WINDOWS\system32\cmd.exe

C:\>ipconfig

Windows IP Configuration


Ethernet adapter VMware Network Adapter VMnet8:

        Connection-specific DNS Suffix  . :
        IP Address. . . . . . . . . . . . : 192.168.145.1
        Subnet Mask . . . . . . . . . . . : 255.255.255.0
        Default Gateway . . . . . . . . . :

Ethernet adapter VMware Network Adapter VMnet1:

        Connection-specific DNS Suffix  . :
        IP Address. . . . . . . . . . . . : 192.168.126.1
        Subnet Mask . . . . . . . . . . . : 255.255.255.0
        Default Gateway . . . . . . . . . :

Ethernet adapter 本地连接:

        Connection-specific DNS Suffix  . :
        IP Address. . . . . . . . . . . . : 192.168.20.98
        Subnet Mask . . . . . . . . . . . : 255.255.255.0
        Default Gateway . . . . . . . . . : 192.168.20.1

C:\>
```

图 4-2

从 PC2 上 ping 所属 VLAN 的网关、VLAN 10 的网关和 PC1 的结果，如图 4-3~图 4-5 所示，说明配置单臂路由后，网络已经全部实现互连互通。

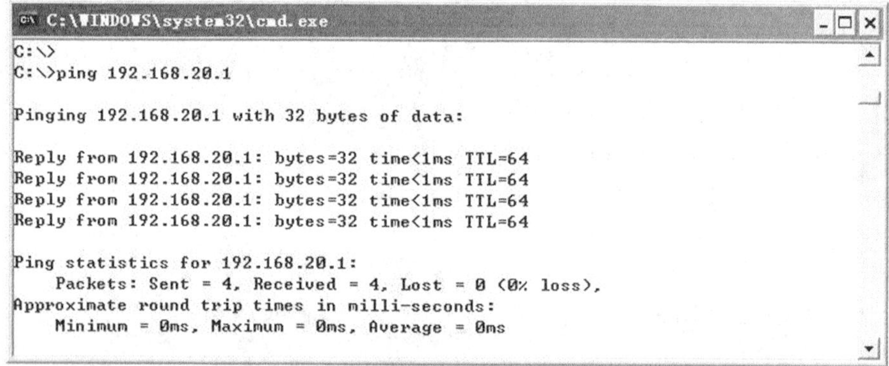

```
C:\WINDOWS\system32\cmd.exe
C:\>
C:\>ping 192.168.20.1

Pinging 192.168.20.1 with 32 bytes of data:

Reply from 192.168.20.1: bytes=32 time<1ms TTL=64
Reply from 192.168.20.1: bytes=32 time<1ms TTL=64
Reply from 192.168.20.1: bytes=32 time<1ms TTL=64
Reply from 192.168.20.1: bytes=32 time<1ms TTL=64

Ping statistics for 192.168.20.1:
    Packets: Sent = 4, Received = 4, Lost = 0 (0% loss),
Approximate round trip times in milli-seconds:
    Minimum = 0ms, Maximum = 0ms, Average = 0ms
```

图 4-3

```
C:\WINDOWS\system32\cmd.exe

C:\>ping 192.168.10.1

Pinging 192.168.10.1 with 32 bytes of data:

Reply from 192.168.10.1: bytes=32 time<1ms TTL=64
Reply from 192.168.10.1: bytes=32 time<1ms TTL=64
Reply from 192.168.10.1: bytes=32 time<1ms TTL=64
Reply from 192.168.10.1: bytes=32 time<1ms TTL=64

Ping statistics for 192.168.10.1:
    Packets: Sent = 4, Received = 4, Lost = 0 (0% loss),
Approximate round trip times in milli-seconds:
    Minimum = 0ms, Maximum = 0ms, Average = 0ms
```

图 4-4

```
C:\>ping 192.168.10.10

Pinging 192.168.10.10 with 32 bytes of data:

Reply from 192.168.10.10: bytes=32 time=1ms TTL=127
Reply from 192.168.10.10: bytes=32 time<1ms TTL=127
Reply from 192.168.10.10: bytes=32 time<1ms TTL=127
Reply from 192.168.10.10: bytes=32 time<1ms TTL=127

Ping statistics for 192.168.10.10:
    Packets: Sent = 4, Received = 4, Lost = 0 (0% loss),
Approximate round trip times in milli-seconds:
    Minimum = 0ms, Maximum = 1ms, Average = 0ms
```

图 4-5

【注意事项】

（1）在给路由器的子接口配置 IP 地址之前，一定要先封装 dot1q 协议。

（2）各个 VLAN 内的主机，要以相应 VLAN 子接口的 IP 地址作为网关。

【参考配置】

```
Router#show running-config
Building configuration...
Current configuration : 586 bytes
!
version RGNOS 10.1.00(4), Release(18443)(Tue Jul 17 20:50:30 CST 2007
-ubu1server)
hostname Router
!
!
interface FastEthernet 0/0
 duplex auto
 speed auto
!
interface FastEthernet 0/0.10
 encapsulation dot1Q 10
 ip address 192.168.10.1 255.255.255.0
!
interface FastEthernet 0/0.20
 encapsulation dot1Q 20
 ip address 192.168.20.1 255.255.255.0
!
interface FastEthernet 0/1
 duplex auto
 speed auto
!
```

```
line con 0
line aux 0
line vty 0 4
 login
!
end
```

L2-SW#show running-config

```
System software version : 1.68 Build Apr 25 2007 Release
Building configuration...
Current configuration : 757 bytes
!
version 1.0
!
hostname L2-SW
vlan 1
!
vlan 10
 name xiaoshou
!
vlan 20
 name jishu
!
interface fastEthernet 0/1
 switchport mode trunk
!
interface fastEthernet 0/6
 switchport access vlan 10
!
interface fastEthernet 0/7
 switchport access vlan 10
!
interface fastEthernet 0/8
 switchport access vlan 10
!
interface fastEthernet 0/9
 switchport access vlan 10
!
interface fastEthernet 0/10
 switchport access vlan 10
!
interface fastEthernet 0/11
 switchport access vlan 20
```

```
!
interface fastEthernet 0/12
 switchport access vlan 20
!
interface fastEthernet 0/13
 switchport access vlan 20
!
interface fastEthernet 0/14
 switchport access vlan 20
!
interface fastEthernet 0/15
 switchport access vlan 20
!
end
```

实验 5　利用三层交换机实现 VLAN 间路由

【实验名称】

利用三层交换机实现 VLAN 间路由。

【实验目的】

掌握如何在三层交换机上配置 SVI 端口，实现 VLAN 间的路由。

【背景描述】

假设某企业有两个主要部门：销售部和技术部，其中销售部门的个人计算机系统分散连接在两台交换机上，他们之间需要相互进行通信，销售部和技术部也需要进行相互通信，现要在交换机上做适当配置来实现这一目标。

【需求分析】

需要在网络内所有的交换机上配置 VLAN，然后在三层交换机上给相应的 VLAN 设置 IP 地址，以实现 VLAN 间的路由。

【实验拓扑】

实验的拓扑图，如图 5-1 所示。

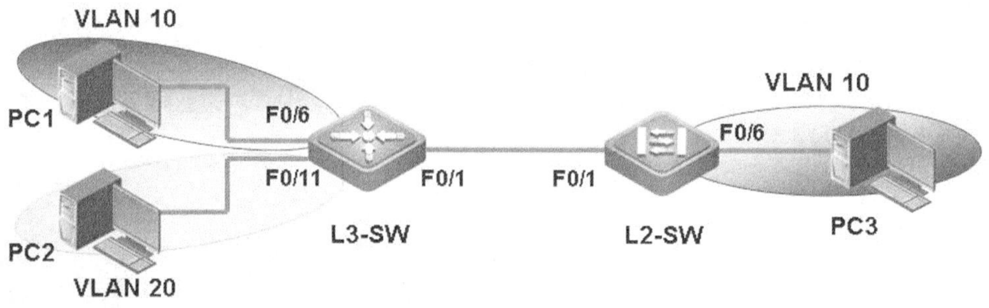

图 5-1

【实验设备】

三层交换机 1 台。

二层交换机 1 台。

【预备知识】

交换机的基本配置方法、VLAN 的工作原理和配置方法、Trunk 的工作原理和配置方法、三层交换的工作原理和配置方法。

【实验原理】

在交换网络中，通过 VLAN 对一个物理网络进行逻辑划分，不同的 VLAN 之间无法直接访问的，必须通过三层的路由设备进行连接。一般利用路由器或三层交换机来实现不同 VLAN 之间的互相访问。三层交换机和路由器具备网络层的功能，能够根据数据的 IP 包头信息，进行选路和转发，从而实现不同网段之间的访问。

直连路由是指：为三层设备的接口配置 IP 地址，并且激活该端口，三层设备会自动产生该接口 IP 所在网段的直连路由信息。

三层交换机实现 VLAN 互访的原理是，利用三层交换机的路由功能，通过识别数据包的 IP 地址，查找路由表进行选路转发。三层交换机利用直连路由可以实现不同 VLAN 之间的互相访问。三层交换机给接口配置 IP 地址，采用 SVI（交换虚拟接口）的方式实现 VLAN 间互连。SVI 是指为交换机中的 VLAN 创建虚拟接口，并且配置 IP 地址。

【实验步骤】

步骤 1　配置两台交换机的主机名。

```
Switch#configure terminal
Enter configuration commands, one per line.  End with CNTL/Z.
Switch(config)#hostname L2-SW
L2-SW(config)#
S3750#configure terminal
Enter configuration commands, one per line.  End with CNTL/Z.
S3750(config)#hostname L3-SW
L3-SW(config)#
```

步骤 2　在三层交换机上划分 VLAN 添加端口，并设置 Trunk。

```
L3-SW(config)#vlan 10
L3-SW(config-vlan)#name xiaoshou
L3-SW(config-vlan)#vlan 20
L3-SW(config-vlan)#name jishu
L3-SW(config-vlan)#exit
L3-SW(config)#
L3-SW(config)#interface range fastEthernet 0/6-10
L3-SW(config-if-range)#switchport mode access
L3-SW(config-if-range)#switchport access vlan 10
L3-SW(config-if-range)#exit
L3-SW(config)#interface range fastEthernet 0/11-15
L3-SW(config-if-range)#switchport mode access
L3-SW(config-if-range)#switchport access vlan 20
L3-SW(config-if-range)#exit
L3-SW(config)#
L3-SW(config)#interface fastEthernet 0/1
L3-SW(config-if)#switchport mode trunk
L3-SW(config-if)#exit
L3-SW(config)#
```

步骤 3　在二层交换机上划分 VLAN 添加端口，并设置 Trunk。

```
L2-SW(config)#vlan 10
L2-SW(config-vlan)#name xiaoshou
```

```
L2-SW(config-vlan)#vlan 20
L2-SW(config-vlan)#name jishu
L2-SW(config-vlan)#exit
L2-SW(config)#
L2-SW(config)#interface range fastEthernet 0/6-10
L2-SW(config-if-range)#switchport mode access
L2-SW(config-if-range)#switchport access vlan 10
L2-SW(config-if-range)#exit
L2-SW(config)#
L2-SW(config)#interface fastEthernet 0/1
L2-SW(config-if)#switchport mode trunk
L2-SW(config-if)#exit
L2-SW(config)#
```

步骤 4　查看 VLAN 和 Trunk 的配置。

```
L2-SW#show vlan
VLAN Name                               Status    Ports
---- ------------------------------ --------- -------------------------------
1    default                            active    Fa0/1 ,Fa0/2 ,Fa0/3
                                                  Fa0/4 ,Fa0/5 ,Fa0/11
                                                  Fa0/12,Fa0/13,Fa0/14
                                                  Fa0/15,Fa0/16,Fa0/17
                                                  Fa0/18,Fa0/19,Fa0/20
                                                  Fa0/21,Fa0/22,Fa0/23
                                                  Fa0/24
10   xiaoshou                           active    Fa0/1 ,Fa0/6 ,Fa0/7
                                                  Fa0/8 ,Fa0/9 ,Fa0/10
20   jishu                              active    Fa0/1
L2-SW#
L2-SW#show interfaces fastEthernet 0/1 switchport
Interface  Switchport Mode      Access  Native   Protected VLAN lists
---------- ---------- --------- -------- --------- --------------------
Fa0/1      Enabled    Trunk     1        1         Disabled All
L3-SW#show vlan
VLAN Name                               Status    Ports
---- ------------------------------ ------------------------------------
  1 VLAN0001                            STATIC    Fa0/1, Fa0/2, Fa0/3, Fa0/4
                                                  Fa0/5, Fa0/16, Fa0/17, Fa0/18
                                                  Fa0/19, Fa0/20, Fa0/21, Fa0/22
                                                  Fa0/23, Fa0/24, Gi0/25, Gi0/26
                                                  Gi0/27, Gi0/28
 10 xiaoshou                            STATIC    Fa0/1, Fa0/6, Fa0/7, Fa0/8
                                                  Fa0/9, Fa0/10
```

```
  20 jishu                          STATIC    Fa0/1, Fa0/11, Fa0/12, Fa0/13
                                              Fa0/14, Fa0/15
L3-SW#
L3-SW#show interfaces fastEthernet 0/1 switchport
Interface                Switchport Mode      Access Native Protected VLAN lists
-------------------- ---------- --------- ------ ------ --------- ----------
FastEthernet 0/1          enabled    TRUNK     1      1     Disabled  ALL
```

步骤 5　验证配置。

PC3 和 PC1 都属于 VLAN 10，它们的 IP 地址都在 C 类网络 192.168.10.0/24 内，PC2 属于 VLAN 20，它的 IP 地址在 C 类网络 192.168.20.0/24 内，此时，不同 VLAN 之间的 PC3 和 PC2 不能 ping 通，如图 5-2 所示。

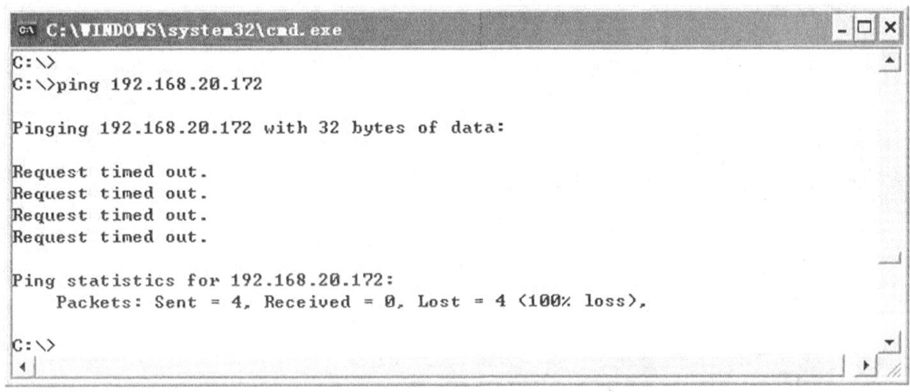

```
C:\WINDOWS\system32\cmd.exe
C:\>
C:\>ping 192.168.20.172

Pinging 192.168.20.172 with 32 bytes of data:

Request timed out.
Request timed out.
Request timed out.
Request timed out.

Ping statistics for 192.168.20.172:
    Packets: Sent = 4, Received = 0, Lost = 4 (100% loss),

C:\>
```

图 5-2

步骤 6　在三层交换机上配置 SVI 端口。

```
L3-SW#configure terminal
Enter configuration commands, one per line.  End with CNTL/Z.
L3-SW(config)#interface vlan 10
! 激活 VLAN 10 的 SVI 端口并配置 IP 地址
L3-SW(config-if)#Dec  2 18:59:30 L3-SW %7:%LINE PROTOCOL CHANGE: Interface VLAN
10, changed state to UP
L3-SW(config-if)#ip address 192.168.10.1 255.255.255.0
L3-SW(config-if)#no shutdown
L3-SW(config-if)#exit
L3-SW(config)#
L3-SW(config)#interface vlan 20
! 激活 VLAN 20 的 SVI 端口并配置 IP 地址
L3-SW(config-if)#Dec  2 19:00:05 L3-SW %7:%LINE PROTOCOL CHANGE: Interface VLAN
20, changed state to UP
L3-SW(config-if)#ip address 192.168.20.1 255.255.255.0
L3-SW(config-if)#no shutdown
L3-SW(config-if)#exit
L3-SW(config)#
```

步骤 7　查看 SVI 端口的配置。

```
L3-SW#show ip route
Codes:  C - connected, S - static,  R - RIP B - BGP
        O - OSPF, IA - OSPF inter area
        N1 - OSPF NSSA external type 1, N2 - OSPF NSSA external type 2
        E1 - OSPF external type 1, E2 - OSPF external type 2
        i - IS-IS, L1 - IS-IS level-1, L2 - IS-IS level-2, ia - IS-IS inter area
        * - candidate default
Gateway of last resort is no set
C    192.168.10.0/24 is directly connected, VLAN 10
C    192.168.10.1/32 is local host.
C    192.168.20.0/24 is directly connected, VLAN 20
C    192.168.20.1/32 is local host.
L3-SW#
```

从中可以看到，VLAN 的虚拟端口上配置的 IP 地址，其网段成为了三层交换机的直连路由。

```
L3-SW#show interfaces vlan 10
Index(dec):4106  (hex):100a
VLAN 10 is UP  , line protocol is UP
Hardware is  VLAN, address is 00d0.f821.a543 (bia 00d0.f821.a543)
Interface address is: 192.168.10.1/24
ARP type: ARPA,ARP Timeout: 3600 seconds
  MTU 1500 bytes, BW 1000000 Kbit
  Encapsulation protocol is Ethernet-II, loopback not set
  Keepalive interval is 10 sec , set
  Carrier delay is 2 sec
  RXload is 1 ,Txload is 1
  Queueing strategy: WFQ
L3-SW#
L3-SW#show interfaces vlan 20
Index(dec):4116  (hex):1014
VLAN 20 is UP  , line protocol is UP
Hardware is  VLAN, address is 00d0.f821.a543 (bia 00d0.f821.a543)
Interface address is: 192.168.20.1/24
ARP type: ARPA,ARP Timeout: 3600 seconds
  MTU 1500 bytes, BW 1000000 Kbit
  Encapsulation protocol is Ethernet-II, loopback not set
  Keepalive interval is 10 sec , set
  Carrier delay is 2 sec
  RXload is 1 ,Txload is 1
  Queueing strategy: WFQ
L3-SW#
```

步骤 8　验证配置。

给 PC3 添加网关 192.168.10.1，如图 5-3 所示，此时再从 PC3 去 ping 不同 VLAN 的主机 PC2，是可以 ping 通的，如图 5-4 所示。

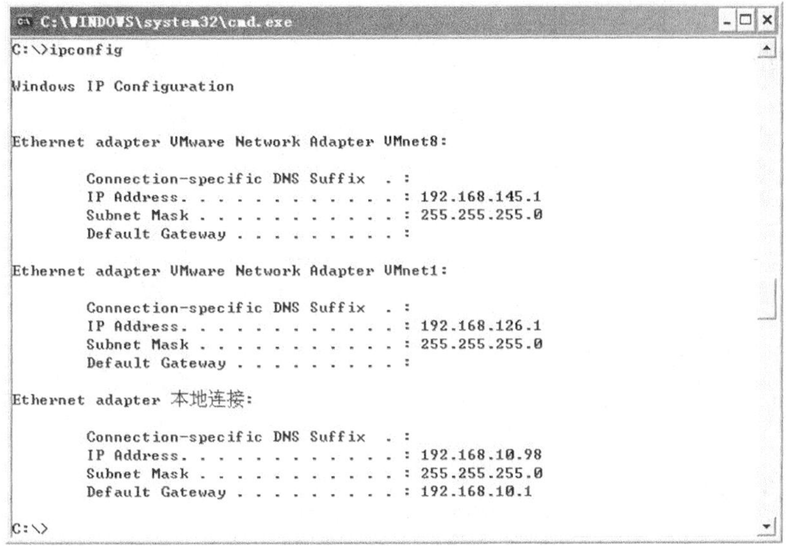

图 5-3

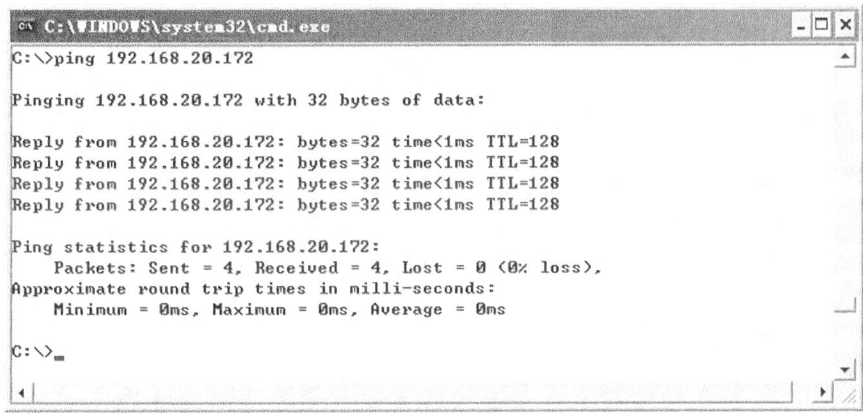

图 5-4

【注意事项】

（1）两台交换机之间相连的端口应该设置为 tag vlan 模式。

（2）为 SVI 端口设置 IP 地址后，一定要使用 no shutdown 命令进行激活，否则无法正常使用。

（3）如果 VLAN 内没有激活的端口，相应 VLAN 的 SVI 端口将无法被激活。

（4）需要设置 PC 的网关为相应 VLAN 的 SVI 接口地址。

【参考配置】

```
L3-SW#show running-config
Building configuration...
Current configuration : 1605 bytes
!
version RGNOS 10.1.00(4), Release(18443)(Tue Jul 17 19:51:54 CST 2007
-ubu6server)
```

```
hostname L3-SW
!
vlan 1
!
vlan 10
 name xiaoshou
!
vlan 20
 name jishu
!
!
interface FastEthernet 0/1
 switchport mode trunk
!
interface FastEthernet 0/2
!
interface FastEthernet 0/3
!
interface FastEthernet 0/4
!
interface FastEthernet 0/5
!
interface FastEthernet 0/6
 switchport access vlan 10
!
interface FastEthernet 0/7
 switchport access vlan 10
!
interface FastEthernet 0/8
 switchport access vlan 10
!
interface FastEthernet 0/9
 switchport access vlan 10
!
interface FastEthernet 0/10
 switchport access vlan 10
!
interface FastEthernet 0/11
 switchport access vlan 20
!
interface FastEthernet 0/12
 switchport access vlan 20
```

```
!
interface FastEthernet 0/13
 switchport access vlan 20
!
interface FastEthernet 0/14
 switchport access vlan 20
!
interface FastEthernet 0/15
 switchport access vlan 20
!
interface FastEthernet 0/16
!
interface FastEthernet 0/17
!
interface FastEthernet 0/18
!
interface FastEthernet 0/19
!
interface FastEthernet 0/20
!
interface FastEthernet 0/21
!
interface FastEthernet 0/22
!
interface FastEthernet 0/23
!
interface FastEthernet 0/24
!
interface GigabitEthernet 0/25
!
interface GigabitEthernet 0/26
!
interface GigabitEthernet 0/27
!
interface GigabitEthernet 0/28
!
interface VLAN 10
 ip address 192.168.10.1 255.255.255.0
!
interface VLAN 20
 ip address 192.168.20.1 255.255.255.0
!
```

```
!
line con 0
line vty 0 4
 login
!
end
```

L2-SW#show running-config

```
System software version : 1.68 Build Apr 25 2007 Release
Building configuration...
Current configuration : 457 bytes
!
version 1.0
!
hostname L2-SW
vlan 1
!
vlan 10
 name xiaoshou
!
vlan 20
 name jishu
!
interface fastEthernet 0/1
 switchport mode trunk
!
interface fastEthernet 0/6
 switchport access vlan 10
!
interface fastEthernet 0/7
 switchport access vlan 10
!
interface fastEthernet 0/8
 switchport access vlan 10
!
interface fastEthernet 0/9
 switchport access vlan 10
!
interface fastEthernet 0/10
 switchport access vlan 10
!
end
```

实验 6　快速生成树配置

【实验名称】

快速生成树配置。

【实验目的】

理解快速生成树协议 RSTP 的工作原理，掌握如何在交换机上配置快速生成树。

【背景描述】

某学校为了开展计算机教学和网络办公，建立了一个计算机教室和一个校办公区，这两处的计算机网络通过两台交换机互连组成内部校园网，为了提高网络的可靠性，网络管理员用 2 条链路将交换机互连，现要在交换机上做适当配置，使网络避免环路。

【需求分析】

两台交换机以双链路互连，需要在启用 RSTP 避免环路的同时，提供链路的冗余备份功能。

【实验拓扑】

实验的拓扑图，如图 6-1 所示。

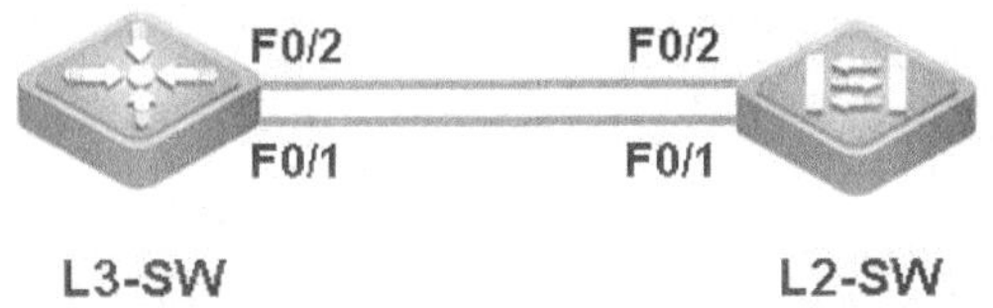

图 6-1

按照拓扑图连接网络时注意，两台交换机都配置完 RSTP 后，再将两台交换机连接起来。如果先连线再配置可能会造成广播风暴，影响交换机的正常工作。

【实验设备】

三层交换机 1 台。

二层交换机 1 台。

【预备知识】

交换机的基本配置方法、Trunk 的工作原理和配置方法、SVI 端口的配置方法、生成树的工作原理以及 RSTP 的配置方法。

【实验原理】

生成树协议（spanning-tree），作用是在交换网络中提供冗余备份链路，并且解决交换网络中的环路问题。

生成树协议是利用 SPA 算法（生成树算法），在存在交换环路的网络中生成一个没有环路的树形网络。运用该算法将交换网络冗余的备份链路逻辑上断开，当主要链路出现故障时，能够自动的切换到备份链路，保证数据的正常转发。

生成树协议目前常见的版本有 STP（生成树协议 IEEE 802.1d）、RSTP（快速生成树协议 IEEE 802.1w）、MSTP（多生成树协议 IEEE 802.1s）。

生成树协议的特点是收敛时间长。当主要链路出现故障以后，到切换到备份链路需要 50 秒的时间。

快速生成树协议（RSTP）在生成树协议的基础上增加了两种端口角色：替换端口（alternate Port）和备份端口（backup Port），分别做为根端口（root Port）和指定端口（designated Port）的冗余端口。当根端口或指定端口出现故障时，冗余端口不需要经过 50 秒的收敛时间，可以直接切换到替换端口或备份端口。从而实现 RSTP 协议小于 1 秒的快速收敛。

【实验步骤】

步骤 1　配置两台交换机的主机名、管理 IP 地址和 Trunk。

```
Switch#configure terminal
Enter configuration commands, one per line.  End with CNTL/Z.
Switch(config)#hostname L2-SW
L2-SW(config)#interface vlan 1
L2-SW(config-if)#ip address 192.168.1.2 255.255.255.0
L2-SW(config-if)#no shutdown
L2-SW(config-if)#exit
L2-SW(config)#
L2-SW(config)#interface fastEthernet 0/1
L2-SW(config-if)#switchport mode trunk
L2-SW(config-if)#exit
L2-SW(config)#
L2-SW(config)#interface fastEthernet 0/2
L2-SW(config-if)#switchport mode trunk
L2-SW(config-if)#exit
L2-SW(config)#
S3750#configure terminal
Enter configuration commands, one per line.  End with CNTL/Z.
S3750(config)#hostname L3-SW
L3-SW(config)#interface vlan 1
L3-SW(config-if)#Dec  2 23:15:26 L3-SW %7:%LINE PROTOCOL CHANGE: Interface VLAN
1, changed state to UP
L3-SW(config-if)#ip address 192.168.1.1 255.255.255.0
L3-SW(config-if)#no shutdown
L3-SW(config-if)#exit
L3-SW(config)#
L3-SW(config)#interface fastEthernet 0/1
L3-SW(config-if)#switchport mode trunk
L3-SW(config-if)#exit
L3-SW(config)#
```

```
L3-SW(config)#interface fastEthernet 0/2
L3-SW(config-if)#switchport mode trunk
L3-SW(config-if)#exit
```

步骤 2　在两台交换机上启用 RSTP。

```
L2-SW(config)#spanning-tree
! 启用生成树协议
L2-SW(config)#spanning-tree mode rstp
! 修改生成树协议的类型为 RSTP
L2-SW(config)#
L3-SW(config)#spanning-tree
Enable spanning-tree.
! 启用生成树协议
L3-SW(config)#spanning-tree mode rstp
! 修改生成树协议的类型为 RSTP
L3-SW(config)#
```

在使用默认参数启用了 RSTP 之后，可以使用 show spanning-tree 命令观察现在两台交换机上生成树的工作状态。

```
L3-SW#show spanning-tree
StpVersion : RSTP
SysStpStatus : ENABLED
MaxAge : 20
HelloTime : 2
ForwardDelay : 15
BridgeMaxAge : 20
BridgeHelloTime : 2
BridgeForwardDelay : 15
MaxHops: 20
TxHoldCount : 3
PathCostMethod : Long
BPDUGuard : Disabled
BPDUFilter : Disabled
BridgeAddr : 00d0.f821.a542
Priority: 32768
TimeSinceTopologyChange : 0d:0h:0m:9s
TopologyChanges : 2
DesignatedRoot : 8000.00d0.f821.a542
RootCost : 0
RootPort : 0
L2-SW#show spanning-tree
StpVersion : RSTP
SysStpStatus : Enabled
BaseNumPorts : 24
```

```
MaxAge : 20
HelloTime : 2
ForwardDelay : 15
BridgeMaxAge : 20
BridgeHelloTime : 2
BridgeForwardDelay : 15
MaxHops : 20
TxHoldCount : 3
PathCostMethod : Long
BPDUGuard : Disabled
BPDUFilter : Disabled
BridgeAddr : 00d0.f88b.ca34
Priority : 32768
TimeSinceTopologyChange : 0d:0h:3m:54s
TopologyChanges : 0
DesignatedRoot : 800000D0F821A542
RootCost : 200000
RootPort : Fa0/1
```

可以看到两台交换机已经正常启用了 RSTP 协议，由于 MAC 地址较小，L3-SW 被选举为根网桥，优先级是 32768；L2-SW 上的根端口是 Fa0/1；两台交换机上计算路径成本的方法都是长整型。

为了在网络中加入其他的交换机后，L3-SW 还是保证能够选举为根网桥，需要提高 L3-SW 的网桥优先级。

步骤 3　指定三层交换机为根网桥，二层交换机的 F0/2 端口为根端口，指定两台交换机的端口路径成本计算方法为短整型。

```
L3-SW(config)#spanning-tree priority ?
  <0-61440>  Bridge priority in increments of 4096
! 查看网桥优先级的可配置范围，在 0~61440 之内，且必须是 4096 的倍数
L3-SW(config)#spanning-tree priority 4096
! 配置网桥优先级为 4096
L3-SW(config)#
L3-SW(config)#interface fastEthernet 0/2
L3-SW(config-if)#spanning-tree port-priority ?
  <0-240>  Port priority in increments of 16
! 查看端口优先级的可配置范围，在 0~240 之内，且必须是 16 的倍数
L3-SW(config-if)#spanning-tree port-priority 96
! 修改 F0/2 端口的优先级为 96
L3-SW(config-if)#exit
L3-SW(config)#spanning-tree pathcost method short
! 修改计算路径成本的方法为短整型
L3-SW(config)#exit
L2-SW(config)#spanning-tree pathcost method short
```

```
！修改计算路径成本的方法为短整型
L2-SW(config)#exit
```

步骤 4　查看生成树的配置。

```
L3-SW#show spanning-tree
StpVersion : RSTP
SysStpStatus : ENABLED
MaxAge : 20
HelloTime : 2
ForwardDelay : 15
BridgeMaxAge : 20
BridgeHelloTime : 2
BridgeForwardDelay : 15
MaxHops: 20
TxHoldCount : 3
PathCostMethod : Short
BPDUGuard : Disabled
BPDUFilter : Disabled
BridgeAddr : 00d0.f821.a542
Priority: 4096
TimeSinceTopologyChange : 0d:0h:0m:34s
TopologyChanges : 7
DesignatedRoot : 1000.00d0.f821.a542
RootCost : 0
RootPort : 0
L3-SW#
L3-SW#show spanning-tree interface fastEthernet 0/1
PortAdminPortFast : Disabled
PortOperPortFast : Disabled
PortAdminLinkType : auto
PortOperLinkType : point-to-point
PortBPDUGuard : disable
PortBPDUFilter : disable
PortState : forwarding
PortPriority : 128
PortDesignatedRoot : 1000.00d0.f821.a542
PortDesignatedCost : 0
PortDesignatedBridge :1000.00d0.f821.a542
PortDesignatedPort : 8001
PortForwardTransitions : 2
PortAdminPathCost : 19
PortOperPathCost : 19
PortRole : designatedPort
```

```
L3-SW#
L3-SW#show spanning-tree interface fastEthernet 0/2
PortAdminPortFast : Disabled
PortOperPortFast : Disabled
PortAdminLinkType : auto
PortOperLinkType : point-to-point
PortBPDUGuard : disable
PortBPDUFilter : disable
PortState : forwarding
PortPriority : 96
PortDesignatedRoot : 1000.00d0.f821.a542
PortDesignatedCost : 0
PortDesignatedBridge :1000.00d0.f821.a542
PortDesignatedPort : 6002
PortForwardTransitions : 4
PortAdminPathCost : 19
PortOperPathCost : 19
PortRole : designatedPort
L3-SW#
```

可以观察到在 L3-SW 中，网桥优先级已经被修改为 4096，Fa0/2 端口的优先级也被修改为 96，在短整型的计算路径成本的方法中，两个端口的路径成本都是 19，现在都处于转发状态。

```
L2-SW#show spanning-tree
StpVersion : RSTP
SysStpStatus : Enabled
BaseNumPorts : 24
MaxAge : 20
HelloTime : 2
ForwardDelay : 15
BridgeMaxAge : 20
BridgeHelloTime : 2
BridgeForwardDelay : 15
MaxHops : 20
TxHoldCount : 3
PathCostMethod : Short
BPDUGuard : Disabled
BPDUFilter : Disabled
BridgeAddr : 00d0.f88b.ca34
Priority : 32768
TimeSinceTopologyChange : 0d:0h:1m:38s
TopologyChanges : 0
DesignatedRoot : 100000D0F821A542
RootCost : 19
```

```
RootPort : Fa0/2
L2-SW#
L2-SW#show spanning-tree interface fastEthernet 0/1
PortAdminPortfast : Disabled
PortOperPortfast : Disabled
PortAdminLinkType : auto
PortOperLinkType : point-to-point
PortBPDUGuard: Disabled
PortBPDUFilter: Disabled
PortState : discarding
PortPriority : 128
PortDesignatedRoot : 100000D0F821A542
PortDesignatedCost : 0
PortDesignatedBridge : 100000D0F821A542
PortDesignatedPort : 8001
PortForwardTransitions : 5
PortAdminPathCost : 0
PortOperPathCost : 19
PortRole : alternatePort
L2-SW#
L2-SW#show spanning-tree interface fastEthernet 0/2
PortAdminPortfast : Disabled
PortOperPortfast : Disabled
PortAdminLinkType : auto
PortOperLinkType : point-to-point
PortBPDUGuard: Disabled
PortBPDUFilter: Disabled
PortState : forwarding
PortPriority : 128
PortDesignatedRoot : 100000D0F821A542
PortDesignatedCost : 0
PortDesignatedBridge : 100000D0F821A542
PortDesignatedPort : 6002
PortForwardTransitions : 3
PortAdminPathCost : 0
PortOperPathCost : 19
PortRole : rootPort
L2-SW#
```

在 L2-SW 中，网桥优先级还是默认的 32768，端口优先级也是默认的 128，路径成本是 19，端口 Fa0/2 被选举为根端口，处于转发状态，而 Fa0/1 则是替换端口，处于丢弃状态。

步骤 5 验证配置。

在三层交换机 L3-SW 上长时间的 ping 二层交换机 L2-SW，其间断开 L2-SW 上的转发端口

Fa0/2，这时观察替换端口能够在多长时间内成为转发端口。

```
L3-SW#ping 192.168.1.2 ntimes 1000
! 使用ping命令的ntimes参数指定ping的次数
Sending 1000, 100-byte ICMP Echoes to 192.168.1.2, timeout is 2 seconds:
  < press Ctrl+C to break >
!!!!!!!!!!!!!!!!!!!!!!!!!!!!!!!!!!!!!!!!!!!!!!!!!!!!!!!!!!!!!!!!!!!!!!!!!!!!!
!!!!!!!!!!!!!!!!!!!!!!!!!!!!!!!!!!!!!!!!!!!!!!!!!!!!!!!!!!!!!!!!!!!!!!!!!!!!!
!!!!!!!!!!!!!!!!!!!!!!!!!!!!!!!!!!!!!!!!!!!!!!!!!!!!!!!!!!!!!!!!!!!!!!!!!!!!!
!!!!!!!!!!!!!!!!!!!!!!!!!!!!!!!!!!!!!!!!!!!!!!!!!!!!!!!!!!!!!!!!!!!!!!!!!!!!!
!!!!!!!!!!!!!!!!!!!!!!!!!!!!!!!!!!!!!!!!!!!!!!!!!!!!!!!!!!!!!!!!!!!!!!!!!!!!!
!!!!!!!!!!!!!!!!!!!!!!!!!!!!!!!!!!!!!!!!!!!!!!!!!!!!!!!!!!!!!!!!!!!!!!!!!!!!!
!!!!!!!!!!!!!!!!!!!!!!!!!!!!!!!!!!!!!!!!!!!!!!!!!!!!!!!!!!!!!!!!!!!!!!!!!!!!!
!!!!!!!!!!!!!!!!!!!!!!!!!!!!!!!Dec  2 23:30:56 L3-SW %7:2008-12-2 23:30:56
topochange:topology is changed.%LINK CHANGED: Interface FastEthernet 0/2,
changed state to down
Dec 2 23:30:57 L3-SW %7:%LINE PROTOCOL CHANGE: Interface FastEthernet 0/2, changed
state to DOWN
Dec  2  23:30:57  L3-SW  %7:2008-12-2  23:30:57  topochange:topology  is
changed.!!!!!!!!!!!!!!!!!!!!!!!!!!!!!!!!!!!!!!!!!!!!!!!!!!!!!!!!!!!!!!!!!!!!!
!!!!!!!!!!!!!!!!!!!!!!!!!!!!!!!!!!!!!!!!!!!!!!!!!!!!!!!!!!!!!!!!!!!!!!!!!!!!!
!!!!!!!!!!!!!!!!!!!!!!!!!!!!!!!!!!!!!!!!!!!!!!!!!!!!!!!!!!!!!!!!!!!!!!!!!!!!!
!!!!!!!!!!!!!!!!!!!!!!!!!!!!!!!!!!!!!!!!!!!!!!!!!!!!!!!!!!!!!!!!!!!!!!!!!!!!!
!!!!!!!!!!!!!!!!!!!!!!!!!!!!!!!!!!!!!!!!!!!!!!!!!!!!!!!!!!!!!!!!!!!!!!!!!!!!!
!!!!!!!!!!!!!!!!!!!!!!!!!!!!!!!!!!!!!!!!!!!!!!!!!!!!!!!!!!!!!!!!
Success rate is 99 percent (998/1000), round-trip min/avg/max = 1/1/10 ms
L3-SW#
```

从中可以看到替换端口变成转发端口的过程中，丢失了 2 个 ping 包，中断时间小于 20ms。

【注意事项】

（1）锐捷交换机 spanning-tree 默认情况下是关闭的，如果网络在物理上存在环路，则必须手工开启 spanning-tree。

（2）锐捷全系列的交换机默认为 MSTP 协议，在配置时注意生成树协议的版本。

【参考配置】

```
L3-SW#show running-config
Building configuration...
Current configuration : 1386 bytes
!
version  RGNOS  10.1.00(4),  Release(18443)(Tue  Jul  17  19:51:54  CST  2007
-ubu6server)
hostname L3-SW
!
vlan 1
```

```
!
spanning-tree
spanning-tree pathcost method short
spanning-tree mode rstp
spanning-tree mst 0 priority 4096
interface FastEthernet 0/1
 switchport mode trunk
!
interface FastEthernet 0/2
 switchport mode trunk
spanning-tree mst 0 port-priority 96
!
interface FastEthernet 0/3
!
interface FastEthernet 0/4
!
interface FastEthernet 0/5
!
interface FastEthernet 0/6
!
interface FastEthernet 0/7
!
interface FastEthernet 0/8
!
interface FastEthernet 0/9
!
interface FastEthernet 0/10
!
interface FastEthernet 0/11
!
interface FastEthernet 0/12
!
interface FastEthernet 0/13
!
interface FastEthernet 0/14
!
interface FastEthernet 0/15
!
interface FastEthernet 0/16
!
interface FastEthernet 0/17
!
```

```
interface FastEthernet 0/18
!
interface FastEthernet 0/19
!
interface FastEthernet 0/20
!
interface FastEthernet 0/21
!
interface FastEthernet 0/22
!
interface FastEthernet 0/23
!
interface FastEthernet 0/24
!
interface GigabitEthernet 0/25
!
interface GigabitEthernet 0/26
!
interface GigabitEthernet 0/27
!
interface GigabitEthernet 0/28
!
interface VLAN 1
 ip address 192.168.1.1 255.255.255.0
!
!
line con 0
line vty 0 4
 login
!
end
```

L2-SW#show running-config

```
System software version : 1.68 Build Apr 25 2007 Release
Building configuration...
Current configuration : 319 bytes
!
version 1.0
!
hostname L2-SW
vlan 1
!
spanning-tree mode rstp
```

```
spanning-tree
spanning-tree pathcost method short
interface fastEthernet 0/1
 switchport mode trunk
!
interface fastEthernet 0/2
 switchport mode trunk
!
interface vlan 1
 no shutdown
 ip address 192.168.1.2 255.255.255.0
!
end
```

实验 7　端口聚合配置

【实验名称】

端口聚合配置。

【实验目的】

理解端口聚合的工作原理，掌握如何在交换机上配置端口聚合。

【背景描述】

假设某企业采用两台交换机组成一个局域网，由于很多数据流量是跨交换机进行转发的，因此需要提高交换机之间的传输带宽，并实现链路冗余备份，为此网络管理员在两台交换机之间要采用两根网线互连，并将相应的两个端口聚合为一个逻辑端口，现要在交换机上做适当配置来实现这一目标。

【需求分析】

需要在两台交换机之间的冗余链路上实现端口聚合，并且在聚合端口上设置 Trunk，以增加网络骨干链路的带宽。

【实验拓扑】

实验的拓扑图，如图 7-1 所示。

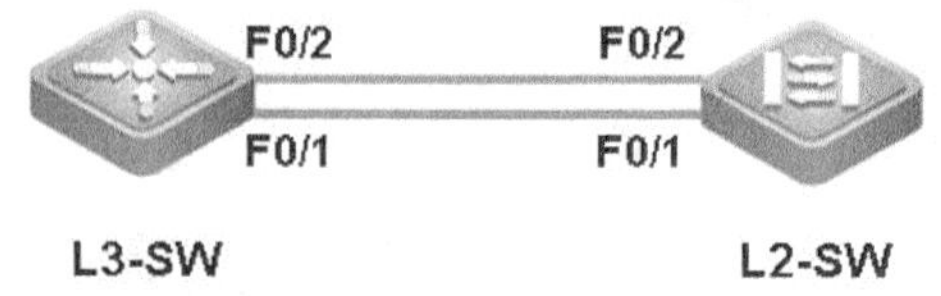

图 7-1

按照拓扑图连接网络时注意，两台交换机都配置完端口聚合后，再将两台交换机连接起来。如果先连线再配置可能会造成广播风暴，影响交换机的正常工作。

【实验设备】

三层交换机 1 台。
二层交换机 1 台。

【预备知识】

交换机的基本配置方法、VLAN 的工作原理和配置方法、Trunk 的工作原理和配置方法、SVI 端口的配置方法、聚合端口的工作原理和配置方法。

【实验原理】

端口聚合（Aggregate-port）又称链路聚合，是指两台交换机之间在物理上将多个端口连接起来，将多条链路聚合成一条逻辑链路。从而增大链路带宽，解决交换网络中因带宽引起的网络瓶颈问题。多条物理链路之间能够相互冗余备份，其中任意一条链路断开，不会影响其他链路的正常转发数据。

端口聚合遵循 IEEE 802.3ad 协议的标准。

【实验步骤】

步骤 1　配置两台交换机的主机名和管理 IP 地址。

```
S3750#configure terminal
Enter configuration commands, one per line.  End with CNTL/Z.
S3750(config)#hostname L3-SW
L3-SW(config)#interface vlan 1
L3-SW(config-if)#Dec  3 01:03:22 L3-SW %7:%LINE PROTOCOL CHANGE: Interface VLAN
1, changed state to UP
L3-SW(config-if)#ip address 192.168.1.1 255.255.255.0
L3-SW(config-if)#no shutdown
L3-SW(config-if)#exit
Switch#configure terminal
Enter configuration commands, one per line.  End with CNTL/Z.
Switch(config)#hostname L2-SW
L2-SW(config)#interface vlan 1
L2-SW(config-if)#ip address 192.168.1.2 255.255.255.0
L2-SW(config-if)#no shutdown
L2-SW(config-if)#exit
```

步骤 2　在两台交换机上配置聚合端口。

```
L3-SW(config)#interface range fastEthernet 0/1-2
L3-SW(config-if-range)#port-group 1
！将端口 Fa0/1~2 加入聚合端口 1，同时创建该聚合端口
L3-SW(config-if-range)#Dec  3 01:03:57 L3-SW %7:%LINE PROTOCOL CHANGE: Interface
AggregatePort 1, changed state to UP
Dec  3 01:03:58 L3-SW %7:%LINK CHANGED: Interface FastEthernet 0/1, changed state
to administratively down
Dec  3 01:03:58 L3-SW %7:%LINE PROTOCOL CHANGE: Interface FastEthernet 0/1,
changed state to DOWN
Dec  3 01:03:58 L3-SW %7:%LINK CHANGED: Interface FastEthernet 0/2, changed state
to administratively down
L3-SW(config-if-range)#exit
L3-SW(config)#
L2-SW(config)#interface range fastEthernet 0/1-2
L2-SW(config-if-range)#port-group 1
！将端口 Fa0/1~2 加入聚合端口 1，同时创建该聚合端口
L2-SW(config-if-range)#exit
L2-SW(config)#
```

步骤 3　将聚合端口设置为 Trunk。

```
L3-SW(config)#interface aggregateport 1
L3-SW(config-if)#switchport mode trunk
L3-SW(config-if)#exit
L3-SW(config)#
L2-SW(config)#interface aggregatePort 1
L2-SW(config-if)#switchport mode trunk
L2-SW(config-if)#exit
L2-SW(config)#
```

步骤 4　设置聚合端口的负载平衡方式。

```
L3-SW(config)#aggregateport load-balance ?
！查看交换机支持的负载平衡方式
  dst-ip       Destination IP address
  dst-mac      Destination MAC address
  ip           Source and destination IP address
  src-dst-mac  Source and destination MAC address
  src-ip       Source IP address
  src-mac      Source MAC address
L3-SW(config)#aggregateport load-balance dst-mac
！设置负载平衡方式为依据目的地址进行，默认是依据源和目的地址
L3-SW(config)#exit
L2-SW(config)#aggregatePort load-balance ?
！查看交换机支持的负载平衡方式
  dst-mac              Destination MAC address
  ip                   Source and destination IP address
  src-mac              Source MAC address
L2-SW(config)#aggregatePort load-balance dst-mac
！设置负载平衡方式为依据目的地址进行，默认是依据源地址
L2-SW(config)#exit
```

步骤 5　查看聚合端口的配置。

```
L3-SW#show aggregatePort load-balance
Load-balance   : Destination MAC
L3-SW#
L3-SW#show aggregatePort summary
AggregatePort MaxPorts SwitchPort Mode   Ports
------------- -------- ---------- ------ ----------------------------------
Ag1           8        Enabled    TRUNK  Fa0/1   ,Fa0/2
L3-SW#
L3-SW#show interfaces aggregateport 1
Index(dec):29 (hex):1d
AggregatePort 1 is UP   , line protocol is UP
Hardware is Aggregate Link AggregatePort
Interface address is: no ip address
```

```
  MTU 1500 bytes, BW 1000000 Kbit
  Encapsulation protocol is Bridge, loopback not set
  Keepalive interval is 10 sec , set
  Carrier delay is 2 sec
  RXload is 1 ,Txload is 1
  Queueing strategy: WFQ
  Switchport attributes:
    interface's description:""
    medium-type is copper
    lastchange time:337 Day: 1 Hour: 3 Minute:56 Second
    Priority is 0
    admin duplex mode is AUTO, oper duplex is Full
    admin speed is AUTO, oper speed is 100Mbps
    flow control admin status is AUTO,flow control oper status is OFF
    broadcast Strom Control is OFF,multicast Strom Control is OFF,unicast Strom
Control is OFF
Aggregate Port Informations:
        Aggregate Number: 1
        Name: "AggregatePort 1"
        Refs: 2
        Members: (count=2)
        FastEthernet 0/1 Link Status: Up
        FastEthernet 0/2 Link Status: Up
L3-SW#
L2-SW#show aggregatePort load-balance
Load-balance   : Destination MAC address
L2-SW#
L2-SW#show aggregatePort summary
AggregatePort MaxPorts SwitchPort Mode   Ports
------------- -------- ---------- ------ ----------------------
Ag1           8        Enabled    Trunk  Fa0/1 , Fa0/2
L2-SW#
L2-SW#show interfaces aggregatePort 1
Interface   : AggregatePort 1
Description :
AdminStatus : up
OperStatus  : up
Hardware    : -
Mtu         : 1500
LastChange  : 0d:0h:0m:0s
AdminDuplex : Auto
OperDuplex  : Full
```

```
AdminSpeed  : Auto
OperSpeed  : 100
FlowControlAdminStatus : Off
FlowControlOperStatus  : Off
Priority   : 0
Broadcast blocked        :DISABLE
Unknown multicast blocked :DISABLE
Unknown unicast blocked   :DISABLE
L2-SW#
```

步骤 6　验证配置。

在三层交换机 L3-SW 上配置另一个用于测试的 VLAN 10，配置 IP 地址为 192.168.10.1/24，然后在二层交换机 L2-SW 上配置默认网关（其作用相当于主机的网关，交换机可将发往其他网段的数据包提交给网关处理），这样 L2-SW 可以 ping 通 192.168.1.1/24 和 192.168.10.1/24，说明聚合端口的 Trunk 配置已经生效。

```
L3-SW(config)#vlan 10
L3-SW(config-vlan)#exit
L3-SW(config)#
L3-SW(config)#interface vlan 10
L3-SW(config-if)#iDec  3 01:16:02 L3-SW %7:%LINE PROTOCOL CHANGE: Interface VLAN
10, changed state to UP
L3-SW(config-if)#ip address 192.168.10.1 255.255.255.0
L3-SW(config-if)#no shutdown
L3-SW(config-if)#exit
L2-SW(config)#ip default-gateway 192.168.1.1
! 设置二层交换机的默认网关
L2-SW(config)#exit
L2-SW#
L2-SW#ping 192.168.1.1
Sending 5, 100-byte ICMP Echos to 192.168.1.1,
timeout is 2000 milliseconds.
!!!!!
Success rate is 100 percent (5/5)
Minimum = 1ms Maximum = 1ms, Average = 1ms
L2-SW#ping 192.168.10.1
Sending 5, 100-byte ICMP Echos to 192.168.10.1,
timeout is 2000 milliseconds.
!!!!!
Success rate is 100 percent (5/5)
Minimum = 1ms Maximum = 1ms, Average = 1ms
```

在三层交换机 L3-SW 上长时间的 ping 二层交换机 L2-SW，然后断开聚合端口中的 Fa0/2 端口。

```
L3-SW#ping 192.168.1.2 ntimes 1000
Sending 1000, 100-byte ICMP Echoes to 192.168.1.2, timeout is 2 seconds:
```

```
  < press Ctrl+C to break >
!!!!!!!!!!!!!!!!!!!!!!!!!!!!!!!!!!!!!!!!!!!!!!!!!!!!!!!!!!!!!!!!!!!!!!!!!!!!!!!!!!!!!!!!
!!!!!!!!!!!!!!!!!!!!!!!!!!!!!!!!!!!!!!!!!!!!!!!!!!!!!!!!!!!!!!!!!!!!!!!!!!!!!!!!!!!!!!!!
!!!!!!!!!!!!!!!!!!!!!!!!!!!!!!!!!!!!!!!!!!!!!!!!!!!!!!!!!!!!!!!!!!!!!!!!!!!!!!!!!!!!!!!!
!!!!!!!!!!!!!!!!!!!!!!!!!!!!!!!!!!!!!!!!!!!!!!!!!!!!!!!!!!!!!!!!!!!!!!!!!!!!!!!!!!!!!!!!
!!!!!!!!!!!!!!!!!!!!!!!!!!!!!!!!!!!!!!!!!!!!!!!!!!!!!!!!!!!!!!!!!!!!!!!!!!!!!!!!!!!!!!!!
!!!!!!!!!!!!!!!!!!!!!!!!!!!!!!!!!!!!!!!!!!!!!!!!!!!!!!!!!!!!!!!!!!!!!!!!!!!!!!!!!!!!!!!!
!!!!!!!!!!!!!!!!!!!!!!!!!!!!!!!!!!!!!!!!!!!!!!!!!!!!!!!!!!!!!!!!!!!!!!!!!!!!!!!!!!!!!!!!
!!!!!!!!!!!!!!!!!!!!!!!!!!!!!!!!!!!!!!!!!!!!!!!!!!!!!!!!!!!!!!!!!!!!!!!!!!!!!!!!!!!!!!!!
!!!!!!!!!!!!!!!!!!!!!!!!!!!!!!!!!!!!!!!!!!!!!!!!!!!!!!!!!!!!!!!!!!!!!!!!!!!!!!!!!!!!!!!!
!!!!!!!!!!!!!!!!!!!!!!!!!!!!!!!!!!!!!!!!!!!!!!!!!!!!!!!!!!!!!!!!!!!!!!!!!!!!!!!!!!!!!!!!
!!!!!!!!!!!!!!!!!!!!!!!!!!!!!!!!!!!!!!!!!!!!!!!!!!!!!!!!!!!!!!!!!!!!!!!!!!!!!!!!!!!!!!!!
!!!!!!!!!!!!!!!!!!!!!!!!!!!!!!!!!!!!!!!!!!!!!!!!!!!!!!!!!!!!!!!!!!!!!!!!!!!!!!!!!!!!!!!!
!!!!!!!!!!!!!!!!!!!!!!!!!!!!!!!!!!!!!!!!!!!!!!!!!!!!!!!!!!!!!!!!!!!!!!!!!!!!!!!!!!!!!!!!
!!!!!!!!!!!!
Success rate is 100 percent (1000/1000), round-trip min/avg/max = 1/1/10 ms
```

可以看到在断开聚合端口中的Fa0/2端口时是没有丢包的。再次实验，此次断开Fa0/1端口。

```
L3-SW#ping 192.168.1.2 ntimes 1000
Sending 1000, 100-byte ICMP Echoes to 192.168.1.2, timeout is 2 seconds:
  < press Ctrl+C to break >
!!!!!!!!!!!!!!!!!!!!!!!!!!!!!!!!!!!!!!!!!!!!!!!!!!!!!!!!!!!!!!!!!!!!!!!!!!!!!!!!!!!!!!!!
!!!!!!!!!!!!!!!!!!!!!!!!!!!!!!!!!!!!!!!!!!!!!!!!!!!!!!!!!!!!!!!!!!!!!!!!!!!!!!!!!!!!!!!!
!!!!!!!!!!!!!!!!!!!!!!!!!!!!!!!!!!!!!!!!!!!!!!!!!!!!!!!!!!!!!!!!!!!!!!!!!!!!!!!!!!!!!!!!
!!!!!!!!!!!!!!!!!!!!!!!!!!!!!!!!!!!!!!!!!!!!!!!!!!!!!!!!!!!!!!!!!!!!!!!!!!!!!!!!!!!!!!!!
!!!!!!!!!!!!!!!!!!!!!!!!!!!!!!!!!!!!!!!!!!!!!!!!!!!!!!!!!!!!!!!!!!!!!!!!!!!!!!!!!!!!!!!!
!!!!!!!!!!!!!!!!!!!!!!!!!!!!!!!!!!!!!!!!!!!!!!!!!!!!!!!!!!!!!!!!!!!!!!!!!!!!!!!!!!!!!!!!
!!!!!!!!!!!!!!!!!!!!!!!!!!!!!!!!!!!!!!!!!!!!!!!!!!!!!!!!!!!!!!!!!!!!!!!!!!!!!!!!!!!!!!!!
!!!!!!!!!!!!!!!!!!!!!!!!!!!!!!!!!!!!!!!!!!!!!!!!!!!!!!!!!!!!!!!!!!!!!!!!!!!!!!!!!!!!!!!!
!!!!!!!!!!!!!!!!!!!!!!!!!!!!!!!!!!!!!!!!!!!!!!!!!!!!!!!!!!!!!!!!!!!!!!!!!!!!!!!!!!!!!!!!.
!!!!!!!!!!!!!!!!!!!!!!!!!!!!!!!!!!!!!!!!!!!!!!!!!!!!!!!!!!!!!!!!!!!!!!!!!!!!!!!!!!!!!!!!!
!!!!!!!!!!!!!!!!!!!!!!!!!!!!!!!!!!!!!!!!!!!!!!!!!!!!!!!!!!!!!!!!!!!!!!!!!!!!!!!!!!!!!!!!
!!!!!!!!!!!!!!!!!!!!!!!!!!!!!!!!!!!!!!!!!!!!!!!!!!!!!!!!!!!!!!!!!!!!!!!!!!!!!!!!!!!!!!!!
!!!!!!!!!!!!!!!!!!!!!!!!!!!!!!!!!!!!!!!!!!!!!!!!!!!!!!!!!!!!!!!!!!!!!!!!!!!!!!!!!!!!!!!!
!!!!!!!!!!
Success rate is 99 percent (999/1000), round-trip min/avg/max = 1/1/10 ms
```

此时发现有一个丢包。这说明在实验中设置的负载均衡方式下，同一对源和目的地址之间的流量只从一个物理端口进行转发，一个端口断开时会将流量切换到另一个端口上，引起了链路短暂的中断。

【注意事项】

（1）只有同类型端口才能聚合为一个AG端口。

（2）所有物理端口必须属于同一个VLAN。

（3）在锐捷交换机上最多支持 8 个物理端口聚合为一个 AG。

（4）在锐捷交换机上最多支持 6 组聚合端口。

【参考配置】

```
L3-SW#show running-config
Building configuration...
Current configuration : 1378 bytes
!
version  RGNOS  10.1.00(4),  Release(18443)(Tue  Jul  17  19:51:54  CST  2007
-ubu6server)
hostname L3-SW
!
vlan 1
!
vlan 10
!
interface FastEthernet 0/1
 port-group 1
!
interface FastEthernet 0/2
 port-group 1
!
interface FastEthernet 0/3
!
interface FastEthernet 0/4
!
interface FastEthernet 0/5
!
interface FastEthernet 0/6
!
interface FastEthernet 0/7
!
interface FastEthernet 0/8
!
interface FastEthernet 0/9
!
interface FastEthernet 0/10
!
interface FastEthernet 0/11
!
interface FastEthernet 0/12
!
```

```
interface FastEthernet 0/13
!
interface FastEthernet 0/14
!
interface FastEthernet 0/15
!
interface FastEthernet 0/16
!
interface FastEthernet 0/17
!
interface FastEthernet 0/18
!
interface FastEthernet 0/19
!
interface FastEthernet 0/20
!
interface FastEthernet 0/21
!
interface FastEthernet 0/22
!
interface FastEthernet 0/23
!
interface FastEthernet 0/24
!
interface GigabitEthernet 0/25
!
interface GigabitEthernet 0/26
!
interface GigabitEthernet 0/27
!
interface GigabitEthernet 0/28
!
interface AggregatePort 1
 switchport mode trunk

!
interface VLAN 1
 ip address 192.168.1.1 255.255.255.0
!
interface VLAN 10
 ip address 192.168.10.1 255.255.255.0
!
```

```
aggregateport load-balance dst-mac
!
!
line con 0
line vty 0 4
 login
!
end
```

L2-SW#show running-config

```
System software version : 1.68 Build Apr 25 2007 Release
Building configuration...
Current configuration : 347 bytes
!
version 1.0
!
hostname L2-SW
vlan 1
!
aggregateport load-balance dst-mac
interface aggregatePort 1
 switchport mode trunk
!
interface fastEthernet 0/1
 port-group 1
!
interface fastEthernet 0/2
 port-group 1
!
interface vlan 1
 no shutdown
 ip address 192.168.1.2 255.255.255.0
!
ip default-gateway 192.168.1.1
end
```

实验 8　路由器的基本操作

【实验名称】

路由器的基本操作。

【实验目的】

理解路由器的工作原理，掌握路由器的基本操作。

【背景描述】

某公司新来一位网管，公司要求其熟悉网络产品，并采用全系列锐捷网络产品。要登录路由器，了解并掌握路由器的命令行操作，进行路由器设备名的配置，配置路由器登录时的描述信息，对路由器的端口配置基本的参数。

【需求分析】

将计算机的 Com 口和路由器的 Console 口通过 Console 线缆连接起来，用 Windows 提供的超级终端工具进行连接，登录路由器的命令行界面进行配置。

【实验拓扑】

实验的拓扑图，如图 8-1 所示。

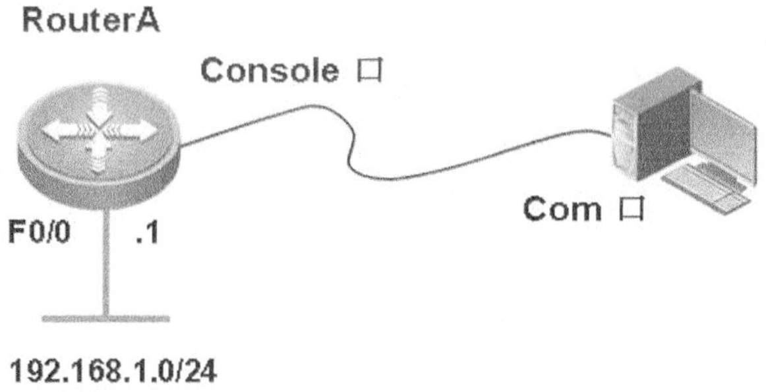

图 8-1

【实验设备】

路由器 1 台。

计算机 1 台。

【预备知识】

路由器的工作原理和基本配置方法。

【实验原理】

路由器的管理方式基本分为带内管理和带外管理两种。通过路由器的 Console 口管理路由器就属于带外管理，不占用路由器的网络接口，但特点是线缆特殊，需要近距离配置。第一次配置路由器时必须利用 Console 进行配置，使其支持 telnet 远程管理。

路由器的命令行操作模式，主要包括：用户模式、特权模式、全局配置模式、端口模式等几种。

- 用户模式：进入路由器后得到的第一个操作模式，该模式下可以简单查看路由器的软、硬件版本信息，并进行简单的测试。用户模式提示符为 Red-Giant>。
- 特权模式：由用户模式进入的下一级模式，该模式下可以对路由器的配置文件进行管理，查看路由器的配置信息，进行网络的测试和调试等。特权模式提示符为 Red-Giant#。
- 全局配置模式：属于特权模式的下一级模式，该模式下可以配置路由器的全局性参数（如主机名、登录信息等）。在该模式中可以进入下一级的配置模式，对路由器具体的功能进行配置。全局模式提示符为 Red-Giant (config)#。
- 端口模式：属于全局配置模式的下一级模式，该模式下可以对路由器的端口进行参数配置。

路由器的基本操作命令包括：

- Exit 命令是退回到上一级操作模式。
- End 命令是直接退回到特权模式。
- 路由器命令行支持获取帮助信息、命令的简写、命令的自动补齐、快捷键功能。配置路由器的设备名称和路由器的描述信息必须在全局配置模式下执行。
- Hostname 配置路由器的设备名称即命令提示符的前部分信息。
- 当用户登录路由器时，需要向用户提示一些必要的信息。网管可以通过设置标题来达到这个目的。可以创建每日通知和登录标题两种类型的标题。
  - Banner motd：配置路由器每日提示信息。
  - Banner login：配置路由器远程登录提示信息，位于每日提示信息之后。
- 锐捷路由器接口 Fastethernet 接口默认情况下是 10Mbps/100Mbps 自适应端口，双工模式也为自　适应。
- 在路由器的物理端口可以灵活配置带宽，但最大值为该端口的实际物理带宽。
- 查看路由器的系统和配置信息命令要在特权模式下执行。
  - Show version：查看路由器的版本信息，可以查看到路由器的硬件版本信息和软件版本信息，用于进行路由器操作系统升级时的依据。
  - Show ip route：查看路由表信息。
  - Show running-config：查看路由器当前生效的配置信息。

【实验步骤】

步骤 1　路由器命令行的基本功能。

```
RSR20>?
！使用？显示当前模式下所有可执行的命令
Exec commands:
  <1-99>              Session number to resume
  disable             Turn off privileged commands
  disconnect          Disconnect an existing network connection
  enable              Turn on privileged commands
  exit                Exit from the EXEC
  help                Description of the interactive help system
  lock                Lock the terminal
  ping                Send echo messages
```

```
  ping6                   ping6
  show                    Show running system information
  start-terminal-service  Start terminal service
  telnet                  Open a telnet connection
  traceroute              Trace route to destination
RSR20>e?
enable  exit
! 显示当前模式下所有以 e 开头的命令
RSR20>en <tab>
! 按键盘的 Tab 键自动补齐命令，路由器支持命令的自动补齐
RSR20>enable
! 使用 enable 命令从用户模式进入特权模式
RSR20#copy ?
! 显示 copy 命令后可执行的参数
flash:          Copy from flash: file system
  running-config  Copy from current system configuration
  startup-config  Copy from startup configuration
  tftp:           Copy from tftp: file system
  xmodem:         Copy from xmodem: file system
RSR20#copy
% Incomplete command.
! 提示命令未完，必须附带可执行的参数
RSR20#conf t
! 路由器支持命令的简写，该命令代表 configure terminal
! 进入路由器的全局配置模式
Enter configuration commands, one per line.  End with CNTL/Z.
RSR20(config)#interface fastEthernet 0/0
! 进入路由器端口 Fa0/0 的接口配置模式
RSR20(config-if)#
RSR20(config-if)#exit
! 使用 exit 命令返回上一级的操作模式
RSR20(config)#interface fastEthernet 0/0
RSR20(config-if)#end
! 使用 end 命令直接返回特权模式
RSR20#
RSR20(config)#interface fastEthernet 0/0
RSR20(config-if)#^Z
! 使用快捷键 ctrl+Z 直接退回到特权模式
RSR20#
RSR20#ping 1.1.1.1
Sending 5, 100-byte ICMP Echoes to 1.1.1.1, timeout is 2 seconds:
  < press Ctrl+C to break >
```

```
..^C
Success rate is 0 percent (0/3)
！在路由器特权模式下执行 ping 1.1.1.1 命令，发现不能 ping 通目标地址，路由器默认情况下需要发送 5 个数据包，若不想等到 5 个数据包均不能 ping 通目标地址时才认为目的地址不可到达，可在数据包未发出 5 个之前按快捷键 Ctrl+C 终止当前操作。
```

步骤 2　配置路由器的名称和每日提示信息。

```
RSR20>enable
RSR20#configure terminal
Enter configuration commands, one per line.  End with CNTL/Z.
RSR20(config)#hostname RouterA
！将路由器的名称设置为 RouterA
RouterA(config)#
RouterA(config)#banner motd &
！设置路由器的每日提示信息，motd 后面的参数为设置的终止符
Enter TEXT message.  End with the character '&'.
Welcome to RouterA, if you are admin, you can config it.
If you are not admin, please EXIT.
&
RouterA(config)#
验证测试：
RouterA#exit
RouterA CON0 is now available
Press RETURN to get started
Welcome to RouterA, if you are admin, you can config it.
If you are not admin, please EXIT.
RouterA>
```

步骤 3　配置路由器的接口并查看接口配置。

```
RouterA#configure terminal
Enter configuration commands, one per line.  End with CNTL/Z.
RouterA(config)#interface fastEthernet 0/0
！进入端口 Fa0/0 的接口配置模式
RouterA(config-if)#ip address 192.168.1.1 255.255.255.0
！配置接口的 IP 地址
RouterA(config-if)#no shutdown
！开启该端口
RouterA(config-if)#end
RouterA#show interfaces  fastEthernet 0/0
！查看端口 Fa0/0 的状态是否为 UP，地址配置和流量统计等信息
Index(dec):1 (hex):1
FastEthernet 0/0 is UP  , line protocol is UP
Hardware is MPC8248 FCC FAST ETHERNET CONTROLLER FastEthernet, address is 00d0.f86b.3832 (bia 00d0.f86b.3832)
```

```
Interface address is: 192.168.1.1/24
ARP type: ARPA,ARP Timeout: 3600 seconds
  MTU 1500 bytes, BW 100000 Kbit
  Encapsulation protocol is Ethernet-II, loopback not set
  Keepalive interval is 10 sec , set
  Carrier delay is 2 sec
  RXload is 1 ,Txload is 1
  Queueing strategy: FIFO
    Output queue 0/40, 0 drops;
    Input queue 0/75, 0 drops
  Link Mode: 100Mbps/Full-Duplex
  5 minutes input rate 1 bits/sec, 0 packets/sec
  5 minutes output rate 1 bits/sec, 0 packets/sec
    1 packets input, 60 bytes, 0 no buffer, 0 dropped
    Received 1 broadcasts, 0 runts, 0 giants
    0 input errors, 0 CRC, 0 frame, 0 overrun, 0 abort
    1 packets output, 42 bytes, 0 underruns , 0 dropped
    0 output errors, 0 collisions, 2 interface resets
```

步骤 4　查看路由器的配置。

```
RouterA#show version
! 查看路由器的版本信息
System description : Ruijie Router(RSR20-04) by Ruijie Network
System start time : 2009-8-16 5:37:38
System hardware version : 1.01
! 硬件版本号
System software version : RGNOS 10.1.00(4), Release(18443)
! 软件版本号
System boot version : 10.2.24515
System serial number : 1234942570135
RouterA#show ip route
! 查看路由表信息
Codes: C - connected, S - static,  R - RIP B - BGP
       O - OSPF, IA - OSPF inter area
       N1 - OSPF NSSA external type 1, N2 - OSPF NSSA external type 2
       E1 - OSPF external type 1, E2 - OSPF external type 2
       i - IS-IS, L1 - IS-IS level-1, L2 - IS-IS level-2, ia - IS-IS inter area
       * - candidate default
Gateway of last resort is no set
C   192.168.1.0/24 is directly connected, FastEthernet 0/0
C   192.168.1.1/32 is local host.
RouterA#show running-config
! 查看路由器当前生效的配置信息
```

```
Building configuration...
Current configuration : 540 bytes
!
version RGNOS 10.1.00(4), Release(18443)(Tue Jul 17 20:50:30 CST 2007
-ubu1server)
hostname RouterA
!
!
interface FastEthernet 0/0
 ip address 192.168.1.1 255.255.255.0
 duplex auto
 speed auto
!
interface FastEthernet 0/1
 duplex auto
 speed auto
!
line con 0
line aux 0
line vty 0 4
 login
!
banner motd ^C
Welcome to RouterA, if you are admin, you can config it.
If you are not admin, please EXIT.
^C
!
end
```

【注意事项】

（1）命令行操作进行自动补齐或命令简写时，要求所简写的字母必须能够区别该命令。如 Red-Giant# conf 可以代表 configure，但 Red-Giant#co 无法代表 configure，因为 co 开头的命令有两个 copy 和 configure，设备无法区别。

（2）注意区别每个操作模式下可执行的命令种类。路由器不可以跨模式执行命令。

（3）配置设备名称的有效字符是 22 个字节。

（4）配置每日提示信息时，注意终止符不能在描述文本中出现。如果输入结束的终止符后再输入字符，则这些字符将被系统丢弃。

（5）Serial 接口正常的端口速率最大是 2.048Mbps。

（6）注意 Show interface 和 show ip interface 之间的区别。

（7）Show running-config 是查看当前生效的配置信息。Show startup-config 是查看保存在 NVRAM 里的配置文件信息。

（8）路由器的配置信息全部加载在 RAM 里生效。路由器在启动过程中是将 NVRAM 里的配置文件加载到 RAM 里生效的。

实验 9　在路由器上配置 Telnet

【实验名称】

在路由器上配置 Telnet。

【实验目的】

掌握如何在路由器上配置 Telnet，以实现路由器的远程登录访问。

【背景描述】

路由器用于连接多个子网时，通常放置的位置都距离较远，查看和修改配置会很麻烦，此时如果可以远程登录到路由器上进行操作，将能够大大降低管理员的工作量。

【需求分析】

需要掌握如何配置路由器的密码，配置 Telnet 服务，以及如何通过 Telnet 远程登录路由器进行操作的方法。

【实验拓扑】

实验的拓扑图，如图 9-1 所示。

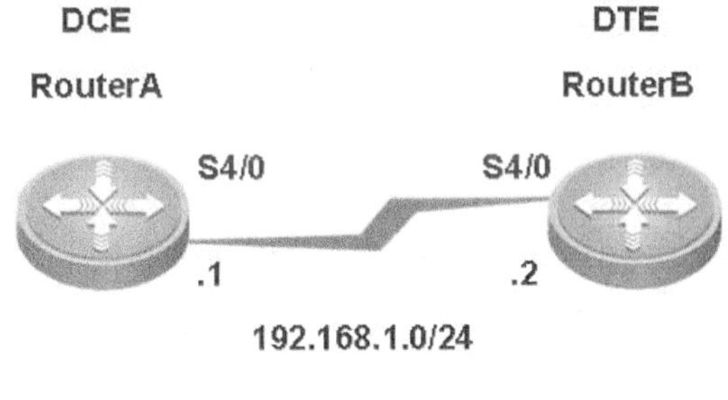

图 9-1

【实验设备】

路由器（带串口）2 台。

V.35 DCE/DTE 线缆 1 对。

【预备知识】

路由器的工作原理和基本配置方法。

【实验原理】

将两台路由器通过串口，用 V.35 DTE/DCE 电缆连接在一起，分别配置 Telnet，可以互相以 Telnet 方式登录对方。

路由器提供广域网接口（serial 高速同步串口），使用 V.35 线缆连接广域网接口链路。在广域网连接时一端为 DCE（数据通信设备）、一端为 DTE（数据终端设备）。要求必须在 DCE 端配置时钟频率（clock rate）才能保证链路的连通。

【实验步骤】

步骤 1　配置路由器的名称、接口 IP 地址和时钟。

```
R3740#configure terminal
Enter configuration commands, one per line.  End with CNTL/Z.
R3740(config)#hostname RouterA
! 配置路由器的名称
RouterA(config)#interface serial 4/0
! 进入串口的接口配置模式
RouterA(config-if)#clock rate 512000
! 设置 DCE 端的时钟频率
RouterA(config-if)#ip address 192.168.1.1 255.255.255.0
! 配置接口 IP 地址
RouterA(config-if)#no shutdown
! 启用端口
RouterA(config-if)#exit
R3740#configure terminal
Enter configuration commands, one per line.  End with CNTL/Z.
R3740(config)#hostname RouterB
RouterB(config)#interface serial 4/0
RouterB(config-if)#ip address 192.168.1.2 255.255.255.0
RouterB(config-if)#no shutdown
RouterB(config-if)#exit
```

步骤 2　配置 Telnet。

```
RouterA(config)#enable password ruijie
! 配置路由器的特权模式密码
RouterA(config)#line vty 0 4
! 进入线程配置模式
RouterA(config-line)#password star
! 配置 Telnet 密码
RouterA(config-line)#login
! 设置 Telnet 登录时进行身份验证
RouterA(config-line)#end
RouterB(config)#enable password ruijie
RouterB(config)#line vty 0 4
RouterB(config-line)#password star
RouterB(config-line)#login
RouterB(config-line)#end
```

步骤 3　测试网络连通性，以 Telnet 方式登录路由器。

```
RouterB#ping 192.168.1.1
Sending 5, 100-byte ICMP Echoes to 192.168.1.1, timeout is 2 seconds:
  < press Ctrl+C to break >
!!!!!
```

```
Success rate is 100 percent (5/5), round-trip min/avg/max = 1/2/10 ms
RouterB#telnet 192.168.1.1
Trying 192.168.1.1, 23...
User Access Verification
Password:
! 提示输入 Telnet 密码，此处输入 ruijie
RouterA>en
Password:
! 提示输入特权模式密码，此处输入 star
RouterA#en
! 远程登录路由器 A，可进行配置
RouterA#
RouterA#conf t
Enter configuration commands, one per line.  End with CNTL/Z.
RouterA(config)#exit
RouterA#
RouterA#
RouterA#exit
! 使用 exit 命令退出 Telnet 登录
RouterB#
RouterA#ping 192.168.1.2
Sending 5, 100-byte ICMP Echoes to 192.168.1.2, timeout is 2 seconds:
  < press Ctrl+C to break >
!!!!!
Success rate is 100 percent (5/5), round-trip min/avg/max = 1/4/10 ms
RouterA#telnet 192.168.1.2
Trying 192.168.1.2, 23...
User Access Verification
Password:
RouterB>en
Password:
RouterB#
RouterB#conf t
Enter configuration commands, one per line.  End with CNTL/Z.
RouterB(config)#exit
RouterB#
RouterB#exit
RouterA#
```

【注意事项】

（1）如果两台路由器通过串口直接互连，则必须在其中一端设置时钟频率（DCE）。

（2）如果没有配置 Telnet 密码，则登录时会提示“Password required, but none set”。

（3）如果没有配置 enable 密码，则远程登录到路由器上后不能进入特权模式，并提示“Password required, but none set”。

【参考配置】

```
RouterA#show running-config
Building configuration...
Current configuration : 582 bytes
!
version  RGNOS  10.1.00(4),  Release(18443)(Tue  Jul  17  21:16:17  CST  2007
-ubu1server)
hostname RouterA
!
!
enable password 7 035122110c3706
!
interface serial 4/0
 ip address 192.168.1.1 255.255.255.0
 clock rate 512000
!
interface serial 4/1
 clock rate 64000
!
interface GigabitEthernet 0/0
 duplex auto
 speed auto
!
interface GigabitEthernet 0/1
 duplex auto
 speed auto
!
!
line con 0
line aux 0
line vty 0 4
 login
 password 7 0133574225
!
end
RouterB#show running-config
Building configuration...
Current configuration : 562 bytes
!
```

```
version  RGNOS  10.1.00(4),  Release(18443)(Tue  Jul  17  21:16:17  CST  2007
-ubu1server)
hostname RouterB
!
!
enable password 7 0251563e120f3b
!
interface serial 4/0
 ip address 192.168.1.2 255.255.255.0
!
interface serial 4/1
 clock rate 64000
!
interface GigabitEthernet 0/0
 duplex auto
 speed auto
!
interface GigabitEthernet 0/1
 duplex auto
 speed auto
!
line con 0
line aux 0
line vty 0 4
 login
 password 7 0059344251
!
end
```

实验 10　静态路由配置

【实验名称】

静态路由配置。

【实验目的】

理解静态路由的工作原理，掌握如何配置静态路由。

【背景描述】

假设校园网分为 2 个区域，每个区域内使用 1 台路由器连接 2 个子网，现要在路由器上做适当配置，实现校园网内各个区域子网之间的相互通信。

【需求分析】

两台路由器通过串口以 V.35 DCE/DTE 电缆连接在一起，并在每个路由器上设置 2 个 Loopback 端口模拟子网及静态路由，实现所有子网间的互通。

【实验拓扑】

实验的拓扑图，如图 10-1 所示。

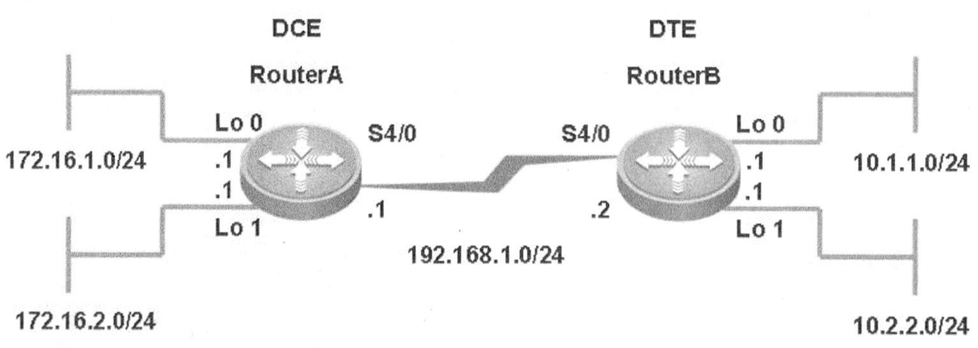

图 10-1

【实验设备】

路由器（带串口）2 台。

V.35 DCE/DTE 电缆 1 对。

【预备知识】

路由器的工作原理和基本配置方法，静态路由的工作原理和配置方法。

【实验原理】

路由器属于网络层设备，能够根据 IP 包头的信息，选择一条最佳路径，将数据包转发出去。实现不同网段的主机之间的互相访问。

路由器是根据路由表进行选路和转发的。而路由表里是由一条条的路由信息组成。路由表的产生方式一般有 3 种。

❑ 直连路由　给路由器接口配置一个 IP 地址，路由器自动产生本接口 IP 所在网段的路由信息。

❑ 静态路由　在拓扑结构简单的网络中，网管可以通过手工的方式配置本路由器未知网段的路由信息，从而实现不同网段之间的连接。

❑ 动态路由协议学习产生的路由　在大规模的网络中，或网络拓扑相对复杂的情况下，通过在路由器上运行动态路由协议，路由器之间互相自动学习产生路由信息。

【实验步骤】

步骤 1　配置路由器的名称、接口 IP 地址和时钟。

```
R3740#configure terminal
Enter configuration commands, one per line.  End with CNTL/Z.
R3740(config)#hostname RouterA
! 配置路由器的名称
RouterA(config)#
RouterA(config)#interface serial 4/0
! 进入端口 S4/0 的接口配置模式
RouterA(config-if)#clock rate 512000
! 设置串口的时钟
RouterA(config-if)#ip address 192.168.1.1 255.255.255.0
! 设置端口的 IP 地址
RouterA(config-if)#no shutdown
! 开启端口
RouterA(config-if)#exit
RouterA(config)#
RouterA(config)#interface loopback 0
! 设置 Loopback 端口用于测试
RouterA(config-if)#Sep 15 01:05:02 RouterA %7:%LINE PROTOCOL CHANGE: Interface
Loopback 0, changed state to UP
RouterA(config-if)#ip address 172.16.1.1 255.255.255.0
RouterA(config-if)#exit
RouterA(config)#
RouterA(config)#interface loopback 1
RouterA(config-if)#Sep 15 01:05:31 RouterA %7:%LINE PROTOCOL CHANGE: Interface
Loopback 1, changed state to UP
RouterA(config-if)#ip address 172.16.2.1 255.255.255.0
RouterA(config-if)#exit
R3740#configure terminal
Enter configuration commands, one per line.  End with CNTL/Z.
R3740 (config)#hostname RouterB
RouterB(config)#
RouterB(config)#interface serial 4/0
RouterB(config-if)#ip address 192.168.1.2 255.255.255.0
```

```
RouterB(config-if)#no shutdown
RouterB(config-if)#exit
RouterB(config)#
RouterB(config)#interface loopback 0
RouterB(config-if)#Aug 22 03:03:36 RouterB %7:%LINE PROTOCOL CHANGE: Interface
Loopback 0, changed state to UP
RouterB(config-if)#ip address 10.1.1.1 255.255.255.0
RouterB(config-if)#exit
RouterB(config)#
RouterB(config)#interface loopback 1
RouterB(config-if)#Aug 22 03:04:03 RouterB %7:%LINE PROTOCOL CHANGE: Interface
Loopback 1, changed state to UP
RouterB(config-if)#ip address 10.2.2.1 255.255.255.0
RouterB(config-if)#exit
```

步骤 2　配置静态路由。

```
RouterA(config)#ip route 10.1.1.0 255.255.255.0 192.168.1.2
! 设置到子网 10.1.1.0 的静态路由，采用下一跳的方式
RouterA(config)#ip route 10.2.2.0 255.255.255.0 s4/0
! 设置到子网 10.2.2.0 的静态路由，采用出站端口的方式
RouterB(config)#ip route 172.16.1.0 255.255.255.0 192.168.1.1
RouterB(config)#ip route 172.16.2.0 255.255.255.0 s4/0
```

步骤 3　查看路由表和接口配置。

```
RouterA#show ip route
Codes:  C - connected, S - static,  R - RIP B - BGP
       O - OSPF, IA - OSPF inter area
       N1 - OSPF NSSA external type 1, N2 - OSPF NSSA external type 2
       E1 - OSPF external type 1, E2 - OSPF external type 2
       i - IS-IS, L1 - IS-IS level-1, L2 - IS-IS level-2, ia - IS-IS inter area
       * - candidate default
Gateway of last resort is no set
S      10.1.1.0/24 [1/0] via 192.168.1.2
S      10.2.2.0/24 is directly connected, serial 4/0
C     172.16.1.0/24 is directly connected, Loopback 0
C     172.16.1.1/32 is local host.
C     172.16.2.0/24 is directly connected, Loopback 1
C     172.16.2.1/32 is local host.
C     192.168.1.0/24 is directly connected, serial 4/0
C     192.168.1.1/32 is local host.
! 可以看到以下一跳方式配置的静态路由和以出站端口方式配置的静态路由，在路由表中的显示方式是不
一样的
RouterA#show interfaces serial 4/0
Index(dec):1 (hex):1
```

```
serial 4/0 is UP    , line protocol is UP
Hardware is Infineon DSCC4 PEB20534 H-10 serial
Interface address is: 192.168.1.1/24
  MTU 1500 bytes, BW 2000 Kbit
  Encapsulation protocol is HDLC, loopback not set
  Keepalive interval is 10 sec , set
  Carrier delay is 2 sec
  RXload is 1 ,Txload is 1
  Queueing strategy: WFQ
    11421118 carrier transitions
    V35 DCE cable
    DCD=up  DSR=up  DTR=up  RTS=up  CTS=up
  5 minutes input rate 19 bits/sec, 0 packets/sec
  5 minutes output rate 19 bits/sec, 0 packets/sec
    95 packets input, 4134 bytes, 0 no buffer, 1 dropped
    Received 69 broadcasts, 0 runts, 0 giants
    0 input errors, 0 CRC, 0 frame, 0 overrun, 0 abort
    94 packets output, 4118 bytes, 0 underruns , 0 dropped
    0 output errors, 0 collisions, 0 interface resets
RouterB#show ip route
Codes:  C - connected, S - static,  R - RIP B - BGP
       O - OSPF, IA - OSPF inter area
       N1 - OSPF NSSA external type 1, N2 - OSPF NSSA external type 2
       E1 - OSPF external type 1, E2 - OSPF external type 2
       i - IS-IS, L1 - IS-IS level-1, L2 - IS-IS level-2, ia - IS-IS inter area
       * - candidate default
Gateway of last resort is no set
C   10.1.1.0/24 is directly connected, Loopback 0
C   10.1.1.1/32 is local host.
C   10.2.2.0/24 is directly connected, Loopback 1
C   10.2.2.1/32 is local host.
S    172.16.1.0/24 [1/0] via 192.168.1.1
S    172.16.2.0/24 is directly connected, serial 4/0
C   192.168.1.0/24 is directly connected, serial 4/0
C   192.168.1.2/32 is local host.
RouterB#show interfaces serial 4/0
Index(dec):1 (hex):1
serial 4/0 is UP    , line protocol is UP
Hardware is Infineon DSCC4 PEB20534 H-10 serial
Interface address is: 192.168.1.2/24
  MTU 1500 bytes, BW 2000 Kbit
  Encapsulation protocol is HDLC, loopback not set
```

```
  Keepalive interval is 10 sec , set
  Carrier delay is 2 sec
  RXload is 1 ,Txload is 1
  Queueing strategy: WFQ
    11421118 carrier transitions
    V35 DTE cable
    DCD=up  DSR=up  DTR=up  RTS=up  CTS=up
  5 minutes input rate 74 bits/sec, 0 packets/sec
  5 minutes output rate 74 bits/sec, 0 packets/sec
    86 packets input, 3942 bytes, 0 no buffer, 0 dropped
    Received 61 broadcasts, 0 runts, 0 giants
    0 input errors, 0 CRC, 0 frame, 0 overrun, 0 abort
    87 packets output, 3964 bytes, 0 underruns , 0 dropped
    0 output errors, 0 collisions, 1 interface resets
```

步骤 4　测试网络连通性。

```
RouterA#ping 10.1.1.1
Sending 5, 100-byte ICMP Echoes to 10.1.1.1, timeout is 2 seconds:
  < press Ctrl+C to break >
!!!!!
Success rate is 100 percent (5/5), round-trip min/avg/max = 1/4/10 ms
RouterA#ping 10.2.2.1
Sending 5, 100-byte ICMP Echoes to 10.2.2.1, timeout is 2 seconds:
  < press Ctrl+C to break >
!!!!!
Success rate is 100 percent (5/5), round-trip min/avg/max = 1/4/10 ms
RouterB#ping 172.16.1.1
Sending 5, 100-byte ICMP Echoes to 172.16.1.1, timeout is 2 seconds:
  < press Ctrl+C to break >
!!!!!
Success rate is 100 percent (5/5), round-trip min/avg/max = 1/4/10 ms
RouterB#ping 172.16.2.1
Sending 5, 100-byte ICMP Echoes to 172.16.2.1, timeout is 2 seconds:
  < press Ctrl+C to break >
!!!!!
Success rate is 100 percent (5/5), round-trip min/avg/max = 1/4/10 ms
```

【注意事项】

（1）如果两台路由器通过串口直接互连，则必须在其中一端设置时钟频率（DCE）。

（2）静态路由必须双向都配置才能互通，在配置时要注意回程路由。

【参考配置】

```
RouterA#show running-config
Building configuration...
```

```
Current configuration : 745 bytes
!
version  RGNOS  10.1.00(4),  Release(18443)(Tue  Jul  17  21:16:17  CST  2007
-ubu1server)
hostname RouterA
!
!
interface serial 4/0
 ip address 192.168.1.1 255.255.255.0
 clock rate 512000
!
interface serial 4/1
 clock rate 64000
!
interface GigabitEthernet 0/0
 duplex auto
 speed auto
!
interface GigabitEthernet 0/1
 duplex auto
 speed auto
!
interface Loopback 0
 ip address 172.16.1.1 255.255.255.0
!
interface Loopback 1
 ip address 172.16.2.1 255.255.255.0
!
ip route  10.1.1.0 255.255.255.0  192.168.1.2
ip route  10.2.2.0 255.255.255.0  serial 4/0
!
line con 0
line aux 0
line vty 0 4
 login
!
end
```

RouterB#show running-config

```
Building configuration...
Current configuration : 725 bytes
!
version  RGNOS  10.1.00(4),  Release(18443)(Tue  Jul  17  21:16:17  CST  2007
```

```
-ubu1server)
hostname RouterB
!
!
interface serial 4/0
 ip address 192.168.1.2 255.255.255.0
!
interface serial 4/1
 clock rate 64000
!
interface GigabitEthernet 0/0
 duplex auto
 speed auto
!
interface GigabitEthernet 0/1
 duplex auto
 speed auto
!
interface Loopback 0
 ip address 10.1.1.1 255.255.255.0
!
interface Loopback 1
 ip address 10.2.2.1 255.255.255.0
!
ip route  172.16.1.0 255.255.255.0  192.168.1.1
ip route  172.16.2.0 255.255.255.0  serial 4/0
!
line con 0
line aux 0
line vty 0 4
 login
!
end
```

实验 11　RIP 路由协议基本配置

【实验名称】

RIP 路由协议基本配置。

【实验目的】

掌握在路由器上如何配置 RIP 路由协议。

【背景描述】

假设某校园网从地理位置上分为 2 个区域，每个区域内分别有一台路由器连接了 2 个子网，需要将两台路由器通过以太网链路连接在一起并进行适当的配置，以实现这 4 个子网之间的互连互通。为了便于管理员在未来每个校园区域扩充子网数量时，不需要同时更改路由器的配置，计划使用 RIP 路由协议实现子网之间的互通。

【需求分析】

两台路由器通过快速以太网端口连接在一起，每个路由器上设置 2 个 Loopback 端口模拟子网，在所有端口运行 RIP 路由协议，实现所有子网间的互通。

【实验拓扑】

实验的拓扑图，如图 11-1 所示。

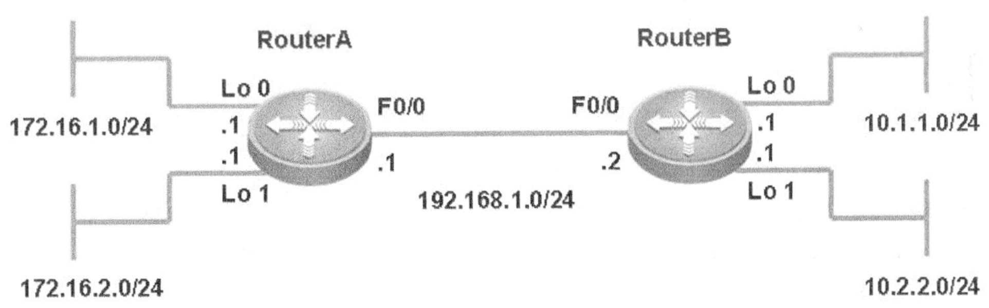

图 11-1

【实验设备】

路由器 2 台。

【预备知识】

路由器的工作原理和基本配置方法，距离矢量路由协议，RIP 工作原理和配置方法。

【实验原理】

RIP（Routing Information Protocols，路由信息协议）是应用较早、使用较普遍的 IGP（Interior Gateway Protocol，内部网关协议），适用于小型同类网络，是典型的距离矢量（distance-vector）协议。

RIP 协议以跳数做为衡量路径开销，RIP 协议里规定最大跳数为 15。

RIP 在构造路由表时会使用到 3 种计时器，分别是更新计时器、无效计时器、刷新计时器。它让每台路由器周期性地向每个相邻的邻居发送完整的路由表。路由表包括每个网络或子网的信息，以及与之相关的度量值。

【实验步骤】

步骤 1　配置两台路由器的主机名、接口 IP 地址。

```
RSR20#configure terminal
Enter configuration commands, one per line.  End with CNTL/Z.
RSR20(config)#hostname RouterA
RouterA(config)#
RouterA(config)#interface fastEthernet 0/0
RouterA(config-if)#ip address 192.168.1.1 255.255.255.0
RouterA(config-if)#no shutdown
RouterA(config-if)#exit
RouterA(config)#
RouterA(config)#interface loopback 0
RouterA(config-if)#Aug 15 23:46:32 RouterA %7:%LINE PROTOCOL CHANGE: Interface
Loopback 0, changed state to UP
RouterA(config-if)#ip address 172.16.1.1 255.255.255.0
RouterA(config-if)#exit
RouterA(config)#
RouterA(config)#interface loopback 1
RouterA(config-if)#Aug 15 23:47:00 RouterA %7:%LINE PROTOCOL CHANGE: Interface
Loopback 1, changed state to UP
RouterA(config-if)#ip address 172.16.2.1 255.255.255.0
RouterA(config-if)#exit
RSR20#configure terminal
Enter configuration commands, one per line.  End with CNTL/Z.
RSR20(config)#hostname RouterB
RouterB(config)#
RouterB(config)#interface fastEthernet 0/0
RouterB(config-if)#ip address 192.168.1.2 255.255.255.0
RouterB(config-if)#no shutdown
RouterB(config-if)#exit
RouterB(config)#
RouterB(config)#interface loopback 0
RouterB(config-if)#Aug  8 21:00:00 RouterB %7:%LINE PROTOCOL CHANGE: Interface
Loopback 0, changed state to UP
RouterB(config-if)#ip address 10.1.1.1 255.255.255.0
RouterB(config-if)#exit
RouterB(config)#
RouterB(config)#interface loopback 1
```

```
RouterB(config-if)#Aug  8 21:00:28 RouterB %7:%LINE PROTOCOL CHANGE: Interface
Loopback 1, changed state to UP
RouterB(config-if)#ip address 10.2.2.1 255.255.255.0
RouterB(config-if)#exit
```

步骤 2　在两台路由器上配置 RIP 路由协议。

```
RouterA(config)#router rip
RouterA(config-router)#network 192.168.1.0
RouterA(config-router)#network 172.16.1.0
RouterA(config-router)#exit
RouterB(config)#router rip
RouterB(config-router)#network 192.168.1.0
RouterB(config-router)#network 10.0.0.0
RouterB(config-router)#exit
```

步骤 3　查看 RIP 配置信息，路由表。

```
RouterA#show ip route
Codes:  C - connected, S - static,  R - RIP B - BGP
       O - OSPF, IA - OSPF inter area
       N1 - OSPF NSSA external type 1, N2 - OSPF NSSA external type 2
       E1 - OSPF external type 1, E2 - OSPF external type 2
       i - IS-IS, L1 - IS-IS level-1, L2 - IS-IS level-2, ia - IS-IS inter area
       * - candidate default
Gateway of last resort is no set
R     10.0.0.0/8 [120/1] via 192.168.1.2, 00:00:17, FastEthernet 0/0
C    172.16.1.0/24 is directly connected, Loopback 0
C    172.16.1.1/32 is local host.
C    172.16.2.0/24 is directly connected, Loopback 1
C    172.16.2.1/32 is local host.
C    192.168.1.0/24 is directly connected, FastEthernet 0/0
C    192.168.1.1/32 is local host.
RouterA#
Routing Protocol is "rip"
  Sending updates every 30 seconds, next due in 21 seconds
  Invalid after 180 seconds, flushed after 120 seconds
  Outgoing update filter list for all interface is: not set
  Incoming update filter list for all interface is: not set
  Default redistribution metric is 1
  Redistributing:
  Default version control: send version 1, receive any version
    Interface                Send  Recv   Key-chain
    FastEthernet 0/0         1     1      2
    Loopback 0               1     1      2
    Loopback 1               1     1      2
```

```
  Routing for Networks:
    172.16.0.0
    192.168.1.0
  Distance: (default is 120)
RouterA#
```
RouterB#show ip route
```
Codes:  C - connected, S - static,  R - RIP B - BGP
        O - OSPF, IA - OSPF inter area
        N1 - OSPF NSSA external type 1, N2 - OSPF NSSA external type 2
        E1 - OSPF external type 1, E2 - OSPF external type 2
        i - IS-IS, L1 - IS-IS level-1, L2 - IS-IS level-2, ia - IS-IS inter area
        * - candidate default
Gateway of last resort is no set
C    10.1.1.0/24 is directly connected, Loopback 0
C    10.1.1.1/32 is local host.
C    10.2.2.0/24 is directly connected, Loopback 1
C    10.2.2.1/32 is local host.
```
R 172.16.0.0/16 [120/1] via 192.168.1.1, 00:00:12, FastEthernet 0/0

C 192.168.1.0/24 is directly connected, FastEthernet 0/0
```
C    192.168.1.2/32 is local host.
```
RouterA#show ip rip database
```
10.0.0.0/8    auto-summary
10.0.0.0/8
    [1] via 192.168.1.2 FastEthernet 0/0  00:09
172.16.0.0/16    auto-summary
172.16.1.0/24
    [1] directly connected, Loopback 0
172.16.2.0/24
    [1] directly connected, Loopback 1
192.168.1.0/24    auto-summary
192.168.1.0/24
    [1] directly connected, FastEthernet 0/0
```
RouterA#show ip rip interface

FastEthernet 0/0 is up, line protocol is up

Routing Protocol: RIP

Receive RIPv1 and RIPv2 packets

Send RIPv1 packets only
```
    Passive interface: Disabled
```
Split horizon: Enabled
```
    V2 Broadcast: Disabled
    Multicast registe: Registed
    Interface Summary Rip:
```

```
      Not Configured
    IP interface address:
      192.168.1.1/24
FastEthernet 0/1 is down, line protocol is down
  RIP is not enabled on this interface
Null 0 is up, line protocol is up
  RIP is not enabled on this interface
Loopback 0 is up, line protocol is up
  Routing Protocol: RIP
    Receive RIPv1 and RIPv2 packets
    Send RIPv1 packets only
    Passive interface: Disabled
    Split horizon: Enabled
    V2 Broadcast: Disabled
    Multicast registe: Registed
    Interface Summary Rip:
      Not Configured
    IP interface address:
      172.16.1.1/24
Loopback 1 is up, line protocol is up
  Routing Protocol: RIP
    Receive RIPv1 and RIPv2 packets
    Send RIPv1 packets only
    Passive interface: Disabled
    Split horizon: Enabled
    V2 Broadcast: Disabled
    Multicast registe: Registed
    Interface Summary Rip:
      Not Configured
    IP interface address:
      172.16.2.1/24
RouterB#show ip rip
Routing Protocol is "rip"
  Sending updates every 30 seconds, next due in 21 seconds
  Invalid after 180 seconds, flushed after 120 seconds
  Outgoing update filter list for all interface is: not set
  Incoming update filter list for all interface is: not set
  Default redistribution metric is 1
  Redistributing:
  Default version control: send version 1, receive any version
    Interface              Send  Recv   Key-chain
    FastEthernet 0/0       1     1      2
```

```
    Loopback 0                1     1     2
    Loopback 1                1     1     2
  Routing for Networks:
    10.0.0.0
    192.168.1.0
  Distance: (default is 120)
RouterB#show ip rip database
10.0.0.0/8    auto-summary
10.1.1.0/24
    [1] directly connected, Loopback 0
10.2.2.0/24
    [1] directly connected, Loopback 1
172.16.0.0/16    auto-summary
172.16.0.0/16
    [1] via 192.168.1.1 FastEthernet 0/0  00:08
192.168.1.0/24    auto-summary
192.168.1.0/24
    [1] directly connected, FastEthernet 0/0
RouterB#show ip rip interface
FastEthernet 0/0 is up, line protocol is up
  Routing Protocol: RIP
    Receive RIPv1 and RIPv2 packets
    Send RIPv1 packets only
    Passive interface: Disabled
    Split horizon: Enabled
    V2 Broadcast: Disabled
    Multicast registe: Registed
    Interface Summary Rip:
      Not Configured
    IP interface address:
      192.168.1.2/24
FastEthernet 0/1 is down, line protocol is down
  RIP is not enabled on this interface
Null 0 is up, line protocol is up
  RIP is not enabled on this interface
Loopback 0 is up, line protocol is up
  Routing Protocol: RIP
    Receive RIPv1 and RIPv2 packets
    Send RIPv1 packets only
    Passive interface: Disabled
    Split horizon: Enabled
    V2 Broadcast: Disabled
```

```
    Multicast registe: Registed
    Interface Summary Rip:
      Not Configured
    IP interface address:
      10.1.1.1/24
Loopback 1 is up, line protocol is up
  Routing Protocol: RIP
    Receive RIPv1 and RIPv2 packets
    Send RIPv1 packets only
    Passive interface: Disabled
    Split horizon: Enabled
    V2 Broadcast: Disabled
    Multicast registe: Registed
    Interface Summary Rip:
      Not Configured
    IP interface address:
      10.2.2.1/24
```

步骤 4　测试网络连通性。

```
RouterA#ping 10.1.1.1
Sending 5, 100-byte ICMP Echoes to 10.1.1.1, timeout is 2 seconds:
  < press Ctrl+C to break >
!!!!!
Success rate is 100 percent (5/5), round-trip min/avg/max = 1/1/1 ms
RouterA#ping 10.2.2.1
Sending 5, 100-byte ICMP Echoes to 10.2.2.1, timeout is 2 seconds:
  < press Ctrl+C to break >
!!!!!
Success rate is 100 percent (5/5), round-trip min/avg/max = 1/2/10 ms
RouterB#ping 172.16.1.1
Sending 5, 100-byte ICMP Echoes to 172.16.1.1, timeout is 2 seconds:
  < press Ctrl+C to break >
!!!!!
Success rate is 100 percent (5/5), round-trip min/avg/max = 1/1/1 ms
RouterB#ping 172.16.2.1
Sending 5, 100-byte ICMP Echoes to 172.16.2.1, timeout is 2 seconds:
  < press Ctrl+C to break >
!!!!!
Success rate is 100 percent (5/5), round-trip min/avg/max = 1/1/1 ms
```

步骤 5　用 debug 命令观察路由器接收和发送路由更新的情况。

下面是一个完整的 RIP 路由器接收更新和发送更新的过程，从中可以看到 RouterB 接收到了 RouterA 发送的更新，其中包含一条路由信息 172.16.0.0（可以看到水平分割原则的作用），然后刷新了路由表。

RouterB 本身发送的更新报文则在 Fa0/0、Lo0 和 Lo1 这 3 个端口采用广播的方式发出，广播地址分别为 192.168.1.255，10.1.1.255，10.2.2.255，使用 UDP 的 520 端口。在水平分割的原则下，每个端口发送的路由信息均不相同。

```
RouterB#debug ip rip
Aug  8 21:06:08 RouterB %7: [RIP] RIP recveived packet, sock=2125 src=192.168.1.1
len=24
Aug  8 21:06:08 RouterB %7: [RIP] Cancel peer remove timer
Aug  8 21:06:08 RouterB %7:[RIP] Peer remove timer shedule...
Aug  8 21:06:08 RouterB %7:      route-entry: family 2 ip 172.16.0.0 metric 1
Aug  8 21:06:08 RouterB %7: [RIP] Received version 1 response packet
Aug  8 21:06:08 RouterB %7: [RIP] Translate mask to 16
Aug  8 21:06:08 RouterB %7: [RIP] Old path is: nhop=192.168.1.1 routesrc=192.168.1.1
intf=1
Aug  8 21:06:08 RouterB %7: [RIP] New path is: nhop=192.168.1.1 routesrc=192.168.1.1
Aug  8 21:06:08 RouterB %7: [RIP] [172.16.0.0/16] RIP route refresh!
Aug  8 21:06:08 RouterB %7: [RIP] [172.16.0.0/16] RIP distance apply from
192.168.1.1!
Aug  8 21:06:08 RouterB %7: [RIP] [172.16.0.0/16] ready to refresh kernel...
Aug  8 21:06:08 RouterB %7: [RIP] NSM refresh: IPv4 RIP Route 172.16.0.0/16 distance=120
metric=1 nexthop_num=1 distance=120 nexhop=192.168.1.1 ifindex=1
Aug  8 21:06:08 RouterB %7: [RIP] [172.16.0.0/16] cancel route timer
Aug  8 21:06:08 RouterB %7: [RIP] [172.16.0.0/16] route timer schedule...
Aug  8 21:06:23 RouterB %7: [RIP] Output timer expired to send reponse
Aug  8 21:06:23 RouterB %7: [RIP] Prepare to send BROADCAST response...
Aug  8 21:06:23 RouterB %7: [RIP] Building update entries on FastEthernet 0/0
Aug  8 21:06:23 RouterB %7: network 10.0.0.0 metric 1
Aug  8 21:06:23 RouterB %7: [RIP] Send packet to 192.168.1.255 Port 520 on FastEthernet
0/0
Aug  8 21:06:23 RouterB %7: [RIP] Prepare to send BROADCAST response...
Aug  8 21:06:23 RouterB %7: [RIP] Building update entries on Loopback 0
Aug  8 21:06:23 RouterB %7: network 10.2.2.0 metric 1
Aug  8 21:06:23 RouterB %7: network 172.16.0.0 metric 2
Aug  8 21:06:23 RouterB %7: network 192.168.1.0 metric 1
Aug  8 21:06:23 RouterB %7: [RIP] Send packet to 10.1.1.255 Port 520 on Loopback 0
Aug  8 21:06:23 RouterB %7: [RIP] Prepare to send BROADCAST response...
Aug  8 21:06:23 RouterB %7: [RIP] Building update entries on Loopback 1
Aug  8 21:06:23 RouterB %7: network 10.1.1.0 metric 1
Aug  8 21:06:23 RouterB %7: network 172.16.0.0 metric 2
Aug  8 21:06:23 RouterB %7: network 192.168.1.0 metric 1
Aug  8 21:06:23 RouterB %7: [RIP] Send packet to 10.2.2.255 Port 520 on Loopback 1
Aug  8 21:06:23 RouterB %7: [RIP] Schedule response send timer
```

【注意事项】

（1）配置 RIP 的 Network 命令时只支持 A、B、C 的主网络号，如果写入子网则自动转为主网络号。

（2）No auto-summary 功能只有在 RIPv2 支持。

【参考配置】

```
RouterA#show running-config
Building configuration...
Current configuration : 612 bytes
!
version RGNOS 10.1.00(4), Release(18443)(Tue Jul 17 20:50:30 CST 2007
-ubu1server)
hostname RouterA
!
interface FastEthernet 0/0
 ip address 192.168.1.1 255.255.255.0
 duplex auto
 speed auto
!
interface FastEthernet 0/1
 duplex auto
 speed auto
!
interface Loopback 0
 ip address 172.16.1.1 255.255.255.0
!
interface Loopback 1
 ip address 172.16.2.1 255.255.255.0
!
router rip
 network 172.16.0.0
 network 192.168.1.0
!
line con 0
line aux 0
line vty 0 4
 login
!
end
RouterB#show running-config
Building configuration...
Current configuration : 606 bytes
```

```
!
version RGNOS 10.1.00(4), Release(18443)(Tue Jul 17 20:50:30 CST 2007
-ubu1server)
hostname RouterB
!
interface FastEthernet 0/0
 ip address 192.168.1.2 255.255.255.0
 duplex auto
 speed auto
!
interface FastEthernet 0/1
 duplex auto
 speed auto
!
interface Loopback 0
 ip address 10.1.1.1 255.255.255.0
!
interface Loopback 1
 ip address 10.2.2.1 255.255.255.0
!
router rip
 network 10.0.0.0
 network 192.168.1.0
!
line con 0
line aux 0
line vty 0 4
 login
!
end
```

实验 12　RIPv2 配置

【实验名称】

RIPv2 配置。

【实验目的】

理解 RIP 两个版本之间的区别，掌握如何配置 RIPv2。

【背景描述】

其校园网在地理位置上分为 2 个区域，每个区域内分别有一台路由器连接了 2 个子网，需要将两台路由器通过以太网链路连接在一起并进行适当的配置，以实现这 4 个子网之间的互连互通。为了便于管理员在未来每个校园区域扩充子网数量时，不需要同时更改路由器的配置，计划使用 RIP 路由协议实现子网之间的互通。

【需求分析】

两台路由器通过快速以太网端口连接在一起，每个路由器上设置 2 个 Loopback 端口模拟子网，在所有端口运行 RIP 路由协议，实现所有子网间的互通。

【实验拓扑】

实验的拓扑图，如图 12-1 所示。

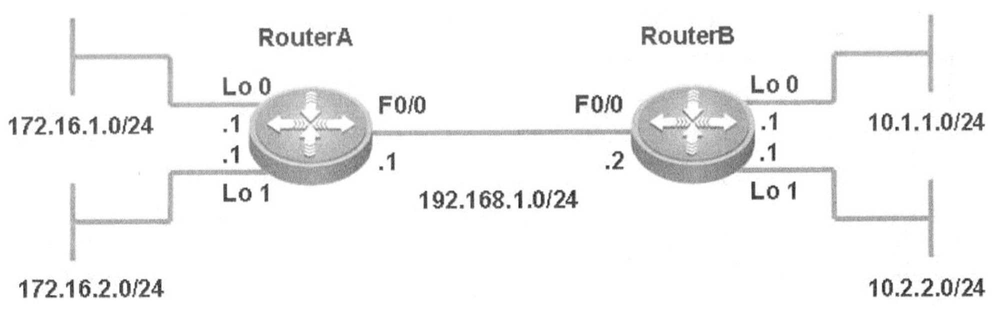

图 12-1

【实验设备】

路由器 2 台。

【预备知识】

路由器的工作原理和基本配置方法、距离矢量路由协议、RIP 工作原理、理解 RIPv1 和 RIPv2 的区别、RIPv2 的配置方法。

【实验原理】

RIP 协议有两个版本 RIPv1 和 RIPv2。

RIPv1 属于有类路由协议，不支持 VLSM（变长子网掩码），RIPv1 是以广播的形式进行路由信息的更新的，更新周期为30秒。

RIPv2 属于无类路由协议，支持 VLSM（变长子网掩码），RIPv2 是以组播的形式进行路由信息的更新的，组播地址是224.0.0.9。RIPv2 还支持基于端口的认证，提高网络的安全性。

【实验步骤】

步骤1　配置两台路由器的主机名、接口IP地址。

```
RSR20#configure terminal
Enter configuration commands, one per line.  End with CNTL/Z.
RSR20(config)#hostname RouterA
RouterA(config)#
RouterA(config)#interface fastEthernet 0/0
RouterA(config-if)#ip address 192.168.1.1 255.255.255.0
RouterA(config-if)#no shutdown
RouterA(config-if)#exit
RouterA(config)#
RouterA(config)#interface loopback 0
RouterA(config-if)#Aug 15 23:46:32 RouterA %7:%LINE PROTOCOL CHANGE: Interface
Loopback 0, changed state to UP
RouterA(config-if)#ip address 172.16.1.1 255.255.255.0
RouterA(config-if)#exit
RouterA(config)#
RouterA(config)#interface loopback 1
RouterA(config-if)#Aug 15 23:47:00 RouterA %7:%LINE PROTOCOL CHANGE: Interface
Loopback 1, changed state to UP
RouterA(config-if)#ip address 172.16.2.1 255.255.255.0
RouterA(config-if)#exit
RSR20#configure terminal
Enter configuration commands, one per line.  End with CNTL/Z.
RSR20(config)#hostname RouterB
RouterB(config)#
RouterB(config)#interface fastEthernet 0/0
RouterB(config-if)#ip address 192.168.1.2 255.255.255.0
RouterB(config-if)#no shutdown
RouterB(config-if)#exit
RouterB(config)#
RouterB(config)#interface loopback 0
RouterB(config-if)#Aug  8 21:00:00 RouterB %7:%LINE PROTOCOL CHANGE: Interface
Loopback 0, changed state to UP
RouterB(config-if)#ip address 10.1.1.1 255.255.255.0
RouterB(config-if)#exit
RouterB(config)#
RouterB(config)#interface loopback 1
```

```
RouterB(config-if)#Aug  8 21:00:28 RouterB %7:%LINE PROTOCOL CHANGE: Interface
Loopback 1, changed state to UP
RouterB(config-if)#ip address 10.2.2.1 255.255.255.0
RouterB(config-if)#exit
```

步骤 2　在两台路由器上启用 RIPv2，但不关闭自动汇总。

```
RouterA(config)#router rip
RouterA(config-router)#network 192.168.1.0
RouterA(config-router)#network 172.16.1.0
RouterA(config-router)#version 2
! 配置使用 RIPv2
RouterA(config-router)#exit
RouterB(config)#router rip
RouterB(config-router)#network 192.168.1.0
RouterB(config-router)#network 10.0.0.0
RouterB(config-router)#version 2
RouterB(config-router)#exit
```

步骤 3　查看路由表。

从路由表中可以看到，仍然只有 B 类主网络 172.16.0.0/16 和 A 类主网络 10.0.0.0/8 出现在路由表之中。虽然 RIPv2 支持 VLSM，但 RouterA 和 RouterB 都是边界路由器，分别是 B 类主网络 172.16.0.0/16 和 C 类主网络 192.168.1.0/24 的边界、A 类主网络 10.0.0.0/8 和 C 类主网络 192.168.1.0/24 的边界，因此在执行自动的路由汇总。

```
RouterA#show ip route
Codes:  C - connected, S - static,  R - RIP B - BGP
       O - OSPF, IA - OSPF inter area
       N1 - OSPF NSSA external type 1, N2 - OSPF NSSA external type 2
       E1 - OSPF external type 1, E2 - OSPF external type 2
       i - IS-IS, L1 - IS-IS level-1, L2 - IS-IS level-2, ia - IS-IS inter area
       * - candidate default
Gateway of last resort is no set
R     10.0.0.0/8 [120/1] via 192.168.1.2, 00:00:10, FastEthernet 0/0
C    172.16.1.0/24 is directly connected, Loopback 0
C    172.16.1.1/32 is local host.
C    172.16.2.0/24 is directly connected, Loopback 1
C    172.16.2.1/32 is local host.
C    192.168.1.0/24 is directly connected, FastEthernet 0/0
C    192.168.1.1/32 is local host.
RouterB#show ip route
Codes:  C - connected, S - static,  R - RIP B - BGP
       O - OSPF, IA - OSPF inter area
       N1 - OSPF NSSA external type 1, N2 - OSPF NSSA external type 2
       E1 - OSPF external type 1, E2 - OSPF external type 2
       i - IS-IS, L1 - IS-IS level-1, L2 - IS-IS level-2, ia - IS-IS inter area
```

```
        * - candidate default
Gateway of last resort is no set
C    10.1.1.0/24 is directly connected, Loopback 0
C    10.1.1.1/32 is local host.
C    10.2.2.0/24 is directly connected, Loopback 1
C    10.2.2.1/32 is local host.
R    172.16.0.0/16 [120/1] via 192.168.1.1, 00:00:02, FastEthernet 0/0
C    192.168.1.0/24 is directly connected, FastEthernet 0/0
C    192.168.1.2/32 is local host.
```

步骤 4　关闭自动路由汇总。

```
RouterA(config)#router rip
RouterA(config-router)#no auto-summary
! 关闭 RIPv2 的自动路由汇总功能
RouterA(config-router)#end
RouterB(config)#router rip
RouterB(config-router)#no auto-summary
RouterB(config-router)#end
```

步骤 5　查看 RIP 配置信息，路由表。

```
RouterA#show ip route
Codes:  C - connected, S - static,  R - RIP B - BGP
        O - OSPF, IA - OSPF inter area
        N1 - OSPF NSSA external type 1, N2 - OSPF NSSA external type 2
        E1 - OSPF external type 1, E2 - OSPF external type 2
        i - IS-IS, L1 - IS-IS level-1, L2 - IS-IS level-2, ia - IS-IS inter area
        * - candidate default
Gateway of last resort is no set
R    10.1.1.0/24 [120/1] via 192.168.1.2, 00:00:03, FastEthernet 0/0
R    10.2.2.0/24 [120/1] via 192.168.1.2, 00:00:03, FastEthernet 0/0
C    172.16.1.0/24 is directly connected, Loopback 0
C    172.16.1.1/32 is local host.
C    172.16.2.0/24 is directly connected, Loopback 1
C    172.16.2.1/32 is local host.
C    192.168.1.0/24 is directly connected, FastEthernet 0/0
C    192.168.1.1/32 is local host.
! 可以看到 RIP 路由表中已经学习到了子网的路由
RouterA#show ip rip
Routing Protocol is "rip"
  Sending updates every 30 seconds, next due in 23 seconds
  Invalid after 180 seconds, flushed after 120 seconds
  Outgoing update filter list for all interface is: not set
  Incoming update filter list for all interface is: not set
  Default redistribution metric is 1
```

```
  Redistributing:
  Default version control: send version 2, receive version 2
    Interface             Send  Recv   Key-chain
    FastEthernet 0/0      2     2
    Loopback 0            2     2
    Loopback 1            2     2
  Routing for Networks:
    172.16.0.0
    192.168.1.0
  Distance: (default is 120)
! 在配置 RIPv2 版本后，RIP 路由器将只接收和发送版本 2 的更新报文
RouterA#show ip rip database
10.0.0.0/8    auto-summary
10.1.1.0/24
    [1] via 192.168.1.2 FastEthernet 0/0   00:28
10.2.2.0/24
    [1] via 192.168.1.2 FastEthernet 0/0   00:28
172.16.0.0/16    auto-summary
172.16.1.0/24
    [1] directly connected, Loopback 0
172.16.2.0/24
    [1] directly connected, Loopback 1
192.168.1.0/24    auto-summary
192.168.1.0/24
    [1] directly connected, FastEthernet 0/0
! RIP 的数据库中保存了子网条目的信息
RouterA#show ip rip interface
FastEthernet 0/0 is up, line protocol is up
  Routing Protocol: RIP
    Receive RIPv2 packets only
    Send RIPv2 packets only
    Passive interface: Disabled
    Split horizon: Enabled
    V2 Broadcast: Disabled
    Multicast registe: Registed
    Interface Summary Rip:
      Not Configured
    IP interface address:
      192.168.1.1/24
FastEthernet 0/1 is down, line protocol is down
  RIP is not enabled on this interface
Null 0 is up, line protocol is up
```

```
  RIP is not enabled on this interface
Loopback 0 is up, line protocol is up
  Routing Protocol: RIP
    Receive RIPv2 packets only
    Send RIPv2 packets only
    Passive interface: Disabled
    Split horizon: Enabled
    V2 Broadcast: Disabled
    Multicast registe: Registed
    Interface Summary Rip:
      Not Configured
    IP interface address:
      172.16.1.1/24
Loopback 1 is up, line protocol is up
  Routing Protocol: RIP
    Receive RIPv2 packets only
    Send RIPv2 packets only
    Passive interface: Disabled
    Split horizon: Enabled
    V2 Broadcast: Disabled
    Multicast registe: Registed
    Interface Summary Rip:
      Not Configured
    IP interface address:
      172.16.2.1/24
RouterB#show ip route
Codes:  C - connected, S - static,  R - RIP B - BGP
        O - OSPF, IA - OSPF inter area
        N1 - OSPF NSSA external type 1, N2 - OSPF NSSA external type 2
        E1 - OSPF external type 1, E2 - OSPF external type 2
        i - IS-IS, L1 - IS-IS level-1, L2 - IS-IS level-2, ia - IS-IS inter area
        * - candidate default
Gateway of last resort is no set
C    10.1.1.0/24 is directly connected, Loopback 0
C    10.1.1.1/32 is local host.
C    10.2.2.0/24 is directly connected, Loopback 1
C    10.2.2.1/32 is local host.
R    172.16.1.0/24 [120/1] via 192.168.1.1, 00:00:22, FastEthernet 0/0
R    172.16.2.0/24 [120/1] via 192.168.1.1, 00:00:22, FastEthernet 0/0
C    192.168.1.0/24 is directly connected, FastEthernet 0/0
C    192.168.1.2/32 is local host.
RouterB#show ip rip
```

```
Routing Protocol is "rip"
  Sending updates every 30 seconds, next due in 22 seconds
  Invalid after 180 seconds, flushed after 120 seconds
  Outgoing update filter list for all interface is: not set
  Incoming update filter list for all interface is: not set
  Default redistribution metric is 1
  Redistributing:
  Default version control: send version 2, receive version 2
    Interface           Send  Recv   Key-chain
    FastEthernet 0/0     2     2
    Loopback 0           2     2
    Loopback 1           2     2
  Routing for Networks:
    10.0.0.0
    192.168.1.0
  Distance: (default is 120)
RouterB#
RouterB#show ip rip dat
RouterB#show ip rip database
10.0.0.0/8    auto-summary
10.1.1.0/24
    [1] directly connected, Loopback 0
10.2.2.0/24
    [1] directly connected, Loopback 1
172.16.0.0/16    auto-summary
172.16.1.0/24
    [1] via 192.168.1.1 FastEthernet 0/0   00:02
172.16.2.0/24
    [1] via 192.168.1.1 FastEthernet 0/0   00:02
192.168.1.0/24    auto-summary
192.168.1.0/24
    [1] directly connected, FastEthernet 0/0
RouterB#show ip rip interface
FastEthernet 0/0 is up, line protocol is up
  Routing Protocol: RIP
    Receive RIPv2 packets only
    Send RIPv2 packets only
    Passive interface: Disabled
    Split horizon: Enabled
    V2 Broadcast: Disabled
    Multicast registe: Registed
    Interface Summary Rip:
```

```
      Not Configured
    IP interface address:
      192.168.1.2/24
FastEthernet 0/1 is down, line protocol is down
  RIP is not enabled on this interface
Null 0 is up, line protocol is up
  RIP is not enabled on this interface
Loopback 0 is up, line protocol is up
  Routing Protocol: RIP
    Receive RIPv2 packets only
    Send RIPv2 packets only
    Passive interface: Disabled
    Split horizon: Enabled
    V2 Broadcast: Disabled
    Multicast registe: Registed
    Interface Summary Rip:
      Not Configured
    IP interface address:
      10.1.1.1/24
Loopback 1 is up, line protocol is up
  Routing Protocol: RIP
    Receive RIPv2 packets only
    Send RIPv2 packets only
    Passive interface: Disabled
    Split horizon: Enabled
    V2 Broadcast: Disabled
    Multicast registe: Registed
    Interface Summary Rip:
      Not Configured
    IP interface address:
      10.2.2.1/24
```

步骤 6 测试网络连通性。

```
RouterA#ping 10.1.1.1
Sending 5, 100-byte ICMP Echoes to 10.1.1.1, timeout is 2 seconds:
  < press Ctrl+C to break >
!!!!!
Success rate is 100 percent (5/5), round-trip min/avg/max = 1/2/10 ms
RouterA#ping 10.2.2.1
Sending 5, 100-byte ICMP Echoes to 10.2.2.1, timeout is 2 seconds:
  < press Ctrl+C to break >
!!!!!
Success rate is 100 percent (5/5), round-trip min/avg/max = 1/1/1 ms
```

```
RouterB#ping 172.16.1.1
Sending 5, 100-byte ICMP Echoes to 172.16.1.1, timeout is 2 seconds:
  < press Ctrl+C to break >
!!!!!
Success rate is 100 percent (5/5), round-trip min/avg/max = 1/1/1 ms
RouterB#ping 172.16.2.1
Sending 5, 100-byte ICMP Echoes to 172.16.2.1, timeout is 2 seconds:
  < press Ctrl+C to break >
!!!!!
Success rate is 100 percent (5/5), round-trip min/avg/max = 1/2/10 ms
```

步骤 7 用 debug 命令观察路由器接收和发送路由更新的情况。

下面是一个完整的 RIPv2 路由器接收更新和发送更新的过程，从中可以看到 RouterB 接收到了 RouterA 发送的更新，其中包含两条路由信息 172.16.1.0 和 172.16.2.0（可以看到水平分割原则的作用），然后刷新了路由表。

RouterB 本身发送的更新报文则在 Fa0/0、Lo0 和 Lo1 这 3 个端口使用组播方式发出，组播地址为 224.0.0.9，使用的是 UDP 520 端口。在水平分割的原则下，每个端口发送的路由信息均不相同。

注意 RIPv2 的更新报文格式和 RIPv1 的不同。

```
RouterB#debug ip rip
RouterB#Aug  8 21:58:08 RouterB %7: [RIP] RIP recveived packet, sock=2125
src=192.168.1.1 len=44
Aug  8 21:58:08 RouterB %7: [RIP] Cancel peer remove timer
Aug  8 21:58:08 RouterB %7:[RIP] Peer remove timer shedule...
Aug  8 21:58:08 RouterB %7: [RIP] Both do not need auth, Auth ok
Aug  8 21:58:08 RouterB %7:      route-entry: family 2 tag 0 ip 172.16.1.0 mask
255.255.255.0 nhop 0.0.0.0 metric 1
Aug  8 21:58:08 RouterB %7:      route-entry: family 2 tag 0 ip 172.16.2.0 mask
255.255.255.0 nhop 0.0.0.0 metric 1
Aug  8 21:58:08 RouterB %7: [RIP] Received version 2 response packet
Aug  8 21:58:08 RouterB %7: [RIP] Old path is: nhop=192.168.1.1
routesrc=192.168.1.1 intf=1
Aug  8 21:58:08 RouterB %7: [RIP] New path is: nhop=192.168.1.1
routesrc=192.168.1.1
Aug  8 21:58:08 RouterB %7: [RIP] [172.16.1.0/24] RIP route refresh!
Aug  8 21:58:08 RouterB %7: [RIP] [172.16.1.0/24] RIP distance apply from
192.168.1.1!
Aug  8 21:58:08 RouterB %7: [RIP] [172.16.1.0/24] ready to refresh kernel...
Aug  8 21:58:08 RouterB %7: [RIP] NSM refresh: IPv4 RIP Route 172.16.1.0/24
distance=120 metric=1 nexthop num=1 distance=120 nexhop=192.168.1.1 ifindex=1
Aug  8 21:58:08 RouterB %7: [RIP] [172.16.1.0/24] cancel route timer
Aug  8 21:58:08 RouterB %7: [RIP] [172.16.1.0/24] route timer schedule...
Aug  8 21:58:08 RouterB %7: [RIP] Old path is: nhop=192.168.1.1
```

```
routesrc=192.168.1.1 intf=1
Aug  8 21:58:08 RouterB %7: [RIP] New path is: nhop=192.168.1.1 routesrc=
192.168.1.1
Aug  8 21:58:08 RouterB %7: [RIP] [172.16.2.0/24] RIP route refresh!
Aug  8 21:58:08 RouterB %7: [RIP] [172.16.2.0/24] RIP distance apply from 192.168.1.1!
Aug  8 21:58:08 RouterB %7: [RIP] [172.16.2.0/24] ready to refresh kernel...
Aug  8 21:58:08 RouterB %7: [RIP] NSM refresh: IPv4 RIP Route 172.16.2.0/24
distance=120 metric=1 nexthop num=1 distance=120 nexhop=192.168.1.1 ifindex=1
Aug  8 21:58:08 RouterB %7: [RIP] [172.16.2.0/24] cancel route timer
Aug  8 21:58:08 RouterB %7: [RIP] [172.16.2.0/24] route timer schedule...
Aug  8 21:58:23 RouterB %7: [RIP] Output timer expired to send reponse
Aug  8 21:58:23 RouterB %7: [RIP] Prepare to send MULTICAST response...
Aug  8 21:58:23 RouterB %7: [RIP] Building update entries on FastEthernet 0/0
Aug  8 21:58:23 RouterB %7: 10.1.1.0/24 via 0.0.0.0 metric 1 tag 0
Aug  8 21:58:23 RouterB %7: 10.2.2.0/24 via 0.0.0.0 metric 1 tag 0
Aug  8 21:58:23 RouterB %7: [RIP] Send packet to 224.0.0.9 Port 520 on FastEthernet 0/0
Aug  8 21:58:23 RouterB %7: [RIP] Prepare to send MULTICAST response...
Aug  8 21:58:23 RouterB %7: [RIP] Building update entries on Loopback 0
Aug  8 21:58:23 RouterB %7: 10.2.2.0/24 via 0.0.0.0 metric 1 tag 0
Aug  8 21:58:23 RouterB %7: 172.16.1.0/24 via 0.0.0.0 metric 2 tag 0
Aug  8 21:58:23 RouterB %7: 172.16.2.0/24 via 0.0.0.0 metric 2 tag 0
Aug  8 21:58:23 RouterB %7: 192.168.1.0/24 via 0.0.0.0 metric 1 tag 0
Aug  8 21:58:23 RouterB %7: [RIP] Send packet to 224.0.0.9 Port 520 on Loopback 0
Aug  8 21:58:23 RouterB %7: [RIP] Prepare to send MULTICAST response...
Aug  8 21:58:23 RouterB %7: [RIP] Building update entries on Loopback 1
Aug  8 21:58:23 RouterB %7: 10.1.1.0/24 via 0.0.0.0 metric 1 tag 0
Aug  8 21:58:23 RouterB %7: 172.16.1.0/24 via 0.0.0.0 metric 2 tag 0
Aug  8 21:58:23 RouterB %7: 172.16.2.0/24 via 0.0.0.0 metric 2 tag 0
Aug  8 21:58:23 RouterB %7: 192.168.1.0/24 via 0.0.0.0 metric 1 tag 0
Aug  8 21:58:23 RouterB %7: [RIP] Send packet to 224.0.0.9 Port 520 on Loopback 1
Aug  8 21:58:23 RouterB %7: [RIP] Schedule response send timer
```

【注意事项】

（1）配置 RIP 的 Network 命令时只支持 A、B、C 的主网络号，如果写入子网则自动转为主网络号。

（2）No auto-summary 功能只有在 RIPv2 支持。

（3）如果配置 no auto-summary 命令后立刻查看路由表，除了能看到子网的路由条目外，还可以看到原本主网络的路由条目，该主网络的路由条目将在无效计时器、刷新计时器超时后才会被清除。

【参考配置】

```
RouterA#show running-config
Building configuration...
```

```
Current configuration : 642 bytes
!
version RGNOS 10.1.00(4), Release(18443)(Tue Jul 17 20:50:30 CST 2007
-ubu1server)
hostname RouterA
!
!
interface FastEthernet 0/0
 ip address 192.168.1.1 255.255.255.0
 duplex auto
 speed auto
!
interface FastEthernet 0/1
 duplex auto
 speed auto
!
interface Loopback 0
 ip address 172.16.1.1 255.255.255.0
!
interface Loopback 1
 ip address 172.16.2.1 255.255.255.0
!
router rip
 version 2
 network 172.16.0.0
 network 192.168.1.0
 no auto-summary
!
line con 0
line aux 0
line vty 0 4
 login
!
end
```

RouterB#show running-config

```
Building configuration...
Current configuration : 636 bytes
!
version RGNOS 10.1.00(4), Release(18443)(Tue Jul 17 20:50:30 CST 2007
-ubu1server)
hostname RouterB
!
```

```
!
interface FastEthernet 0/0
 ip address 192.168.1.2 255.255.255.0
 duplex auto
 speed auto
!
interface FastEthernet 0/1
 duplex auto
 speed auto
!
interface Loopback 0
 ip address 10.1.1.1 255.255.255.0
!
interface Loopback 1
 ip address 10.2.2.1 255.255.255.0
!
router rip
 version 2
 network 10.0.0.0
 network 192.168.1.0
 no auto-summary
!
line con 0
line aux 0
line vty 0 4
 login
!
End
```

实验 13　OSPF 基本配置

【实验名称】

OSPF 基本配置。

【实验目的】

掌握在路由器上配置 OSPF 单区域。

【背景描述】

假设校园网通过 1 台三层交换机连到校园网的出口路由器上，路由器再和校园外的另 1 台路由器连接，现在要进行适当配置，实现校园网内部主机与校园网外部主机的相互通信。

本实验以两台路由器、1 台三层交换机为例。S3550 上划分有 VLAN10 和 VLAN50，其中，VLAN10 用于连接 RA，VLAN50 用于连接校园网主机。

【需求分析】

需要在路由器和交换机上配置 OSPF 路由协议，使全网互通，从而实现信息的共享和传递。

【实验拓扑】

实验的拓扑图，如图 13-1 所示。

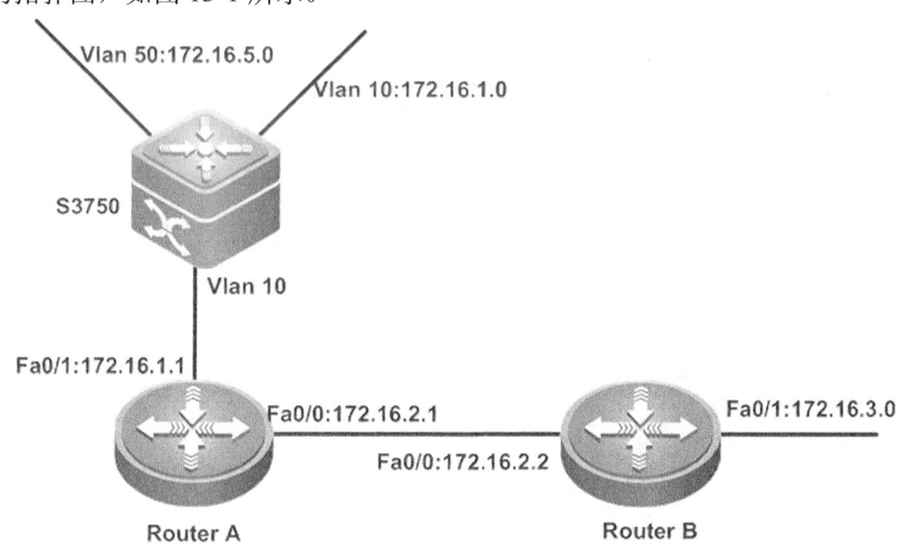

图 13-1

【实验设备】

三层交换机 1 台。

路由器 2 台。

交叉线或直连线 3 条。

【预备知识】

路由器基本配置知识、OSPF。

【实验原理】

OSPF（Open Shortest Path First，开放式最短路径优先）协议，是目前网络中应用最广泛的路由协议之一。属于内部网关路由协议，能够适应各种规模的网络环境，是典型的链路状态（link-state）协议。

OSPF 路由协议通过向全网扩散本设备的链路状态信息，使网络中的每台设备最终同步于一个具有全网链路状态的数据库（LSDB），然后路由器采用 SPF 算法，以自己为根，计算到达其他网络的最短路径，最终形成全网路由信息。

OSPF 属于无类路由协议，支持 VLSM（变长子网掩码）。OSPF 是以组播的形式进行链路状态的通告的。

在大模型的网络环境中，OSPF 支持区域的划分，将网络进行合理地规划。划分区域时，必须存在 area0（骨干区域）。其他区域和骨干区域直接相连，或通过虚链路的方式连接。

【实验步骤】

步骤 1　在路由器和三层交换机配置 IP 地址。

```
switch#configure terminal
switch(config)#hostname S3750
S3750(config)#vlan 10
S3750(config-vlan)#exit
S3750(config)#vlan 50
S3750(config-vlan)#exit
S3750(config)#interface f0/1
S3750(config-if)#switchport access vlan 10
S3750(config-if)#exit
S3750(config)#interface f0/2
S3750(config-if)#switchport access vlan 50
S3750(config-if)#exit
S3750(config)#interface vlan 10
S3750(config-if)#ip address 172.16.1.2 255.255.255.0
S3750(config-if)#no shutdown
S3750(config-if)#exit
S3750(config)#interface vlan 50
S3750(config-if)#ip address 172.16.5.1 255.255.255.0
S3750(config-if)#no shutdown
S3750(config-if)#exit
RouterA(config)# interface fastethernet 0/1
RouterA(config-if)# ip address 172.16.1.1 255.255.255.0
RouterA(config-if)# no shutdown
RouterA(config-if)#exit
RouterA(config)# interface fastethernet 0/0
```

```
RouterA(config-if)# ip address 172.16.2.1 255.255.255.0
RouterB(config-if)# no shutdown
RouterB(config)# interface fastethernet 0/1
RouterB(config-if)# ip address 172.16.3.1 255.255.255.0
RouterB(config-if)# no shutdown
RouterB(config-if)#exit
RouterB(config)# interface fastethernet 0/0
RouterB(config-if)# ip address 172.16.2.2 255.255.255.0
RouterB(config-if)# no shutdown
```

步骤 2　配置 OSPF 路由协议。

```
S3750(config)#router ospf
S3750(config-router)#network 172.16.5.0 0.0.0.255 area 0
S3750(config-router)#network 172.16.1.0 0.0.0.255 area 0
S3750(config-router)#end
RouterA(config)# router ospf
RouterA(config-router)#network 172.16.1.0 0.0.0.255 area 0
RouterA(config-router)#network 172.16.2.0 0.0.0.255 area 0
RouterA(config-router)#end
RouterB(config)#router ospf
RouterB(config-router)#network 172.16.2.0 0.0.0.255 area 0
RouterB(config-router)#network 172.16.3.0 0.0.0.255 area 0
RouterB(config-router)#end
```

步骤 3　验证测试。

```
S3750#show vlan
VLAN Name                              Status    Ports
---- -------------------------------- --------- -----------------------------------
   1 VLAN0001                          STATIC    Fa0/3, Fa0/4, Fa0/5, Fa0/6
                                                 Fa0/7, Fa0/8, Fa0/9, Fa0/10
                                                 Fa0/11, Fa0/12, Fa0/13,
Fa0/14   Fa0/15, Fa0/16,
                                                 Fa0/17, Fa0/18, Fa0/22
Fa0/19, Fa0/20, Fa0/21,
                                                 Fa0/23, Fa0/24, Gi0/25,
Gi0/26,Gi0/27, Gi0/28
  10 VLAN0010                          STATIC    Fa0/1
  50 VLAN0050                          STATIC    Fa0/2
S3750#show ip interface brief
Interface                   IP-Address(Pri)   OK?     Status
VLAN 10                     172.16.1.2/24     YES     UP
VLAN 50                     172.16.5.1/24     YES     UP

RA#show ip interface brief
```

```
Interface              IP-Address(Pri)    OK?    Status
FastEthernet 0/0       172.16.2.1/24      YES    UP
FastEthernet 0/1       172.16.1.1/24      YES    UP
```

RB#show ip interface brief

```
Interface              IP-Address(Pri)    OK?    Status
FastEthernet 0/0       172.16.2.2/24      YES    UP
FastEthernet 0/1       172.16.1.3/24      YES    UP
Loopback 0             no address         YES    DOWN
```

S3750#show ip route

```
Codes:  C - connected, S - static,  R - RIP B - BGP
        O - OSPF, IA - OSPF inter area
        N1 - OSPF NSSA external type 1, N2 - OSPF NSSA external type 2
        E1 - OSPF external type 1, E2 - OSPF external type 2
        i - IS-IS, L1 - IS-IS level-1, L2 - IS-IS level-2, ia - IS-IS inter area
        * - candidate default
Gateway of last resort is no set
C    172.16.1.0/24 is directly connected, VLAN 10
C    172.16.1.2/32 is local host.
O    172.16.2.0/24 [110/2] via 172.16.1.1, 00:14:09, VLAN 10
O    172.16.3.0/24 [110/3] via 172.16.1.1, 00:04:39, VLAN 10
C    172.16.5.0/24 is directly connected, VLAN 50
C    172.16.5.1/32 is local host.
```

RA#show ip route

```
Codes:  C - connected, S - static,  R - RIP B - BGP
        O - OSPF, IA - OSPF inter area
        N1 - OSPF NSSA external type 1, N2 - OSPF NSSA external type 2
        E1 - OSPF external type 1, E2 - OSPF external type 2
        i - IS-IS, L1 - IS-IS level-1, L2 - IS-IS level-2, ia - IS-IS inter area
        * - candidate default
Gateway of last resort is no set
C    172.16.1.0/24 is directly connected, FastEthernet 0/1
C    172.16.1.1/32 is local host.
C    172.16.2.0/24 is directly connected, FastEthernet 0/0
C    172.16.2.1/32 is local host.
O    172.16.3.0/24 [110/2] via 172.16.2.2, 00:05:21, FastEthernet 0/0
O    172.16.5.0/24 [110/2] via 172.16.1.2, 00:14:51, FastEthernet 0/1
```

RB#show ip route

```
Codes: C - connected, S - static,  R - RIP B - BGP
        O - OSPF, IA - OSPF inter area
        N1 - OSPF NSSA external type 1, N2 - OSPF NSSA external type 2
```

```
       E1 - OSPF external type 1, E2 - OSPF external type 2
       i - IS-IS, L1 - IS-IS level-1, L2 - IS-IS level-2, ia - IS-IS inter area
       * - candidate default
Gateway of last resort is no set
O      172.16.1.0/24 [110/2] via 172.16.2.1, 00:05:58, FastEthernet 0/0
C    172.16.2.0/24 is directly connected, FastEthernet 0/0
C    172.16.2.2/32 is local host.
C    172.16.3.0/24 is directly connected, FastEthernet 0/1
C    172.16.3.1/32 is local host.
O      172.16.5.0/24 [110/3] via 172.16.2.1, 00:15:22, FastEthernet 0/0
RA#show ip ospf neighbor
OSPF process 1:
Neighbor ID     Pri   State        Dead Time   Address          Interface
172.16.5.1        1   Full/DR     00:00:38    172.16.1.2      FastEthernet 0/1
172.16.2.2        1   Full/DR     00:00:36    172.16.2.2      FastEthernet 0/0
RA#show ip ospf interface fastEthernet 0/0
FastEthernet 0/0 is up, line protocol is up
  Internet Address 172.16.2.1/24, Ifindex 1, Area 0.0.0.0, MTU 1500
  Matching network config: 172.16.2.0/24
  Process ID 1, Router ID 172.167.1.1, Network Type BROADCAST, Cost: 1
  Transmit Delay is 1 sec, State BDR, Priority 1
  Designated Router (ID) 172.16.2.2, Interface Address 172.16.2.2
  Backup Designated Router (ID) 172.167.1.1, Interface Address 172.16.2.1
  Timer intervals configured, Hello 10, Dead 40, Wait 40, Retransmit 5
    Hello due in 00:00:05
  Neighbor Count is 1, Adjacent neighbor count is 1
  Crypt Sequence Number is 82589
  Hello received 114 sent 115, DD received 4 sent 5
  LS-Req received 1 sent 1, LS-Upd received 5 sent 9
  LS-Ack received 6 sent 4, Discarded 0
```

【注意事项】

（1）在申明直连网段时，注意要写该网段的反掩码。

（2）在申明直连网段时，必须指明所属的区域。

【参考配置】

```
S3750#show running-config
Building configuration...
Current configuration : 1399 bytes
!
version  RGNOS  10.1.00(4),  Release(18443)(Tue  Jul  17  19:51:54  CST  2007
-ubu6server)
hostname S3750
```

```
!
vlan 1
!
vlan 10
!
vlan 50
!
interface FastEthernet 0/1
 switchport access vlan 10
!
interface FastEthernet 0/2
 switchport access vlan 50
!
interface FastEthernet 0/3
!
interface FastEthernet 0/4
!
interface FastEthernet 0/5
!
interface FastEthernet 0/6
!
interface FastEthernet 0/7
!
interface FastEthernet 0/8
!
interface FastEthernet 0/9
!
interface FastEthernet 0/10
!
interface FastEthernet 0/11
!
interface FastEthernet 0/12
!
interface FastEthernet 0/13
!
interface FastEthernet 0/14
!
interface FastEthernet 0/15
!
interface FastEthernet 0/16
!
interface FastEthernet 0/17
```

```
!
interface FastEthernet 0/18
!
interface FastEthernet 0/19
!
interface FastEthernet 0/20
!
interface FastEthernet 0/21
!
interface FastEthernet 0/22
!
interface FastEthernet 0/23
!
interface FastEthernet 0/24
!
interface GigabitEthernet 0/25
!
interface GigabitEthernet 0/26
!
interface GigabitEthernet 0/27
!
interface GigabitEthernet 0/28
!
interface VLAN 10
 ip address 172.16.1.2 255.255.255.0
!
interface VLAN 50
 ip address 172.16.5.1 255.255.255.0
!
router ospf 1
 network 172.16.1.0 0.0.0.255 area 0
 network 172.16.5.0 0.0.0.255 area 0
!
line con 0
line vty 0 4
 login
!
end
```

RB#show running-config

```
Building configuration...
Current configuration : 579 bytes
```

```
!
version  RGNOS  10.1.00(4),  Release(18443)(Tue  Jul  17  20:50:30  CST  2007
-ubu1server)
hostname RB
!
interface FastEthernet 0/0
 ip address 172.16.2.2 255.255.255.0
 duplex auto
 speed auto
!
interface FastEthernet 0/1
 ip address 172.16.3.1 255.255.255.0
 duplex auto
 speed auto
!
interface Loopback 0
!
router ospf 1
 network 172.16.2.0 0.0.0.255 area 0
 network 172.16.3.0 0.0.0.255 area 0
!
line con 0
line aux 0
line vty 0 4
 login
!
end
```

RA#show running-config

```
Building configuration...
Current configuration : 554 bytes
!
version  RGNOS  10.1.00(4),  Release(18443)(Tue  Jul  17  20:50:30  CST  2007
-ubu1server)
hostname RA
!
interface FastEthernet 0/0
 ip address 172.16.2.1 255.255.255.0
 duplex auto
 speed auto
!
interface FastEthernet 0/1
```

```
 ip address 172.16.1.1 255.255.255.0
 duplex auto
 speed auto
!
router ospf 1
 network 172.16.1.0 0.0.0.255 area 0
 network 172.16.2.0 0.0.0.255 area 0
!
line con 0
line aux 0
line vty 0 4
 login
!
end
```

实验 14　OSPF 单区域配置

【实验名称】

OSPF 单区域配置。

【实验目的】

配置 OSPF 单区域实验，实现简单的 OSPF 配置。

【背景描述】

拓扑图中有 3 台路由器，共有 5 个网段，并且是无类的子网。

【需求分析】

在以下拓扑图中，使用 OSPF 路由协议学习路由信息，并且使用的是单区域，所有的路由器都在区域 0 中。

【实验拓扑】

实验的拓扑图，如图 14-1 所示。

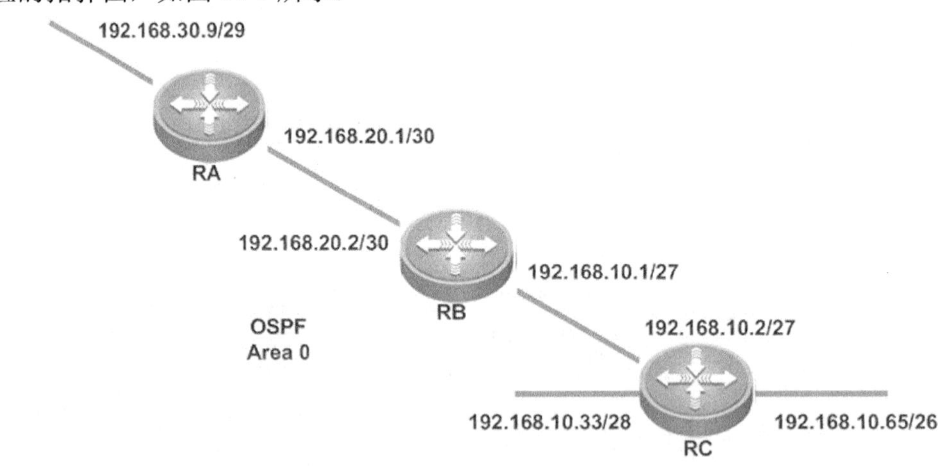

图 14-1

【实验设备】

路由器 3 台。

【预备知识】

路由器基本配置知识、OSPF。

【实验原理】

在路由器上启用 OSFP 进程，所有的路由信息通过 OSFP 路由协议传递。

【实验步骤】

步骤 1　在路由器上配置 IP 地址。

```
RA#config  t
RA(config)# interface FastEthernet 0/0
RA(config-if)#ip address 192.168.20.1 255.255.255.252
RA(config)#interface Loopback 0
RA(config-if)#ip address 192.168.30.9 255.255.255.248
RB#config  t
RB(config)# interface FastEthernet 0/0
RB(config-if)#ip address 192.168.20.2 255.255.255.252
RB(config)#interface FastEthernet 0/1
RB(config-if)#ip address 192.168.10.1 255.255.255.224
RC#config  t
RC(config)# interface FastEthernet 0/0
RC(config-if)#ip address 192.168.10.2 255.255.255.224
RC(config)#interface Loopback 0
RC(config-if)#ip address 192.168.10.33 255.255.255.240
RC(config)#interface Loopback 1
RC(config-if)#ip address 192.168.10.65 255.255.255.192
```

步骤 2　配置 OSPF。

```
RA(config)#router ospf 10
RA(config-router)#network 192.168.30.8 0.0.0.7 area 0
    RA(config-router)#network 192.168.20.0 0.0.0.3 area 0
RB(config)# router ospf 10
RB(config-router)#network 192.168.10.0 0.0.0.31 area 0
RB(config-router)#network 192.168.20.0 0.0.0.3 area 0

RC(config)# router ospf 10
RC(config-router)#network 192.168.10.0 0.0.0.31 area 0
RC(config-router)#network 192.168.10.32 0.0.0.15 area 0
    RC(config-router)#network 192.168.10.64 0.0.0.63 area 0
```

步骤 3　验证测试。

```
RA#show ip interface brief
Interface                  IP-Address(Pri)     OK?       Status
FastEthernet 0/0          192.168.20.1/30      YES       UP
FastEthernet 0/1           no address          YES       DOWN
Loopback 0                 192.168.30.9/29     YES       UP
RB#show ip interface brief
Interface                  IP-Address(Pri)      OK?       Status
FastEthernet 0/0          192.168.20.2/30      YES       UP
FastEthernet 0/1          192.168.10.1/27      YES       UP
RC#show ip interface brief
Interface                  IP-Address(Pri)      OK?       Status
FastEthernet 0/0            192.168.10.2/27      YES       UP
```

```
FastEthernet 0/1               no address        YES      DOWN
Loopback 0                     192.168.10.33/28   YES      UP
Loopback 1                     192.168.10.65/26   YES      UP
```

用命令 show ip route 和 sh ip ospf neighbor 来验证配置

```
RA#sh ip route
Codes:  C - connected, S - static,  R - RIP B - BGP
        O - OSPF, IA - OSPF inter area
        N1 - OSPF NSSA external type 1, N2 - OSPF NSSA external type 2
        E1 - OSPF external type 1, E2 - OSPF external type 2
        i - IS-IS, L1 - IS-IS level-1, L2 - IS-IS level-2, ia - IS-IS inter area
        * - candidate default
Gateway of last resort is no set
O    192.168.10.0/27 [110/2] via 192.168.20.2, 00:01:32, FastEthernet 0/0
C    192.168.30.8/29 is directly connected, Loopback 0
C    192.168.30.9/32 is local host.
O    192.168.10.33/32 [110/2] via 192.168.20.2, 00:01:32, FastEthernet 0/0
O    192.168.10.65/32 [110/2] via 192.168.20.2, 00:01:32, FastEthernet 0/0
C    192.168.20.0/30 is directly connected, FastEthernet 0/0
C    192.168.20.1/32 is local host.
RB#show ip route
Codes:  C - connected, S - static,  R - RIP B - BGP
        O - OSPF, IA - OSPF inter area
        N1 - OSPF NSSA external type 1, N2 - OSPF NSSA external type 2
        E1 - OSPF external type 1, E2 - OSPF external type 2
        i - IS-IS, L1 - IS-IS level-1, L2 - IS-IS level-2, ia - IS-IS inter area
        * - candidate default
Gateway of last resort is no set
C    192.168.10.0/27 is directly connected, FastEthernet 0/1
C    192.168.10.1/32 is local host.
O    192.168.10.33/32 [110/1] via 192.168.10.2, 00:02:25, FastEthernet 0/1
O    192.168.10.65/32 [110/1] via 192.168.10.2, 00:02:14, FastEthernet 0/1
C    192.168.20.0/30 is directly connected, FastEthernet 0/0
C    192.168.20.2/32 is local host.
O    192.168.30.9/32 [110/1] via 192.168.20.1, 00:05:16, FastEthernet 0/0
RC#show ip route
Codes:  C - connected, S - static,  R - RIP B - BGP
        O - OSPF, IA - OSPF inter area
        N1 - OSPF NSSA external type 1, N2 - OSPF NSSA external type 2
        E1 - OSPF external type 1, E2 - OSPF external type 2
        i - IS-IS, L1 - IS-IS level-1, L2 - IS-IS level-2, ia - IS-IS inter area
        * - candidate default
Gateway of last resort is no set
```

```
C    192.168.10.0/27 is directly connected, FastEthernet 0/0
C    192.168.10.2/32 is local host.
C    192.168.10.32/28 is directly connected, Loopback 0
C    192.168.10.33/32 is local host.
C    192.168.10.64/26 is directly connected, Loopback 1
C    192.168.10.65/32 is local host.
O    192.168.20.0/30 [110/2] via 192.168.10.1, 00:01:23, FastEthernet 0/0
O    192.168.30.9/32 [110/2] via 192.168.10.1, 00:01:23, FastEthernet 0/0
RB#show ip ospf neighbor
OSPF process 1:
Neighbor ID     Pri   State      Dead Time   Address         Interface
192.168.10.65     1   Full/BDR  00:00:38    192.168.10.2    FastEthernet 0/1
192.168.30.9      1   Full/DR    00:00:33    192.168.20.1    FastEthernet 0/0
RB#show ip ospf neighbor detail
 Neighbor 192.168.10.65, interface address 192.168.10.2
   In the area 0.0.0.0 via interface FastEthernet 0/1
   Neighbor priority is 1, State is Full, 5 state changes
   DR is 192.168.10.1, BDR is 192.168.10.2
   Options is 0x42 (*|O|-|-|-|-|E|-)
   Dead timer due in 00:00:37
   Neighbor is up for 00:04:33
   Database Summary List 0
   Link State Request List 0
   Link State Retransmission List 0
   Crypt Sequence Number is 0
   Thread Inactivity Timer on
   Thread Database Description Retransmission off
   Thread Link State Request Retransmission off
   Thread Link State Update Retransmission off
 Neighbor 192.168.30.9, interface address 192.168.20.1
   In the area 0.0.0.0 via interface FastEthernet 0/0
   Neighbor priority is 1, State is Full, 6 state changes
   DR is 192.168.20.1, BDR is 192.168.20.2
   Options is 0x42 (*|O|-|-|-|-|E|-)
   Dead timer due in 00:00:30
   Neighbor is up for 00:07:50
   Database Summary List 0
   Link State Request List 0
   Link State Retransmission List 0
   Crypt Sequence Number is 0
   Thread Inactivity Timer on
   Thread Database Description Retransmission off
```

```
   Thread Link State Request Retransmission off
 Thread Link State Update Retransmission off
```

【注意事项】

在做本实验前，注意子网掩码的换算。

【参考配置】

```
RA#show running-config
Building configuration...
Current configuration : 587 bytes
!
version RGNOS 10.1.00(4), Release(18443)(Tue Jul 17 20:50:30 CST 2007
-ubu1server)
hostname RA
!
interface FastEthernet 0/0
 ip address 192.168.20.1 255.255.255.252
 duplex auto
 speed auto
!
interface FastEthernet 0/1
 duplex auto
 speed auto
!
interface Loopback 0
 ip address 192.168.30.9 255.255.255.248
!
router ospf 1
 network 192.168.20.0 0.0.0.3 area 0
 network 192.168.30.8 0.0.0.7 area 0
!
line con 0
line aux 0
line vty 0 4
 login
!
end
RB#show running-config
Building configuration...
Current configuration : 563 bytes
!
version RGNOS 10.1.00(4), Release(18443)(Tue Jul 17 20:50:30 CST 2007
-ubu1server)
```

```
hostname RB
!
interface FastEthernet 0/0
 ip address 192.168.20.2 255.255.255.252
 duplex auto
 speed auto
!
interface FastEthernet 0/1
 ip address 192.168.10.1 255.255.255.224
 duplex auto
 speed auto
!
router ospf 1
 network 192.168.10.0 0.0.0.31 area 0
 network 192.168.20.0 0.0.0.3 area 0
!
line con 0
line aux 0
line vty 0 4
 login
!
end
```

RC#show running-config

```
Building configuration...
Current configuration : 699 bytes
!
version RGNOS 10.1.00(4), Release(18443)(Tue Jul 17 20:50:30 CST 2007
-ubu1server)
hostname RC
!
interface FastEthernet 0/0
 ip address 192.168.10.2 255.255.255.224
 duplex auto
 speed auto
!
interface FastEthernet 0/1
 duplex auto
 speed auto
!
interface Loopback 0
 ip address 192.168.10.33 255.255.255.240
!
```

```
interface Loopback 1


 ip address 192.168.10.65 255.255.255.192
!
router ospf 1
 network 192.168.10.0 0.0.0.31 area 0
 network 192.168.10.32 0.0.0.15 area 0
 network 192.168.10.64 0.0.0.63 area 0
!
line con 0
line aux 0
line vty 0 4
 login
!
end
```

实验 15　广域网协议的封装

【实验名称】

广域网协议的封装。

【实验目的】

掌握广域网协议的封装类型和封装方法。

【背景描述】

假设你是某公司的网络管理员，两个分公司之间希望能够申请一条广域网专线进行连接。公司现有锐捷路由器两台，希望你了解该设备的广域网接口所支持的协议，以确定选择哪一种广域网链路。

【需求分析】

查看路由器广域网接口支持的数据链路层协议，并进行正确的封装。

【实验拓扑】

实验的拓扑图，如图 15-1 所示。

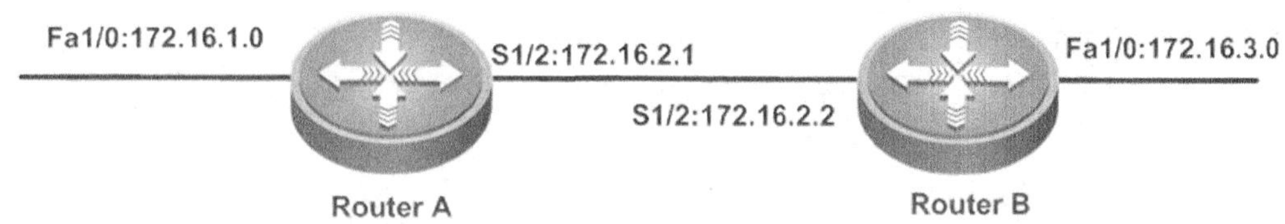

图 15-1

【预备知识】

路由器基本配置知识、广域网知识。

【实验设备】

路由器 2 台。

【实验原理】

常见的广域网专线技术有 DDN 专线、PSTN/ISDN 专线、帧中继专线和 X.25 专线等。数据链路层提供各种专线技术的协议，主要有 PPP、HDLC、X.25、Frame-relay 及 ATM 等。

【实验步骤】

步骤 1　路由器的基本配置。

```
Router A(config)#interface serial 4/0
Router A(config-if)#ip address 172.16.2.1 255.255.255.0
Router B(config)#interface serial 4/0
Router B(config-if)#ip address 172.16.2.2 255.255.255.0
```

步骤 2　封装 HDLC。

```
Router A(config)#interface serial 4/0
Router A (config-if)#encapsulation hdlc
Router B(config)#interface serial 4/0
Router B(config-if)#encapsulation hdlc
```

验证广域网接口的封装类型：

```
Router A#show interfaces serial 4/0
Index(dec):1 (hex):1
serial 4/0 is UP, line protocol is UP
Hardware is Infineon DSCC4 PEB20534 H-10 serial
Interface address is: 172.16.2.2/24
  MTU 1500 bytes, BW 2000 Kbit
  Encapsulation protocol is HDLC, loopback not set
  Keepalive interval is 10 sec, set
  Carrier delay is 2 sec
  RXload is 1,Txload is 1
  Queueing strategy: WFQ
    11421118 carrier transitions
    V35 DTE cable
    DCD=up  DSR=up  DTR=up  RTS=up  CTS=up
  5 minutes input rate 17 bits/sec, 0 packets/sec
  5 minutes output rate 17 bits/sec, 0 packets/sec
    57 packets input, 1664 bytes, 0 no buffer, 0 dropped
    Received 52 broadcasts, 0 runts, 0 giants
    0 input errors, 0 CRC, 0 frame, 0 overrun, 0 abort
    68 packets output, 2726 bytes, 0 underruns, 0 dropped
    0 output errors, 0 collisions, 0 interface resets
```

注意锐捷路由器广域网接口默认封装的就是 HDLC。

步骤 3　封装 PPP。

```
Router A#configure terminal
Enter configuration commands, one per line.  End with CNTL/Z.
Router A (config)#interface serial 4/0
Router A (config-if)#encapsulation ppp
Router B#configure terminal
Enter configuration commands, one per line.  End with CNTL/Z.
Router B (config)#interface serial 4/0
Router B (config-if)#encapsulation ppp
```

验证广域网接口的封装类型：

```
Router A#show interfaces serial 4/0
Index(dec):1 (hex):1
serial 4/0 is UP, line protocol is UP
Hardware is Infineon DSCC4 PEB20534 H-10 serial
Interface address is: 172.16.2.1/24
```

```
MTU 1500 bytes, BW 2000 Kbit
Encapsulation protocol is PPP, loopback not set
Keepalive interval is 10 sec, set
Carrier delay is 2 sec
RXload is 1,Txload is 1
LCP Open
Open: ipcp
Queueing strategy: WFQ
  11421118 carrier transitions
  V35 DCE cable
  DCD=up  DSR=up  DTR=up  RTS=up  CTS=up
5 minutes input rate 30 bits/sec, 0 packets/sec
5 minutes output rate 19 bits/sec, 0 packets/sec
  123 packets input, 3638 bytes, 0 no buffer, 28 dropped
  Received 68 broadcasts, 0 runts, 0 giants
  0 input errors, 0 CRC, 0 frame, 0 overrun, 0 abort
  89 packets output, 2312 bytes, 0 underruns, 0 dropped
  0 output errors, 0 collisions, 7 interface resets
```

【注意事项】

封装广域网协议时，要求 V.35 线缆的两个端口上的封装协议一致，否则将无法建立链路。

【参考配置】

无。

实验 16　PPP PAP 认证

【实验名称】

PPP PAP 认证。

【实验目的】

掌握 PPP PAP 认证的过程及配置。

【背景描述】

假设你是公司的网络管理员，公司为了满足不断增长的业务需求，申请了专线接入，当客户端路由器与 ISP 进行链路协商时，需要验证身份，配置路由器以保证链路的建立，并考虑其安全性。

【需求分析】

在链路协商时保证安全验证。链路协商时，用户名、密码以明文的方式传输。

【实验拓扑】

实验的拓扑图，如图 16-1 所示。

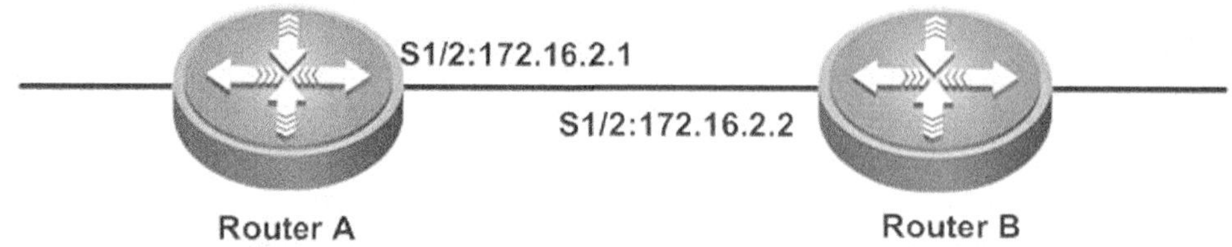

图 16-1

【预备知识】

路由器基本配置知识、PPP PAP 知识。

【实验设备】

路由器（带串口）2 台。

V.35 线缆（DTE/DCE）1 对。

【实验原理】

PPP 协议位于 OSI 七层模型的数据链路层，PPP 协议按照功能划分为两个子层：LCP、NCP。LCP 主要负责链路的协商、建立、回拨、认证、数据的压缩、多链路捆绑等功能。NCP 主要负责与上层的协议进行协商，为网络层协议提供服务。

PPP 的认证功能是指在建立 PPP 链路的过程中进行密码的验证，若验证通过就建立连接，若验证不通过则拆除链路。

PPP 协议支持两种认证方式：PAP 和 CHAP。PAP（Password Authentication Protocol，密码

验证协议）是指验证双方通过两次握手完成验证过程，它是一种用于对试图登录到点对点协议服务器上的用户进行身份验证的方法。由被验证方主动发出验证请求，其中包含了验证的用户名和密码。验证方验证后做出回复：通过验证或验证失败。在验证过程中，用户名和密码以明文的方式在链路上传输。

【实验步骤】

步骤 1　路由器的基本配置。

```
Router(config)#hostname Router A
Router A(config)#interface serial 4/0
Router A(config-if)#ip address 172.16.2.1 255.255.255.0
Router A(config-if)# encapsulation ppp
Router(config)#hostname Router B
Router B(config)#interface serial 4/0
Router B(config-if)#ip address 172.16.2.2 255.255.255.0
Router B(config-if)#encapsulation ppp
```

步骤 2　配置 PAP 认证。

```
Router A(config)#interface serial 4/0
Router A(config-if)#ppp pap sent-username RouterA password 0 123
Router B(config)#username RouterA password 123
Router B(config)#interface serial 4/0
Router B(config-if)#ppp authentication pap
```

步骤 3　验证 PAP 认证。

```
Router A#show interfaces serial 4/0
Index(dec):1 (hex):1
serial 4/0 is UP, line protocol is UP
Hardware is Infineon DSCC4 PEB20534 H-10 serial
Interface address is: 172.16.2.1/24
  MTU 1500 bytes, BW 2000 Kbit
  Encapsulation protocol is PPP, loopback not set
  Keepalive interval is 10 sec, set
  Carrier delay is 2 sec
  RXload is 1,Txload is 1
  LCP Open
  Open: ipcp
  Queueing strategy: WFQ
    11421118 carrier transitions
    V35 DCE cable
    DCD=up  DSR=up  DTR=up  RTS=up  CTS=up
  5 minutes input rate 54 bits/sec, 0 packets/sec
  5 minutes output rate 46 bits/sec, 0 packets/sec
    677 packets input, 14796 bytes, 0 no buffer, 28 dropped
    Received 68 broadcasts, 0 runts, 0 giants
```

```
    0 input errors, 0 CRC, 0 frame, 0 overrun, 0 abort
    655 packets output, 11719 bytes, 0 underruns, 5 dropped
    0 output errors, 0 collisions, 18 interface resets
```

使用 debug ppp authentication 命令验证配置。

```
Router B#debug ppp authentication
Router B#conf t
Enter configuration commands, one per line.  End with CNTL/Z.
Router B(config)#interface serial 4/0
Router B(config-if)#shutdown
Router B(config-if)#Sep  1 23:33:37 RouterB %7:%LINK CHANGED: Interface serial 4/0, changed state to administratively down
Sep  1 23:33:37 RouterB %7:%LINE PROTOCOL CHANGE: Interface serial 4/0, changed state to DOWN
Router B(config-if)#no shutdown
Router B(config-if)#Sep  1 23:33:43 RouterB %7:PPP: ppp clear author(), protocol = LCP
Sep  1 23:33:43 RouterB %7:%LINK CHANGED: Interface serial 4/0, changed state to up
Sep  1 23:33:45 RouterB %7:PPP: serial 4/0 [I] PAP-REQ id 2 len 12
Sep  1 23:33:45 RouterB %7:PPP: Authenticating peer serial 4/0
Sep  1 23:33:45 RouterB %7:PPP: serial 4/0 PAP authentication OK!
Sep  1 23:33:45 RouterB %7:PPP: serial 4/0 [O] PAP SUCCESS id 2 len 1
Sep  1 23:33:45 RouterB %7::PPP: serial 4/0 authentication OK, begin networkphase!
Sep  1 23:33:45 RouterB %7:PPP: ppp clear author(), protocol = IPCP
Sep  1 23:33:46 RouterB %7:%LINE PROTOCOL CHANGE: Interface serial 4/0, changed state to UP
```

【注意事项】

封装广域网协议时，要求 V.35 线缆的两个端口的封装协议保持一致，否则无法建立链路。

【参考配置】

```
Router A#show running-config
Building configuration...
Current configuration : 593 bytes
!
version RGNOS 10.1.00(4), Release(18443)(Tue Jul 17 21:16:17 CST 2007 -ubu1server)
hostname Router A
!
interface serial 4/0
 encapsulation PPP
 ppp pap sent-username RouterA password 7 001b7210
 ip address 172.16.2.1 255.255.255.0
```

```
 clock rate 64000
!
interface serial 4/1
 clock rate 64000
!
interface GigabitEthernet 0/0
 duplex auto
 speed auto
!
interface GigabitEthernet 0/1
 duplex auto
 speed auto
!
line con 0
line aux 0
line vty 0 4
 login
!
end
```

Router B#show running-config

```
Building configuration...
Current configuration : 580 bytes
!
version  RGNOS  10.1.00(4),  Release(18443)(Tue  Jul  17  21:16:17  CST  2007
-ubu1server)
hostname Router B
!
username RouterA password 0 123
!
interface serial 4/0
 encapsulation PPP
 ppp authentication pap
 ip address 172.16.2.2 255.255.255.0
!
interface serial 4/1
 clock rate 64000
!
interface GigabitEthernet 0/0
 duplex auto
 speed auto
!
interface GigabitEthernet 0/1
```

```
 duplex auto
 speed auto
!
line con 0
line aux 0
line vty 0 4
 login
!
end
```

实验 17　PPP CHAP 认证

【实验名称】

PPP CHAP 认证。

【实验目的】

掌握 PPP CHAP 认证的过程及配置。

【背景描述】

假设你是某公司的网络管理员，公司为了满足不断增长的业务需求，申请了专线接入，当客户端路由器与 ISP 进行链路协商时，需要验证身份，配置路由器以保证链路的建立，并考虑其安全性。

【需求分析】

在链路协商时保证安全验证。链路协商时 MD5 以密文的方式传输。

【实验拓扑】

实验的拓扑图，如图 17-1 所示。

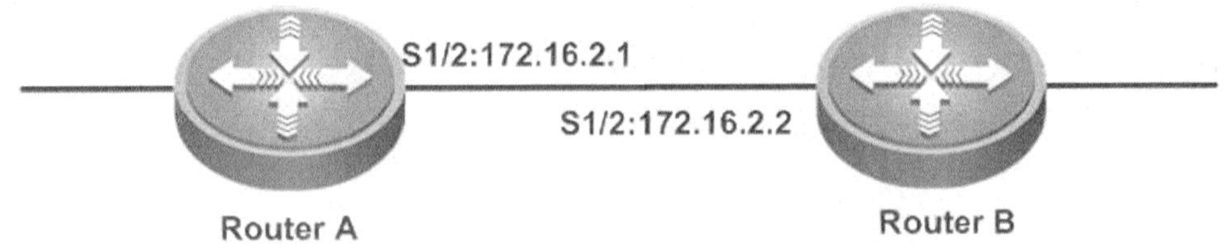

图 17-1

【预备知识】

路由器基本配置知识、PPP CHAP 知识。

【实验设备】

路由器（带串口）2 台。

V.35 线缆（DTE/DCE）1 对。

【实验原理】

PPP 协议位于 OSI 七层模型的数据链路层，PPP 协议按照功能划分为两个子层：LCP、NCP。LCP 主要负责链路的协商、建立、回拨、认证、数据的压缩、多链路捆绑等功能。NCP 主要负责与上层的协议进行协商，为网络层协议提供服务。

PPP 的认证功能是指在建立 PPP 链路的过程中进行密码的验证，若验证通过就建立连接，若验证不通过则拆除链路。

CHAP（Challenge Handshake Authentication Protocol，挑战式握手验证协议）是指验证双方通过三次握手完成验证过程，此方式比 PAP 更安全。CHAP 由验证方主动发出挑战报文，由被验证方应答。在整个验证过程中，链路上传递的信息都进行了加密处理。

【实验步骤】

步骤 1　路由器的基本配置。

```
Router(config)#hostname Router A
Router A(config)#interface serial 4/0
Router A(config-if)#ip address 172.16.2.1 255.255.255.0
Router A(config-if)#encapsulation ppp
Router(config)#hostname Router B
Router B(config)#interface serial 4/0
Router B(config-if)#ip address 172.16.2.2 255.255.255.0
Router B(config-if)#encapsulation ppp
```

步骤 2　配置 CHAP 认证。

```
Router A(config)# username RouterB password 0 123
Router B(config)#username RouterA password 0123
Router B(config)#interface serial 4/0
Router B(config-if)#ppp authentication chap
```

步骤 3　验证 CHAP 认证。

```
Router A#show interfaces serial 4/0
Index(dec):1 (hex):1
serial 4/0 is UP, line protocol is UP
Hardware is Infineon DSCC4 PEB20534 H-10 serial
Interface address is: 172.16.2.1/24
  MTU 1500 bytes, BW 2000 Kbit
  Encapsulation protocol is PPP, loopback not set
  Keepalive interval is 10 sec, set
  Carrier delay is 2 sec
  RXload is 1,Txload is 1
  LCP Open
  Open: ipcp
  Queueing strategy: WFQ
    11421118 carrier transitions
    V35 DCE cable
    DCD=up  DSR=up  DTR=up  RTS=up  CTS=up
  5 minutes input rate 45 bits/sec, 0 packets/sec
  5 minutes output rate 44 bits/sec, 0 packets/sec
    889 packets input, 18810 bytes, 0 no buffer, 28 dropped
    Received 68 broadcasts, 0 runts, 0 giants
    0 input errors, 0 CRC, 0 frame, 0 overrun, 0 abort
    848 packets output, 15203 bytes, 0 underruns, 5 dropped
    0 output errors, 0 collisions, 28 interface resets
```

使用 debug ppp authentication 命令验证配置。

```
Router A#debug ppp authentication
Router A#configure terminal
```

```
Enter configuration commands, one per line.  End with CNTL/Z.
Router A(config)#interface serial 4/0
Router A(config-if)#shutdown
Router A(config-if)#Aug  9 01:46:10 RouterA %7:%LINK CHANGED: Interface serial
4/0, changed state to administratively down
Aug  9 01:46:10 RouterA %7:%LINE PROTOCOL CHANGE: Interface serial 4/0, changed
state to DOWN
Router A(config-if)#no shutdown
Router A(config-if)#Aug  9 01:46:22 RouterA %7:PPP: ppp clear author(), protocol
= LCP
Aug  9 01:46:22 RouterA %7:%LINK CHANGED: Interface serial 4/0, changed state to
up
RouterA(config-if)#Aug  9 01:46:38 RouterA %7:PPP: serial 4/0 [I] CHAP CHALLENGE
id 17 len 24
Aug  9 01:46:38 RouterA %7:PPP: serial 4/0 recv CHAP challenge from RouterB
Aug  9 01:46:38 RouterA %7:PPP: serial 4/0 Search Password in local.
Aug  9 01:46:38 RouterA %7:PPP: serial 4/0 [I] CHAP CHALLENGE id 18 len 24
Aug  9 01:46:38 RouterA %7:PPP: serial 4/0 recv CHAP challenge from RouterB
Aug  9 01:46:38 RouterA %7:PPP: serial 4/0 Search Password in local.
Aug  9 01:46:38 RouterA %7:PPP: serial 4/0 [I] CHAP SUCCESS id 18 len 0
Aug  9 01:46:38 RouterA %7::PPP: serial 4/0 authentication OK, begin networkphase!
Aug  9 01:46:38 RouterA %7:PPP: ppp clear author(), protocol = IPCP
Aug  9 01:46:39 RouterA %7:%LINE PROTOCOL CHANGE: Interface serial 4/0, changed
state to UP
```

【注意事项】

封装广域网协议时，要求 V.35 线缆的两个端口的封装协议保持一致，否则无法建立链路。

【参考配置】

```
Router A#show running-config
Building configuration...
Current configuration : 574 bytes
!
version RGNOS 10.1.00(4), Release(18443)(Tue Jul 17 21:16:17 CST 2007
-ubu1server)
hostname Router A
!
username RouterB password 0 123
!
interface serial 4/0
 encapsulation PPP
 ip address 172.16.2.1 255.255.255.0
 clock rate 64000
```

```
!
interface serial 4/1
 clock rate 64000
!
interface GigabitEthernet 0/0
 duplex auto
 speed auto
!
interface GigabitEthernet 0/1
 duplex auto
 speed auto
!
line con 0
line aux 0
line vty 0 4
 login
!
end
```

Router B#show running-config

```
Building configuration...
Current configuration : 581 bytes
!
version  RGNOS  10.1.00(4),  Release(18443)(Tue  Jul  17  21:16:17  CST  2007
-ubu1server)
hostname Router B
!
username RouterA password 0 123
!
interface serial 4/0
 encapsulation PPP
 ppp authentication chap
 ip address 172.16.2.2 255.255.255.0
!
interface serial 4/1
 clock rate 64000
!
interface GigabitEthernet 0/0
 duplex auto
 speed auto
!
interface GigabitEthernet 0/1
 duplex auto
```

```
 speed auto
!
line con 0
line aux 0
line vty 0 4
 login
!
end
```

实验 18　交换机的端口安全

【实验名称】

交换机的端口安全。

【实验目的】

掌握交换机的端口安全功能，控制用户的安全接入。

【背景描述】

假设你是某公司的网络管理员，公司要求对网络进行严格控制。为了防止公司内部用户的 IP 地址冲突，防止公司内部的网络攻击和破坏行为。为每一位员工分配了固定的 IP 地址，并且只允许公司员工的主机使用网络，不得随意连接其他主机。例如，某员工分配的 IP 地址是 172.16.1.55/24，主机 MAC 地址是 00-06-1B-DE-13-B4，该主机连接在 1 台 2126G 上。

【需求分析】

针对交换机的所有端口，配置最大连接数为 1，针对 PC1 主机的接口进行 IP＋MAC 地址绑定。

【实验拓扑】

实验的拓扑图，如图 18-1 所示。

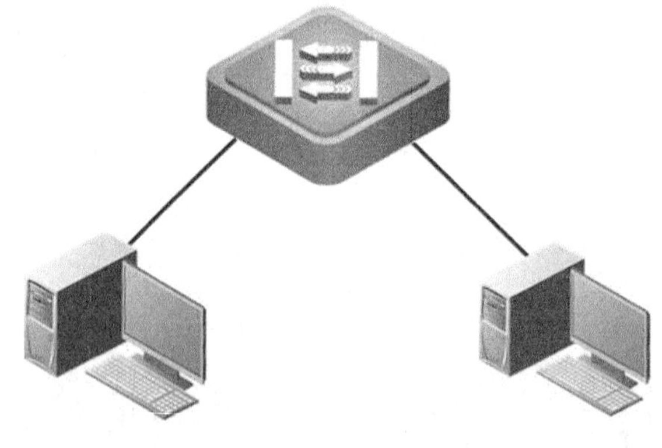

图 18-1

【预备知识】

路由器基本配置知识、端口安全知识。

【实验设备】

交换机 1 台。

PC 机 1 台。

直连网线 1 条。

【实验原理】

交换机端口安全功能是指针对交换机的端口进行安全属性的配置，从而控制用户的安全接入。交换机端口安全主要有两种类型：一是限制交换机端口的最大连接数，二是针对交换机端口进行 MAC 地址、IP 地址的绑定。

限制交换机端口的最大连接数可以控制交换机端口下连的主机数，并防止用户进行恶意的 ARP 欺骗。

交换机端口的地址绑定，可以针对 IP 地址、MAC 地址、IP＋MAC 进行灵活的绑定。可以实现对用户的严格控制。保证用户的安全接入和防止常见的内网网络攻击。如 ARP 欺骗，IP 或 MAC 地址欺骗，IP 地址攻击等。

配置了交换机的端口安全功能后，当实际应用超出了配置的要求时，将产生一个安全违例，产生安全违例的处理方式有以下 3 种。

- protect：当安全地址个数满后，安全端口将丢弃未知名地址（不是该端口的安全地址中的任何一个）的包。
- restrict：当违例产生时，将发送一个 Trap 通知。
- shutdown：当违例产生时，将关闭端口并发送一个 Trap 通知。

当端口因为违例而被关闭后，在全局配置模式下使用命令 errdisable recovery 来将接口从错误状态中恢复过来。

【实验步骤】

步骤 1　配置交换机端口的最大连接数限制。

```
S3750#configure terminal
S3750(config)#interface range fastethernet 0/1-23
S3750(config-if-range)#switchport port-security
S3750(config-if-range)#switchport port-secruity maximum 1
S3750(config-if-range)#switchport port-secruity violation shutdown
```

步骤 2　验证交换机端口的最大连接数限制。

```
S3750#sh port-security
Secure Port           MaxSecureAddr(count) CurrentAddr(count) Security Action
----------------------------------------------------------------
     FastEthernet 0/1          1              0        Shutdown
     FastEthernet 0/2          1              0        Shutdown
     FastEthernet 0/3          1              0        Shutdown
     FastEthernet 0/4          1              0        Shutdown
     FastEthernet 0/5          1              0        Shutdown
     FastEthernet 0/6          1              0        Shutdown
     FastEthernet 0/7          1              0        Shutdown
     FastEthernet 0/8          1              0        Shutdown
     FastEthernet 0/9          1              0        Shutdown
     FastEthernet 0/10          1              0        Shutdown
     FastEthernet 0/11          1              0        Shutdown
     FastEthernet 0/12          1              0        Shutdown
     FastEthernet 0/13          1              0        Shutdown
     FastEthernet 0/14          1              0        Shutdown
     FastEthernet 0/15          1              0        Shutdown
     FastEthernet 0/16          1              0        Shutdown
```

```
    FastEthernet 0/17          1              0         Shutdown
    FastEthernet 0/18          1              0         Shutdown
    FastEthernet 0/19          1              0         Shutdown
    FastEthernet 0/20          1              0         Shutdown
    FastEthernet 0/21          1              0         Shutdown
    FastEthernet 0/22          1              0         Shutdown
    FastEthernet 0/23          1              0         Shutdown
S3750#show port-security interface fastEthernet 0/1
Interface : FastEthernet 0/1
Port Security : Enabled
Port status : up
Violation mode : Shutdown
Maximum MAC Addresses : 1
Total MAC Addresses : 0
Configured MAC Addresses : 0
Aging time : 0 mins
SecureStatic address aging : Disabled
```

步骤 3　配置交换机端口的 MAC 与 IP 地址绑定。

查看主机的 IP 和 MAC 地址信息。

在主机上打开 CMD 命令提示符窗口，执行 ipconfig /all 命令，如图 18-2 所示。

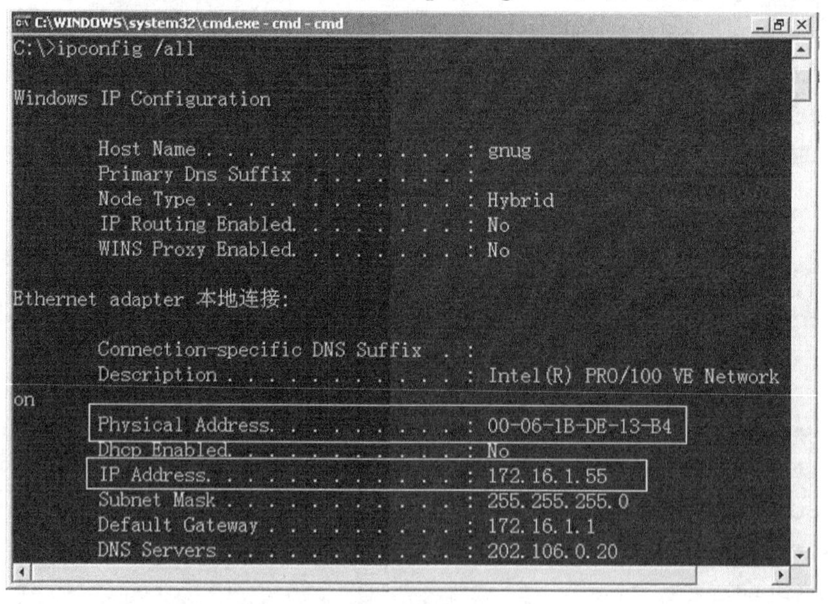

图 18-2

配置交换机端口的地址绑定。

```
S3750#configure terminal
S3750(config)#interface  fastethernet 0/3
S3750(config-if)#switchport port-security
S3750(config-if)#switchport port-security mac-address 0006.1bde.13b4 ip-address
172.16.1.55
```

步骤 4　查看地址安全绑定配置。

```
S3750#sh port-security address all
Vlan Port  Arp-Check Mac Address    IP Address   Type    Remaining Age(mins)
---- ------------------------- -------------- --------- -------------------
1 FastEthernet 0/3 Disabled 0006.1bde.13b4    172.16.1.55 Configured    -
S3750#sh port-security address interface fa0/3
Vlan Mac Address     IP Address      Type        Port            Remaining Age(mins)
---- ----------------- ---------- ---------------------- -------------------
  1  0006.1bde.13b4    172.16.1.55 Configured       FastEthernet 0/3          -
```

步骤 5　配置交换机端口的 IP 地址绑定。

```
S3750(config)#int fastEthernet 0/2
S3750(config-if)#switchport port-security ip-address 10.1.1.1
S3750#show port-security address all
Vlan Port  Arp-Check Mac Address    IP Address  Type     Remaining Age(mins)
---- ------------------------- -------------- --------- -------------------
1   FastEthernet 0/2   Disabled             10.1.1.1 Configured        -
1  FastEthernet 0/3   Disabled 0006.1bde.13b4  172.16.1.55 Configured    -
```

【注意事项】

（1）交换机端口安全功能只能在 ACCESS 接口进行配置。

（2）交换机最大连接数限制的取值范围是 1~128，默认是 128。

（3）交换机最大连接数限制的默认处理方式是 protect。

【参考配置】

```
S3750#sh running-config
Building configuration...
Current configuration : 3759 bytes
!
version  RGNOS  10.1.00(4),  Release(18443)(Tue  Jul  17  19:51:54  CST  2007
-ubu6server)
!
vlan 1
!
interface FastEthernet 0/1
 switchport port-security maximum 1
 switchport port-security violation shutdown
 switchport port-security
!
interface FastEthernet 0/2
 switchport port-security ip-address 10.1.1.1
 switchport port-security maximum 1
 switchport port-security violation shutdown
 switchport port-security
```

```
!
interface FastEthernet 0/3
 switchport port-security mac-address 0006.1bde.13b4 ip-address 172.16.1.55
 switchport port-security maximum 1
 switchport port-security violation shutdown
 switchport port-security
!
interface FastEthernet 0/4
 switchport port-security maximum 1
 switchport port-security violation shutdown
 switchport port-security
!
interface FastEthernet 0/5
 switchport port-security maximum 1
 switchport port-security violation shutdown
 switchport port-security
!
interface FastEthernet 0/6
 switchport port-security maximum 1
 switchport port-security violation shutdown
 switchport port-security
!
interface FastEthernet 0/7
 switchport port-security maximum 1
 switchport port-security violation shutdown
 switchport port-security
!
interface FastEthernet 0/8
 switchport port-security maximum 1
 switchport port-security violation shutdown
 switchport port-security
!
interface FastEthernet 0/9
 switchport port-security maximum 1
 switchport port-security violation shutdown
 switchport port-security
!
interface FastEthernet 0/10
 switchport port-security maximum 1
 switchport port-security violation shutdown
 switchport port-security
!
```

```
interface FastEthernet 0/11
 switchport port-security maximum 1
 switchport port-security violation shutdown
 switchport port-security
!
interface FastEthernet 0/12
 switchport port-security maximum 1
 switchport port-security violation shutdown
 switchport port-security
!
interface FastEthernet 0/13
 switchport port-security maximum 1
 switchport port-security violation shutdown
 switchport port-security
!
interface FastEthernet 0/14
 switchport port-security maximum 1
 switchport port-security violation shutdown
 switchport port-security
!
interface FastEthernet 0/15
 switchport port-security maximum 1
 switchport port-security violation shutdown
 switchport port-security
!
interface FastEthernet 0/16
 switchport port-security maximum 1
 switchport port-security violation shutdown
 switchport port-security
!
interface FastEthernet 0/17
 switchport port-security maximum 1
 switchport port-security violation shutdown
 switchport port-security
!
interface FastEthernet 0/18
 switchport port-security maximum 1
 switchport port-security violation shutdown
 switchport port-security
!
interface FastEthernet 0/19
 switchport port-security maximum 1
```

```
 switchport port-security violation shutdown
 switchport port-security
!
interface FastEthernet 0/20
 switchport port-security maximum 1
 switchport port-security violation shutdown
 switchport port-security
!
interface FastEthernet 0/21
 switchport port-security maximum 1
 switchport port-security violation shutdown
 switchport port-security
!
interface FastEthernet 0/22
 switchport port-security maximum 1
 switchport port-security violation shutdown
 switchport port-security
!
interface FastEthernet 0/23
 switchport port-security maximum 1
 switchport port-security violation shutdown
 switchport port-security
!
interface FastEthernet 0/24
!
interface GigabitEthernet 0/25
!
interface GigabitEthernet 0/26
!
interface GigabitEthernet 0/27
!
interface GigabitEthernet 0/28
!
line con 0
line vty 0 4
 login
!
!
end
```

实验 19 利用 IP 标准访问列表进行网络流量的控制

【实验名称】

利用 IP 标准访问列表进行网络流量的控制。

【实验目的】

掌握路由器上编号的标准 IP 访问列表规则及配置。

【背景描述】

假设你是某公司的网络管理员，公司的经理部、财务部门和销售部门分属于 3 个不同的网段，3 个部门之间用路由器进行信息传递，为了安全起见，公司领导要求销售部门不能对财务部门进行访问，但经理部可以对财务部门进行访问。

经理部的网段为 172.16.2.0，销售部门的网段为 172.16.1.0、财务部门的网段为 172.16.4.0。

【需求分析】

只允许 172.16.2.0 网段与 172.16.4.0 网段的主机进行通信，不允许 172.16.1.0 网段去访问 172.16.4.0 网段的主机。

【实验拓扑】

实验的拓扑图，如图 19-1 所示。

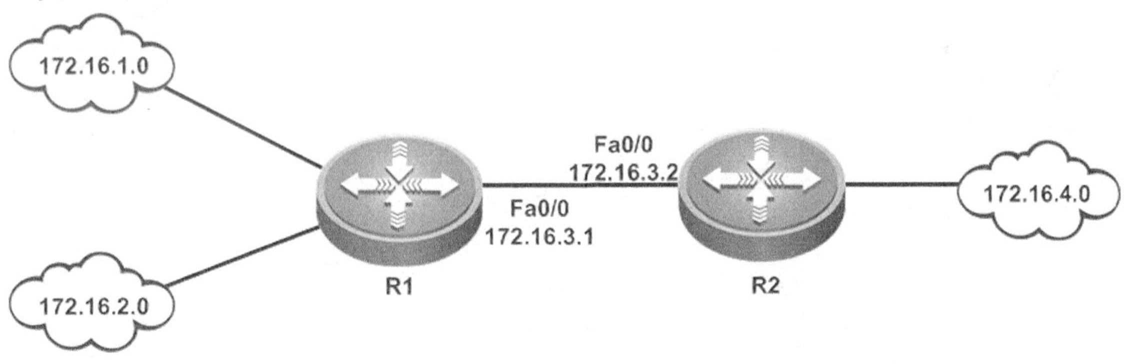

图 19-1

【预备知识】

路由器基本配置知识、访问控制列表知识。

【实验设备】

路由器 2 台。

V.35 线缆 1 条。

直连线或交叉线 3 条。

【实验原理】

IP ACL 是 IP 访问控制列表或 IP 访问列表，它对流经路由器或交换机的数据包根据一定的规则进行过滤，从而提高网络的可管理性和安全性。

IP ACL 分为两种：标准 IP 访问列表和扩展 IP 访问列表。

标准 IP 访问列表可以根据数据包的源 IP 地址定义规则，进行数据包的过滤。

扩展 IP 访问列表可以根据数据包的源 IP、目的 IP、源端口、目的端口、协议来定义规则，进行数据包的过滤。

IP ACL 基于接口进行规则的应用，分为入栈应用和出栈应用。入栈应用是指数据由外部经该接口进入路由器时进行数据包的过滤。出栈应用是指路由器从该接口向外转发数据时进行数据包的过滤。

IP ACL 的配置有两种方式：按照编号的访问列表和按照命名的访问列表。标准 IP 访问列表的编号范围是 1~99、1300~1999，扩展 IP 访问列表的编号范围是 100~199、2000~2699。

【实验步骤】

步骤 1　路由器的基本配置。

```
R1(config)#
R1(config)# interface  loopback 0
R1 (config-if)#ip add 172.16.1.1 255.255.255.0
R1 (config-if)#no shutdown
R1 (config-if)# interface  loopback 1
R1 (config-if)#ip add 172.16.2.1 255.255.255.0
R1 (config-if)#no shutdown
R1 (config-if)#interface FastEthernet0/0
R1 (config-if)#ip add 172.16.3.1 255.255.255.0
R1 (config-if)#no shutdown
R1 (config-if)#end
R2(config)# interface FastEthernet 0/0
R2 (config-if)#ip add 172.16.3.1 255.255.255.0
R2 (config-if)#no shutdown
R2 (config-if)#exit
R2 (config-if)#interface FastEthernet 0/1
R2 (config-if)#ip add 172.16.4.1 255.255.255.0
R2 (config-if)#no shutdown
R2 (config-if)#end
```

步骤 2　配置路由。

```
R1(config)#ip route 0.0.0.0 0.0.0.0 172.16.3.2
R2(config)#ip route 0.0.0.0 0.0.0.0 172.16.3.1
```

步骤 3　配置标准 IP 访问控制列表。

```
R2(config)#access-list 10 deny 172.16.1.0 0.0.0.255
R2(config)#access-list 10 permit 172.16.2.0 0.0.0.255
R2(config)# interface FastEthernet 0/1
R2(config-if)#ip access-group 10 out
```

步骤 4　验证测试。

在没有配置 ACL 时，可以使用原地址为 172.16.1.1，目标地址为 172.16.4.10（此为连接到

R2 接口 fa0/1 的一台主机），进行 ping 通信，如下所示。

```
R1#ping
Protocol [ip]:
Target IP address: 172.16.4.10
Repeat count [5]:
Datagram size [100]:
Timeout in seconds [2]:
Extended commands [n]: y
Source address:172.16.1.1
Time to Live [1, 64]:
Type of service [0, 31]:
Data Pattern [0xABCD]:0xabcd
Sending 5, 100-byte ICMP Echoes to 172.16.4.1, timeout is 2 seconds:
  < press Ctrl+C to break >
!!!!!
Success rate is 100 percent (5/5), round-trip min/avg/max = 1/1/1 ms
```

配置 ACL 后的测试，如下所示。

```
R1#ping
Protocol [ip]:
Target IP address: 172.16.4.10
Repeat count [5]:
Datagram size [100]:
Timeout in seconds [2]:
Extended commands [n]: y
Source address:172.16.1.1
Time to Live [1, 64]:
Type of service [0, 31]:
Data Pattern [0xABCD]:0xabcd
Sending 5, 100-byte ICMP Echoes to 172.16.4.10, timeout is 2 seconds:
  < press Ctrl+C to break >
.....
Success rate is 0 percent (0/5)
R1#ping
Protocol [ip]:
Target IP address: 172.16.4.10
Repeat count [5]:
Datagram size [100]:
Timeout in seconds [2]:
Extended commands [n]: y
Source address:172.16.2.1
Time to Live [1, 64]:
Type of service [0, 31]:
```

```
Data Pattern [0xABCD]:0xabcd
Sending 5, 100-byte ICMP Echoes to 172.16.4.10, timeout is 2 seconds:
  < press Ctrl+C to break >
!!!!!
Success rate is 100 percent (5/5), round-trip min/avg/max = 1/2/10 ms
```

ping（172.16.2.0 网段的主机不能 ping 通 172.16.4.0 网段的主机；172.16.1.0 网段的主机能 ping 通 172.16.4.0 网段的主机）。

```
R2#show access-lists
ip access-list standard 10
 10 deny 172.16.1.0 0.0.0.255
 20 permit 172.16.2.0 0.0.0.255
 35 packets filtered
R2#sh ip access-group interface fa0/1
ip access-group 10 out
Applied On interface FastEthernet 0/1
```

【注意事项】

（1）访问控制列表中的网络掩码是反掩码。

（2）标准控制列表要应用在尽量靠近目的地址的接口上。

【参考配置】

```
R1#show running-config
Building configuration...
Current configuration : 590 bytes
!
version RGNOS 10.1.00(4), Release(18443)(Tue Jul 17 20:50:30 CST 2007
-ubu1server)
hostname R1
!
interface FastEthernet 0/0
 ip address 172.16.3.1 255.255.255.0
 duplex auto
 speed auto
!
interface FastEthernet 0/1
 duplex auto
 speed auto
!
interface Loopback 0
 ip address 172.16.1.1 255.255.255.0
!
interface Loopback 1
 ip address 172.16.2.1 255.255.255.0
```

```
!
ip route  0.0.0.0 0.0.0.0  172.16.3.2
!
line con 0
line aux 0
line vty 0 4
 login
!
end
```

R2#show running-config

```
Building configuration...
Current configuration : 627 bytes
!
version  RGNOS  10.1.00(4),  Release(18443)(Tue  Jul  17  20:50:30  CST  2007
-ubu1server)
hostname R2
!
ip access-list standard 10
 10 deny 172.16.1.0 0.0.0.255
 20 permit 172.16.2.0 0.0.0.255
!
interface FastEthernet 0/0
 ip address 172.16.3.2 255.255.255.0
 duplex auto
 speed auto
!
interface FastEthernet 0/1
 ip access-group 10 out
 ip address 172.16.4.1 255.255.255.0
 duplex auto
 speed auto
!
ip route  0.0.0.0 0.0.0.0  172.16.3.1
!
line con 0
line aux 0
line vty 0 4
 login
!
end
```

实验 20　利用 IP 扩展访问列表实现应用服务的访问限制

【实验名称】

利用 IP 扩展访问列表实现应用服务的访问限制。

【实验目的】

掌握在交换机上编号的扩展 IP 访问列表规则及配置。

【背景描述】

假设你是学校的网络管理员，在 3750-24 交换机上连着学校的提供 WWW 和 FTP 的服务器，另外，还连接着学生宿舍楼和教工宿舍楼，学校规定学生只能对服务器进行 FTP 访问，不能进行 WWW 访问，教工则没有此限制。

【需求分析】

不允许 VLAN30 的主机去访问 VLAN10 网络中的 web 服务，其他的不受限制。

【实验拓扑】

实验的拓扑图，如图 20-1 所示。

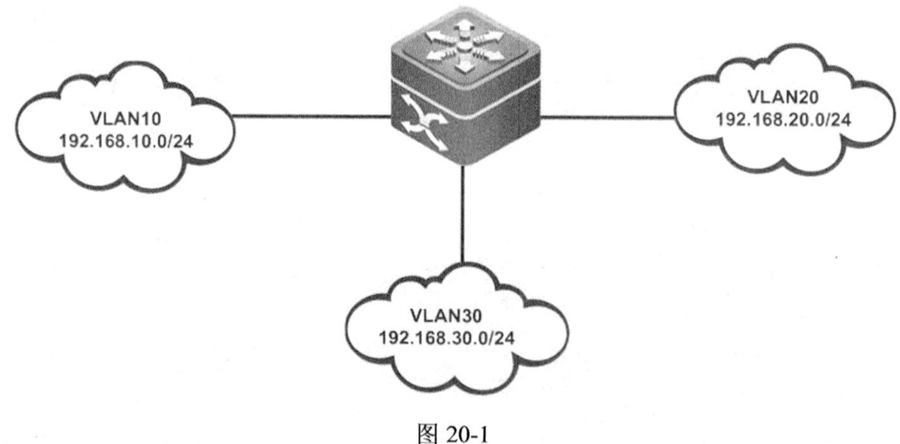

图 20-1

【预备知识】

路由器基本配置知识、访问控制列表知识。

【实验设备】

S3750 交换机 1 台。

PC 3 台。

直连线 3 条。

【实验原理】

IP ACL 是 IP 访问控制列表或 IP 访问列表，它对流经路由器或交换机的数据包根据一定的规则进行过滤，从而提高网络的可管理性和安全性。

IP ACL 分为两种：标准 IP 访问列表和扩展 IP 访问列表。

标准 IP 访问列表可以根据数据包的源 IP 地址定义规则，进行数据包的过滤。

扩展 IP 访问列表可以根据数据包的源 IP、目的 IP、源端口、目的端口、协议来定义规则，进行数据包的过滤。

IP ACL 基于接口进行规则的应用，分为入栈应用和出栈应用。入栈应用是指数据由外部经该接口进入路由器时进行数据包的过滤。出栈应用是指路由器从该接口向外转发数据时进行数据包的过滤。

IP ACL 的配置有两种方式：按照编号的访问列表和按照命名的访问列表。标准 IP 访问列表的编号范围是 1~99、1300~1999，扩展 IP 访问列表的编号范围是 100~199、2000~2699。

【实验步骤】

步骤 1　路由器的基本配置。

```
S3750 (config)#vlan 10
S3750 (config-vlan)#name server
S3750 (config)#vlan 20
S3750 (config-vlan)#name teachers
S3750 (config)#vlan 30
S3750 (config-vlan)#name students
S3750 (config)#interface f0/1
S3750 (config-if)#switchport mode access
S3750 (config-if)#switchport access vlan 10
S3750 (config)#interface f0/2
S3750 (config-if)#switchport mode access
S3750 (config-if)#switchport access vlan 20
S3750 (config)#interface f0/3
S3750 (config-if)#switchport mode access
S3750 (config-if)#switchport access vlan 30
S3750 (config)#int vlan10
S3750 (config-if)#ip add 192.168.10.1 255.255.255.0
S3750 (config-if)#no shutdown
S3750 (config-if)#int vlan 20
S3750 (config-if)#ip add 192.168.20.1 255.255.255.0
S3750 (config-if)#no shutdown
S3750 (config-if)#int vlan 30
S3750 (config-if)#ip add 192.168.30.1 255.255.255.0
S3750 (config-if)#no shutdown
```

步骤 2　配置编号的扩展 IP 访问控制列表。

```
S3750 (config)#access-list 110 deny tcp 192.168.30.0 0.0.0.255  192.168.10.0
0.0.0.255 eq www
S3750 (config)#access-list 110 permit ip any any
S3750 (config)#int vlan 30
S3750 4(config-if)#ip access-group 110 in
```

步骤 3　验证测试。

```
S3750#show access-lists
ip access-list extended 110
 10 deny tcp 192.168.30.0 0.0.0.255 192.168.10.0 0.0.0.255 eq www
 20 permit ip any any
```

可以自己搭建一个 Web 服务，使用客户端进行访问测试。

【注意事项】

（1）访问控制列表要在接口下应用。

（2）要注意，deny 某个网段后要 peimit 其他网段。

【参考配置】

```
S3750#show running-config
Building configuration...
Current configuration : 1563 bytes
!
version RGNOS 10.1.00(4), Release(18443)(Tue Jul 17 19:51:54 CST 2007
-ubu6server)
!
vlan 1
!
vlan 10
!
vlan 20
!
vlan 30
!
ip access-list extended 110
 10 deny tcp 192.168.30.0 0.0.0.255 192.168.10.0 0.0.0.255 eq www
 20 permit ip any any
!
interface FastEthernet 0/1
 switchport access vlan 10
!
interface FastEthernet 0/2
 switchport access vlan 20
!
interface FastEthernet 0/3
 switchport access vlan 30
!
interface FastEthernet 0/4
!
interface FastEthernet 0/5
```

```
!
interface FastEthernet 0/6
!
interface FastEthernet 0/7
!
interface FastEthernet 0/8
!
interface FastEthernet 0/9
!
interface FastEthernet 0/10
!
interface FastEthernet 0/11
!
interface FastEthernet 0/12
!
interface FastEthernet 0/13
!
interface FastEthernet 0/14
!
interface FastEthernet 0/15
!
interface FastEthernet 0/16
!
interface FastEthernet 0/17
!
interface FastEthernet 0/18
!
interface FastEthernet 0/19
!
interface FastEthernet 0/20
!
interface FastEthernet 0/21
!
interface FastEthernet 0/22
!
interface FastEthernet 0/23
!
interface FastEthernet 0/24
!
interface GigabitEthernet 0/25
!
interface GigabitEthernet 0/26
```

```
!
interface GigabitEthernet 0/27
!
interface GigabitEthernet 0/28
!
interface VLAN 10
 ip address 192.168.10.1 255.255.255.0
!
interface VLAN 20
 ip address 192.168.20.1 255.255.255.0
!
interface VLAN 30
 ip access-group 110 in
 ip address 192.168.30.1 255.255.255.0
!
line con 0
line vty 0 4
 login
!
end
```

实验 21　利用访问列表进行 VTY 访问限制

【实验名称】

利用访问列表进行 VTY 访问限制。

【实验目的】

掌握在路由器上使用 ACL 进行 VTY 访问控制，增强路由器的安全性。

【背景描述】

假设你是某大公司的 IT 管理员，公司在全国各地有很多分公司，你需要在路由器上配置 VTY 进行远程登录，但由于远程登录的不安全性，为加强其安全性能，所以需要指定某个 IP 地址可以访问。

【需求分析】

指定只有 IP 地址为 172.16.1.10 和 172.16.1.11 的主机才可以远程登录路由器。

【实验拓扑】

实验的拓扑图，如图 21-1 所示。

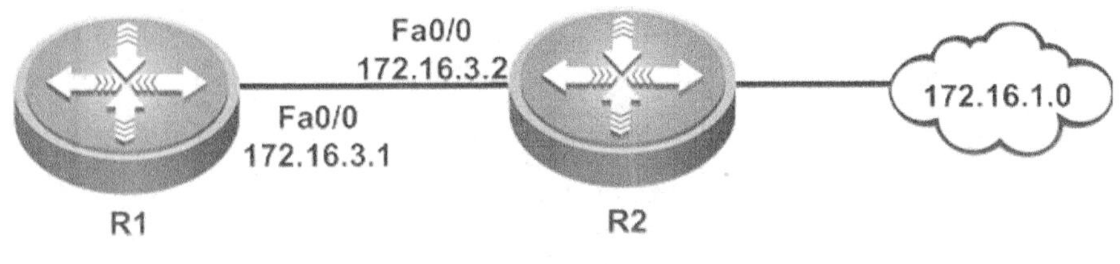

图 21-1

【预备知识】

路由器基本配置知识、访问控制列表知识。

【实验设备】

路由器 2 台。

PC 3 台。

直连线 3 条。

【实验原理】

IP ACL 是 IP 访问控制列表或 IP 访问列表，它对流经路由器或交换机的数据包根据一定的规则进行过滤，从而提高网络的可管理性和安全性。

IP ACL 分为两种：标准 IP 访问列表和扩展 IP 访问列表。

标准 IP 访问列表可以根据数据包的源 IP 地址定义规则，进行数据包的过滤。

扩展 IP 访问列表可以根据数据包的源 IP、目的 IP、源端口、目的端口、协议来定义规则，进行数据包的过滤。

IP ACL 基于接口进行规则的应用，分为入栈应用和出栈应用。入栈应用是指数据由外部经该接口进入路由器时进行数据包的过滤。出栈应用是指路由器从该接口向外转发数据时进行数据包的过滤。

IP ACL 的配置有两种方式：按照编号的访问列表和按照命名的访问列表。标准 IP 访问列表的编号范围是 1~99、1300~1999，扩展 IP 访问列表的编号范围是 100~199、2000~2699。

访问控制列表有以下作用。

安全控制：允许一些分合匹配规则的数据包通过访问的同时而拒绝另一部分不符合匹配规则的数据包。举个例子来说，财务部的数据库服务器上的数据应该是比较机密的，不是任何人都可以访问的，这个时候就需要用到访问控制列表，在此列表中，可以定义哪些主机能够访问财务部数据库服务器，当此列表外的主机访问此服务器时，就会被路由器过滤掉。

流量过滤：此功能是防止一些不必要的数据包通过路由器，提高网络带宽的利用率。

数据流量标识：当公司有两条或两条以上的网络链路时，此功能可使用访问控制列表与路由策略等来实现分工，让不同的数据包选择不同的链路。

【实验步骤】

步骤 1　路由器的基本配置。

```
R1 (config)# interface FastEthernet 0/0
R1 (config-if)#ip address 172.16.3.1 255.255.255.0
R2 (config)#interface FastEthernet 0/0
R2 (config-if)#ip address 172.16.3.2 255.255.255.0
R2 (config)#interface FastEthernet 0/1
R2 (config-if)#ip address 172.16.1.1 255.255.255.0
```

步骤 2　配置远程登录。

```
R1 (config)#enable password ruijie
R1 (config)#line vty 0 4
R1(config-line)#password ruijie
R1(config-line)#login
```

步骤 3　配置路由。

```
R1 (config)#ip route  0.0.0.0 0.0.0.0  172.16.3.2
R2 (config)#ip route  0.0.0.0 0.0.0.0  172.16.3.1
```

步骤 4　配置访问控制列表。

```
R1 (config)#access-list 10 permit host 172.16.1.10
R1 (config)#access-list 10 permit host 172.16.1.11
R1 (config)#access-list 10 deny any
R1 (config)#line vty 0 4
R1(config-line)# access-class 10 in
```

步骤 5　验证测试。

将 fa0/1 接口 IP 地址设置为 172.16.1.1，使用此地址为原地址登录远程路由器 R1，如下所示。

```
R2(config)#interface fastEthernet 0/1
R2(config-if)#ip address 172.16.1.1 255.255.255.0
R2(config-if)#no shutdown
R2(config-if)#end
R2#Aug 21 06:32:10 R2 %5:Configured from console by console
```

```
R2#telnet 172.16.3.1 /source-interface fastEthernet 0/1
Trying 172.16.3.1, 23...
% Destination unreachable; gateway or host down
```

将 fa0/1 接口 IP 地址设置为 172.16.1.10，使用此地址为原地址登录远程路由器 R1，如下所示。

```
R2#configure terminal
Enter configuration commands, one per line.  End with CNTL/Z.
R2(config)#interface fastEthernet 0/1
R2(config-if)#ip address 172.16.1.10 255.255.255.0
R2(config-if)#no shutdown
R2(config-if)#end
R2#Aug 21 06:33:49 R2 %5:Configured from console by console
R2#telnet 172.16.3.1 /source-interface fastEthernet 0/1
Trying 172.16.3.1, 23...
User Access Verification
Password:
R1>en
R1>enable
Password:
R1#
R1#
```

将 fa0/1 接口 IP 地址设置为 172.16.1.11，使用此地址为原地址登录远程路由器 R1，如下所示。

```
R2#configure terminal
Enter configuration commands, one per line.  End with CNTL/Z.
R2(config)#interface fastEthernet 0/1
R2(config-if)#ip address 172.16.1.11 255.255.255.0
R2(config-if)#no shutdown
R2(config-if)#end
R2#Aug 21 06:34:55 R2 %5:Configured from console by console
R2#telnet 172.16.3.1 /source-interface fastEthernet 0/1
Trying 172.16.3.1, 23...
User Access Verification
Password:
R1>enable
Password:
R1#
R1#
R1#show access-lists
ip access-list standard 10
 10 permit host 172.16.1.10
 20 permit host 172.16.1.11
 30 deny any
```

【注意事项】

（1）访问控制列表要在接口下应用。

（2）要注意，deny 某个网段后要 peimit 其他网段。

【参考配置】

```
R1#show running-config
Building configuration...
Current configuration : 657 bytes
!
version RGNOS 10.1.00(4), Release(18443)(Tue Jul 17 20:50:30 CST 2007
-ubu1server)
hostname R1
!
ip access-list standard 10
 10 permit host 172.16.1.10
 20 permit host 172.16.1.11
 30 deny any
!
enable password 7 13170743072817
!
interface FastEthernet 0/0
 ip address 172.16.3.1 255.255.255.0
 duplex auto
 speed auto
!
interface FastEthernet 0/1
 duplex auto
 speed auto
!
ip route 0.0.0.0 0.0.0.0 172.16.3.2
!
line con 0
line aux 0
line vty 0 4
 access-class 10 in
 login
 password 7 035122110c3706
!
end

R2#show running-config
Building configuration...
```

```
Current configuration : 503 bytes
!
version  RGNOS  10.1.00(4),  Release(18443)(Tue  Jul  17  20:50:30  CST  2007
-ubu1server)
hostname R2
!
interface FastEthernet 0/0
 ip address 172.16.3.2 255.255.255.0
 duplex auto
 speed auto
!
interface FastEthernet 0/1
 ip address 172.16.1.11 255.255.255.0
 duplex auto
 speed auto
!
ip route  0.0.0.0 0.0.0.0  172.16.3.1
!
line con 0
line aux 0
line vty 0 4
 login
!
end
```

实验 22　利用动态 NAPT 实现局域网访问互联网

【实验名称】

利用动态 NAPT 实现局域网访问互联网。

【实验目的】

掌握内网中所有主机连接到 Internet 时，通过端口号区分的复用内部全局地址转换。

【背景描述】

假设你是某公司的网络管理员，公司只向 ISP 申请了一个公网 IP 地址，希望全公司的主机都能访问外网，请你实现。

【需求分析】

允许内部所有主机在公网地址缺乏的情况下可以访问外部网络。

【实验拓扑】

实验的拓扑图，如图 22-1 所示。

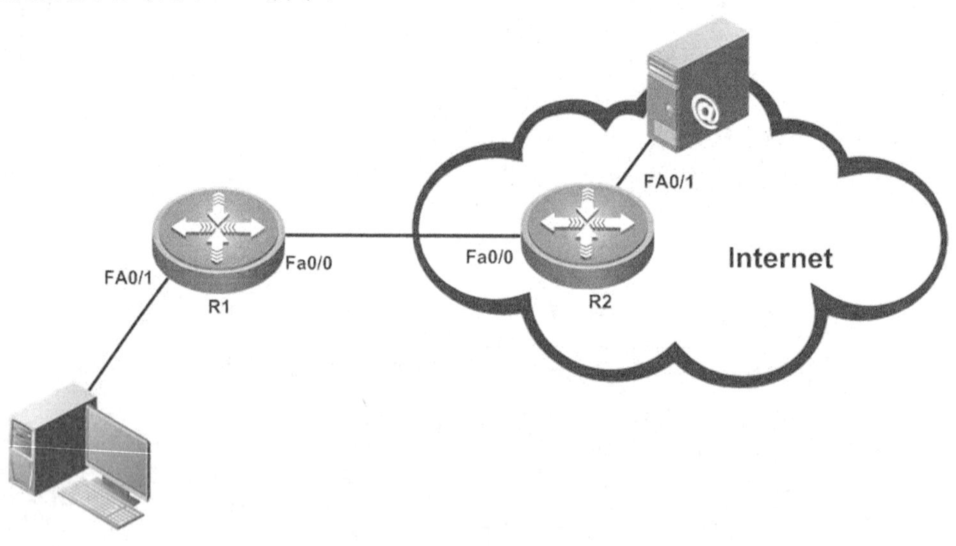

图 22-1

【预备知识】

路由器基本配置知识、NAT 知识。

【实验设备】

路由器（带串口）2 台。

V.35 线缆（DTE/DCE）1 对。

PC 2 台。

直连线或交叉线 2 条。

【实验原理】

NAT（网络地址转换或网络地址翻译），是指将网络地址从一个地址空间转换为另一个地址空间的行为。

NAT 将网络划分为内部网络（inside）和外部网络（outside）两部分。局域网主机利用 NAT 访问网络时，是将局域网内部的本地地址转换为全局地址（互联网合法 IP 地址）后转发数据包。

NAT 分为两种类型：NAT（网络地址转换）和 NAPT（网络地址端口转换）。实现 NAT 转换后，一个本地 IP 地址对应一个全局地址。实现 NAPT 转换后，多个本地 IP 地址对应一个全局 IP 地址。在目前的网络中，由于公网的 IP 地址紧缺，而局域网中的主机数较多，因此，一般使用动态的 NAPT 来实现局域网多台主机共用一个或少数几个公网 IP 访问互联网。

【实验步骤】

步骤 1　路由器的基本配置。

```
R1(config)#
R1 (config)#interface fastEthernet 0/1
R1 (config-if)#ip address 172.16.1.1 255.255.255.0
R1 (config-if)#no shutdown
R1 (config-if)#exit
R1 (config)#interface fastEthernet 0/0
R1 (config-if)#ip address 200.1.8.7 255.255.255.0
R1 (config-if)#no shutdown
R1 (config-if)#exit
R2(config)#interface fastEthernet 1/0
R2 (config-if)#ip address 63.19.6.1 255.255.255.0
R2 (config-if)#no shutdown
R2 (config-if)#exit
R2 (config)#interface fastEthernet 0/0
R2 (config-if)#ip address 200.1.8.8 255.255.255.0
R2 (config-if)#no sh
R2 (config-if)#end
```

步骤 2　配置默认路由。

```
R1(config)#ip route 0.0.0.0 0.0.0.0 200.1.8.8
R2 (config)# ip route 0.0.0.0 0.0.0.0 200.1.8.7
```

步骤 3　配置动态 NAPT 映射。

```
R1(config)#interface fastEthernet 0/1
R1 (config-if)#ip nat inside
R1 (config-if)#exit
R1 (config)#interface fastEthernet 0/0
R1 (config-if)#ip nat outside
R1 (config-if)#exit
R1 (config)#ip nat pool to internet 200.1.8.7 200.1.8.7 netmask 255.255.255.0
R1 (config)#access-list 10 permit 172.16.1.0 0.0.0.255
R1 (config)#ip nat inside source list 10 pool to_internet  overload
```

步骤 4　验证测试。

（1）在路由器 R2 上配置 telnet 服务。

（2）在 PC 机上用 telnet 测试访问 63.19.6.1 路由器。

（3）在路由器 lan-router 查看 NAPT 映射关系。

```
R1#sh ip nat statistics
Total translations: 1, max entries permitted: 30000
 Peak translations: 1 @ 00:02:50 ago
Outside interfaces: FastEthernet 0/0
Inside interfaces: FastEthernet 0/1
Rule statistics:
[ID: 1] inside source dynamic
 hit: 21
 match (after routing):
  ip packet with source-ip match access-list 10
 action :
  translate ip packet's source-ip use pool to internet
R1#show ip nat translations
Pro Inside global      Inside local       Outside local      Outside global
tcp 200.1.8.7:1025     172.16.1.10:1025   63.19.6.1:23       63.19.6.1:23
```

【注意事项】

（1）不要把 inside 和 outside 应用的接口弄错。

（2）要加上能使数据包向外转发的路由，比如默认路由。

（3）尽量不要用广域网接口地址作为映射的全局地址，本例子中特定仅有一个公网地址，实际工作中不推荐。

【参考配置】

```
R1#show running-config
Building configuration...
Current configuration : 724 bytes
!
version RGNOS 10.1.00(4), Release(18443)(Tue Jul 17 20:50:30 CST 2007
-ubu1server)
hostname R1
!
ip access-list standard 10
 10 permit 172.16.1.0 0.0.0.255
!
!
interface FastEthernet 0/0
 ip nat outside
 ip address 200.1.8.7 255.255.255.0
 duplex auto
```

```
 speed auto
!
interface FastEthernet 0/1
 ip nat inside
 ip address 172.16.1.1 255.255.255.0
 duplex auto
 speed auto
!
ip nat pool to internet 200.1.8.7 200.1.8.7 netmask 255.255.255.0
ip nat inside source list 10 pool to internet overload
!
ip route  0.0.0.0 0.0.0.0  200.1.8.8
!
line con 0
line aux 0
line vty 0 4
 login
!
end
```

R2#show running-config

```
Building configuration...
Current configuration : 550 bytes
!
version  RGNOS  10.1.00(4),  Release(18443)(Tue  Jul  17  20:50:30  CST  2007
-ubu1server)
hostname R2
!
enable password 7 10444074
!
!
interface FastEthernet 0/0
 ip address 200.1.8.8 255.255.255.0
 duplex auto
 speed auto
!
interface FastEthernet 0/1
 ip address 63.19.6.1 255.255.255.0
 duplex auto
 speed auto
!
ip route  0.0.0.0 0.0.0.0  200.1.8.7
!
```

```
line con 0
line aux 0
line vty 0 4
 login
 password 7 097e4741
!
end
```

实验 23　利用 NAT 实现外网主机访问内网服务器

【实验名称】

利用 NAT 实现外网主机访问内网服务器。

【实验目的】

掌握 NAT 源地址转换和目的地址转换的区别，掌握如何向外网发布内网的服务器。

【背景描述】

假设你是某公司的网络管理员，公司只向 ISP 申请了一个公网 IP 地址，公司的网站在内网中，要求在互联网上也可以访问到公司的网站，现需要你来实现此功能。172.16.8.5 是 Telent 服务器的 IP 地址。

【需求分析】

内网服务器能够转换成外网的公网 IP，被互联网访问。

【实验拓扑】

实验的拓扑图，如图 23-1 所示。

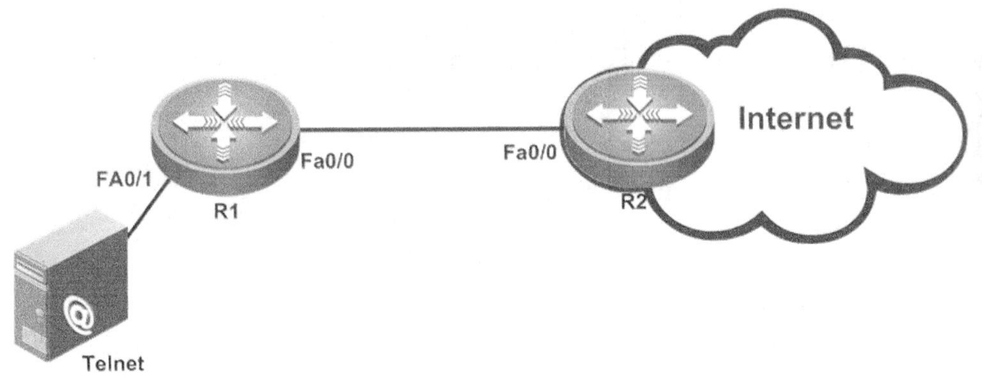

图 23-1

【预备知识】

路由器基本配置知识、NAT 知识。

【实验设备】

路由器（带串口）2 台。

V.35 线缆（DTE/DCE）1 对。

PC 1 台。

直连线 1 条。

【实验原理】

NAT（网络地址转换或网络地址翻译），是指将网络地址从一个地址空间转换为另一个地址空间的行为。

NAT 将网络划分为内部网络（inside）和外部网络（outside）两部分。局域网主机利用 NAT 访问网络时，是将局域网内部的本地地址转换为了全局地址（互联网上的合法 IP 地址）后转发数据包。

NAT 分为两种类型：NAT（网络地址转换）和 NAPT（网络地址端口转换）。实现 NAT 转换后，一个本地 IP 地址对应一个全局地址；实现 NAPT 转换后，多个本地 IP 地址对应一个全局 IP 地址。由于在目前的网络中，公网 IP 地址紧缺，而局域网内的主机数较多，因此，一般使用动态的 NAPT 来实现局域网内多台主机共用一个或少数几个公网 IP 访问互联网。

【实验步骤】

步骤 1　路由器的基本配置。

```
R1(config)#
R1 (config)#interface FastEthernet 0/1
R1 (config-if)#ip address 172.16.8.1 255.255.255.0
R1 (config-if)#no shutdown
R1 (config-if)#exit
R1 (config)#interface interface FastEthernet 0/0
R1 (config-if)#ip address 200.1.8.7 255.255.255.0
R1 (config-if)#no shutdown
R1 (config-if)#exit
R2 (config)#interface fastEthernet 0/1
R2 (config-if)#ip address 63.19.6.1 255.255.255.0
R2 (config-if)#no shutdown
R2 (config-if)#exit
R2 (config)#interface interface FastEthernet 0/0
R2 (config-if)#ip address 200.1.8.8 255.255.255.0
R2 (config-if)#no sh
R2 (config-if)#end
```

步骤 2　配置默认路由。

```
R1(config)#ip route 0.0.0.0 0.0.0.0 200.1.8.7
R2 (config)# ip route 0.0.0.0 0.0.0.0200.1.8.8
```

步骤 3　配置 NAT。

```
R1(config)#interface fastEthernet 0/1
R1(config-if)#ip nat inside
R1(config-if)#exit
R1(config)#interface fastEthernet 0/0
R1(config-if)#ip nat outside
R1(config-if)#exit
R1(config)# ip nat inside source static tcp 172.16.8.5 23 200.1.8.5 23
```

步骤 4　验证测试。

（1）在内网主机上配置 telnet 服务。

（2）在外网的 1 台主机上通过 telnet 登录 200.1.8.5。

```
R1#show ip nat translations
Pro Inside global      Inside local       Outside local      Outside global
tcp 200.1.8.5:23       172.16.8.5:23      63.19.6.1:1033     63.19.6.1:1033
R1#show ip nat statistics
Total translations: 1, max entries permitted: 30000
 Peak translations: 2 @ 00:11:25 ago
Outside interfaces: FastEthernet 0/0
Inside interfaces: FastEthernet 0/1
Rule statistics:
[ID: 5] inside source static
 hit: 2
 match (before routing):
  tcp packet with destination-ip 200.1.8.5 destination-port 23
 action :
  translate ip packet's destination-ip use ip 172.16.8.5 with port set to 23
```

【注意事项】

（1）不要把 inside 和 outside 应用的接口弄错。

（2）配置目标地址转换后，需要利用静态 NAPT 配置静态的端口地址转换。

【参考配置】

```
R1#sh running-config
Building configuration...
Current configuration : 919 bytes
!
version  RGNOS  10.1.00(4),  Release(18443)(Tue  Jul  17  20:50:30  CST  2007
-ubu1server)
hostname R1
!
ip access-list standard 3
 10 permit host 200.1.8.5
!
interface FastEthernet 0/0
 ip nat outside
ip address 200.1.8.7 255.255.255.0
 duplex auto
 speed auto
!
interface FastEthernet 0/1
 ip nat inside
 ip address 172.16.8.1 255.255.255.0
 duplex auto
 speed auto
```

```
!
ip nat inside source static tcp 172.16.8.5 23 200.1.8.5 23
!
ip route  0.0.0.0 0.0.0.0  200.1.8.8
!
line con 0
line aux 0
line vty 0 4
 login
!
end
```

R2#sh running-config

```
Building configuration...
Current configuration : 550 bytes
!
version  RGNOS  10.1.00(4),  Release(18443)(Tue  Jul  17  20:50:30  CST  2007
-ubu1server)
hostname R2
!
enable password 7 151b5f72
!
!
interface FastEthernet 0/0
 ip address 200.1.8.8 255.255.255.0
 duplex auto
 speed auto
!
interface FastEthernet 0/1
 ip address 63.19.6.1 255.255.255.0
 duplex auto
 speed auto
!
ip route  0.0.0.0 0.0.0.0  200.1.8.7
!
line con 0
line aux 0
line vty 0 4
 login
 password 7 0549546d
!
end
```

实验 24　建立开放式的无线接入服务

【实验名称】

建立开放式的无线接入服务。

【实验目的】

掌握最基础的开放式无线接入服务的配置方法。

【背景描述】

小李在某国有企业担任网络管理员职务，不久前采购了一套智能无线交换产品，用于办公区域和会议室的无线覆盖。由于该企业员工对电脑的操作水平比较低，只会打开无线网卡搜寻 AP 信号，不会配置 IP 地址，使用无线网络也只是进行简单的网页浏览和收发邮件。因此，小李需要建立一个开放式的无需认证的无线网络。

【需求分析】

客户需要一个不使用加密、认证的无线网络。无线客户端通过 DHCP 方式获取 IP 地址。

【实验拓扑】

实验的拓扑图，如图 24-1 所示。

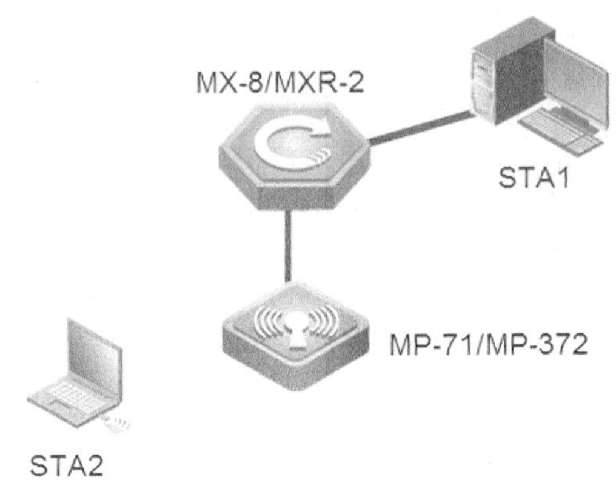

图 24-1

【实验设备】

RG-WG54U 1 块。

PC 2 台。

MP-71/MP-372 1 台。

MX-8/MXR-2 1 台。

RingMaster 服务器 1 台。

【预备知识】

无线局域网基本知识、智能无线交换网络工作原理、智能无线交换产品基本操作。

【实验步骤】

步骤 1　配置无线交换机的基本参数。

无线交换机的默认 IP 地址是 192.168.100.1/24，因此将 STA-1 的 IP 地址配置为 192.168.100.2/24，并打开浏览器，登录到 https://192.168.100.1，弹出如图 24-2 所示的界面，选择“是（Y）”按钮。

系统的默认管理用户名是 admin，密码为空，如图 24-3 所示。

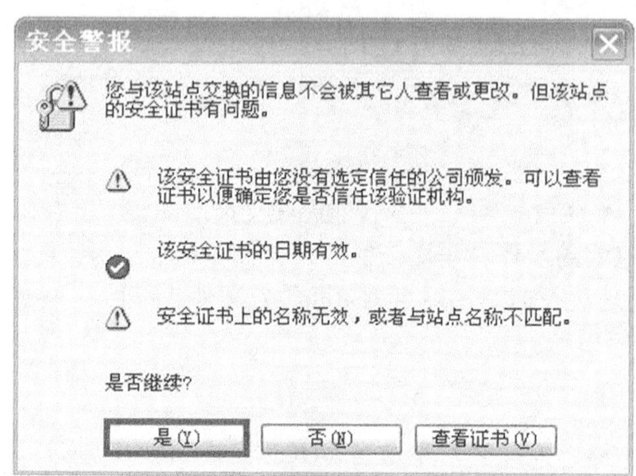

图 24-2

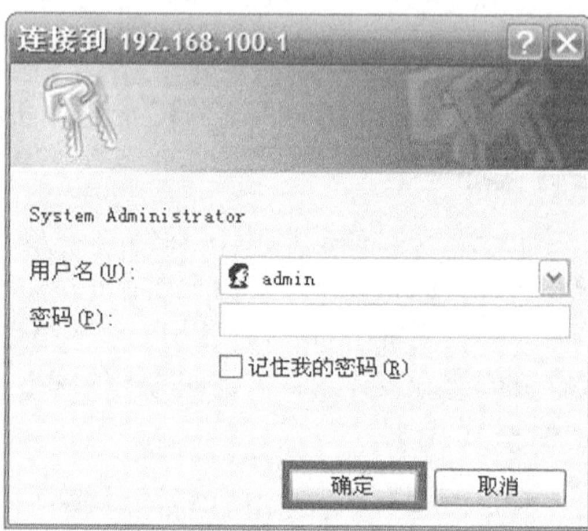

图 24-3

输入用户名和密码后，就进入了无线交换机的 Web 配置页面，单击“Start”按钮，进入快速配置指南页面，如图 24-4 所示。

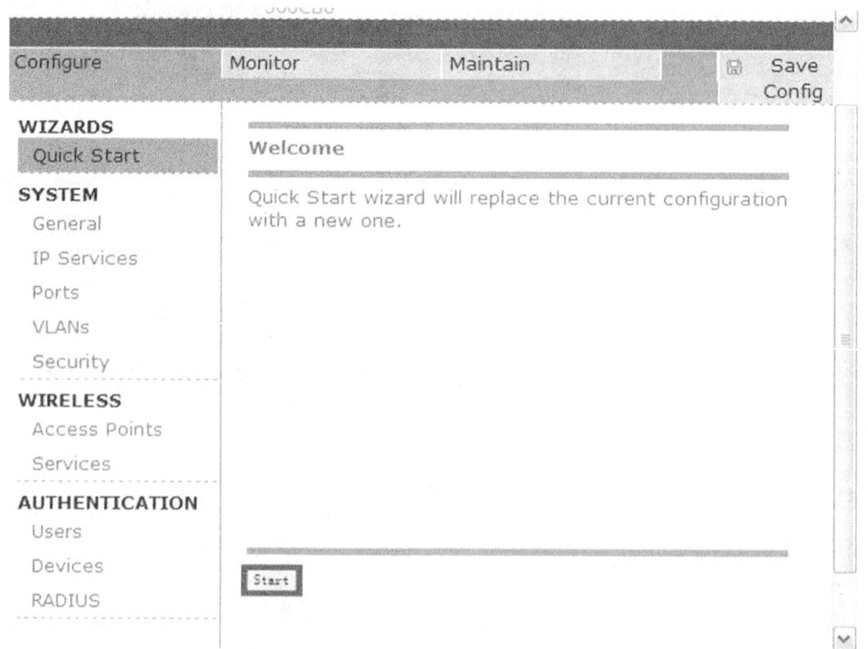

图 24-4

选择管理无线交换机的工具“RingMaster”网管软件，如图 24-5 所示。

设置系统的管理密码，如图 24-7 所示。

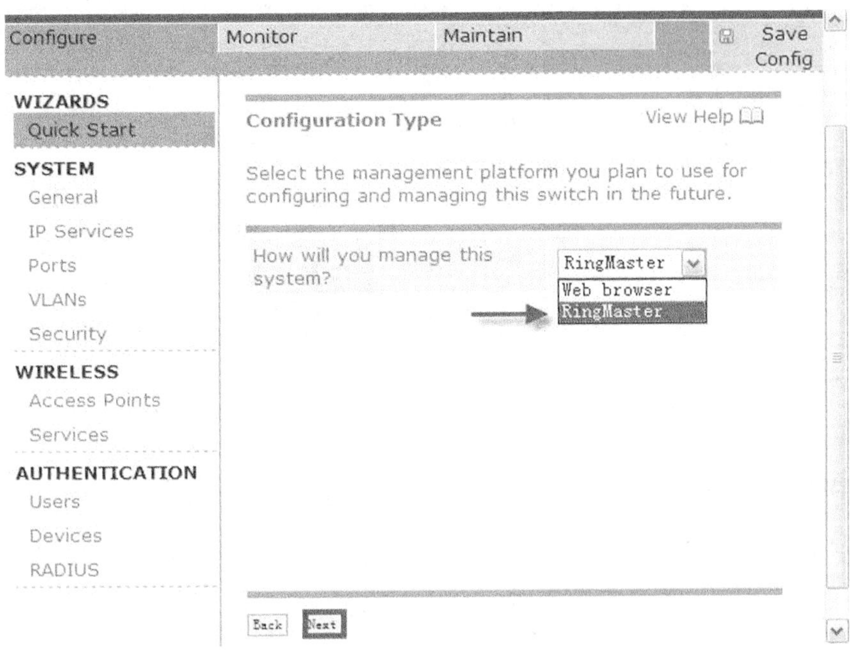

图 24-5

配置无线交换机的 IP 地址、子网掩码以及默认网关，如图 24-6 所示。

Configure
Monitor
Maintain
Save Config
WIZARDS
Quick Start
SYSTEM
General
IP Services
Ports
VLANs
Security
WIRELESS
Access Points
Services
AUTHENTICATION
Users
Devices
RADIUS
IP Configuration
View Help
Specify the following parameters to connect this switch to the network.
IP address 192.168.100.254
IP mask bits 24
Default router 192.168.100.1
Back
Next

图 24-6

设置系统的管理密码，如图 24-7 所示。

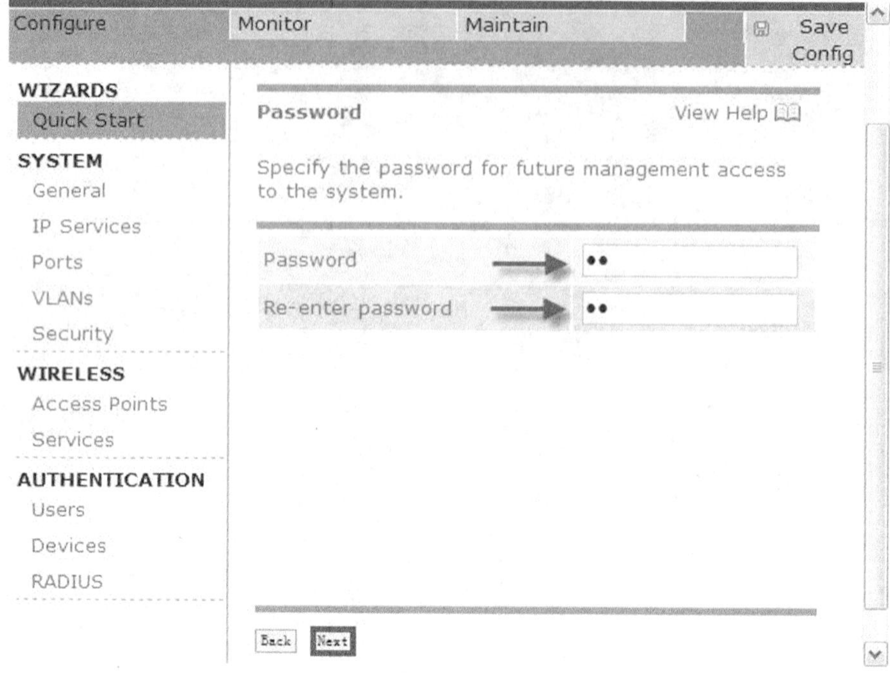

图 24-7

设置系统时间及时区，如图 24-8 所示。

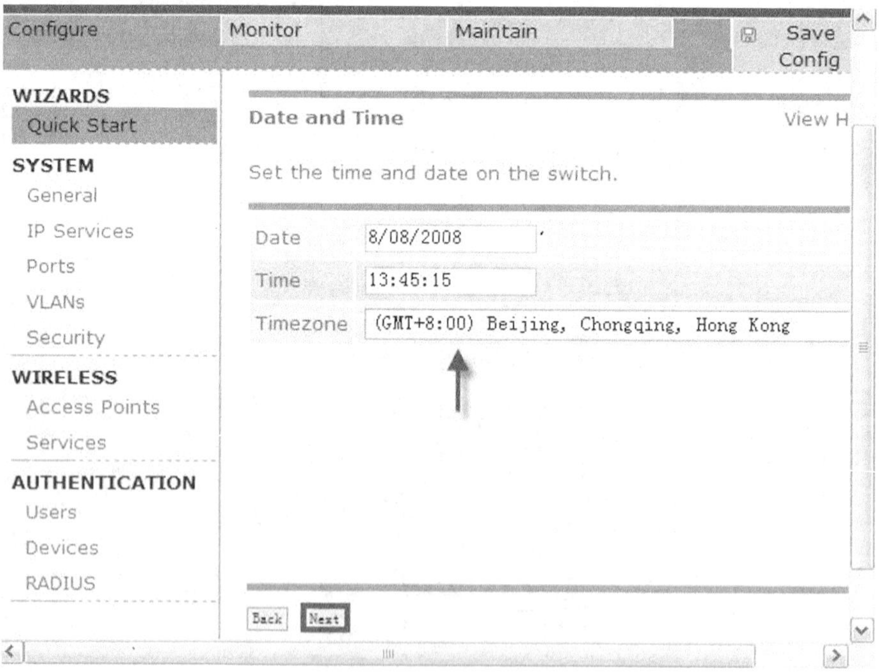

图 24-8

确认无线交换机的基本配置，如图 24-9 所示。

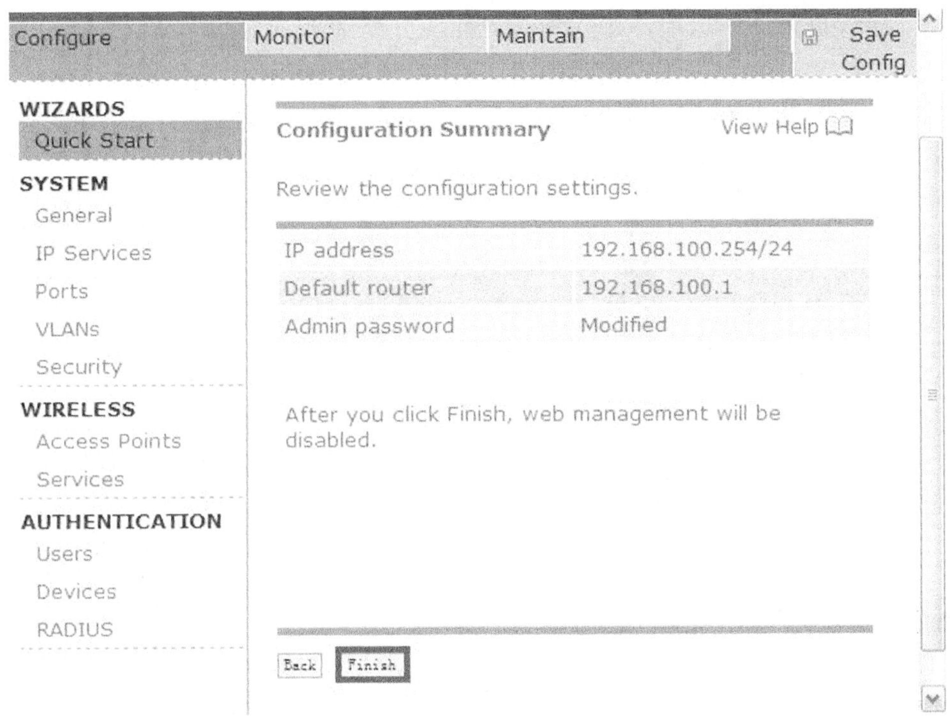

图 24-9

完成无线交换机的基本配置。

步骤 2　通过 RingMaster 网管软件来进行无线交换机的高级配置。

运行 RingMaster 软件，地址为 127.0.0.1，端口为 443，用户名和密码默认为空，如图 24-10 所示。

图 24-10

选择“Configuration”选项，进入配置界面，并添加被管理的无线交换机，如图 24-11 所示。

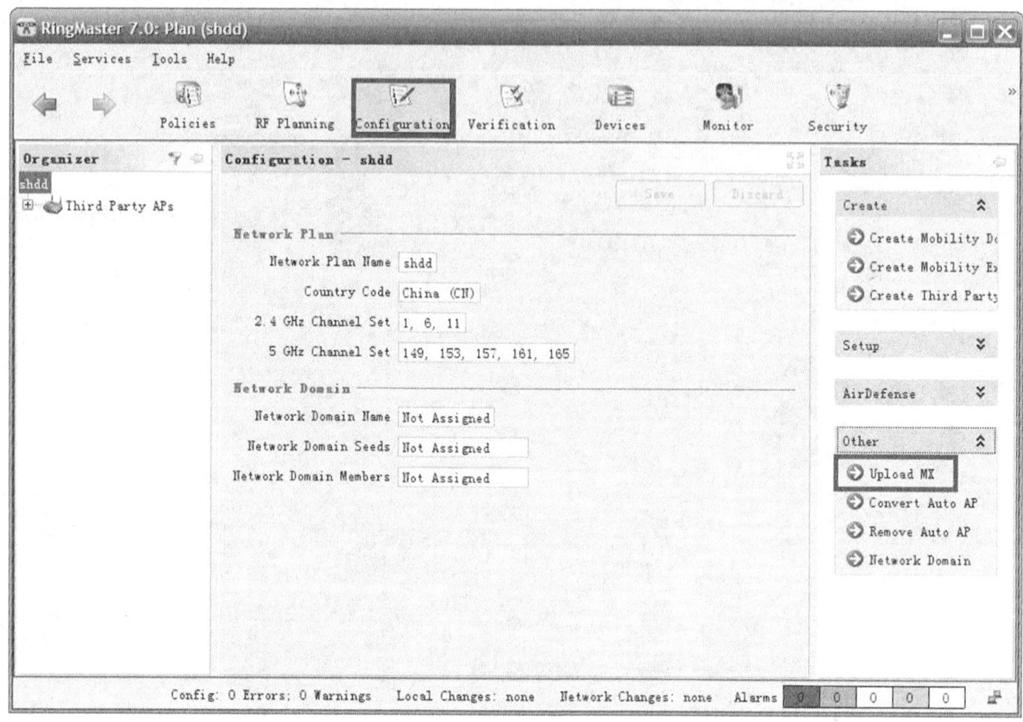

图 24-11

输入被管理的无线交换机的 IP 地址和 Enable 密码，如图 24-12 至图 24-14 所示。

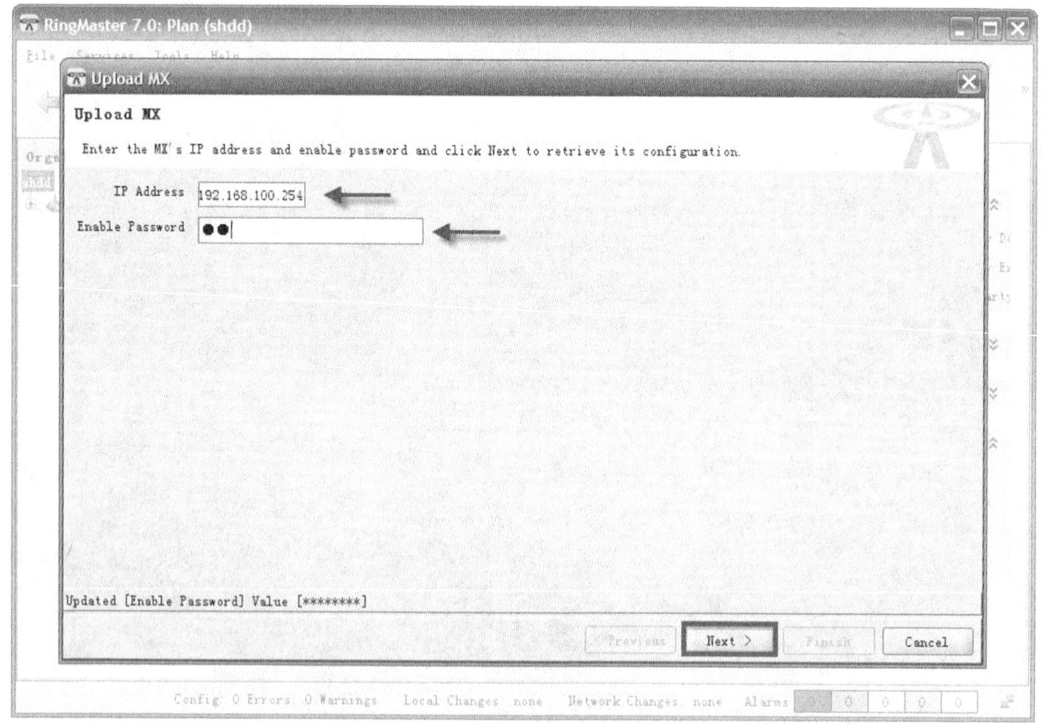

图 24-12

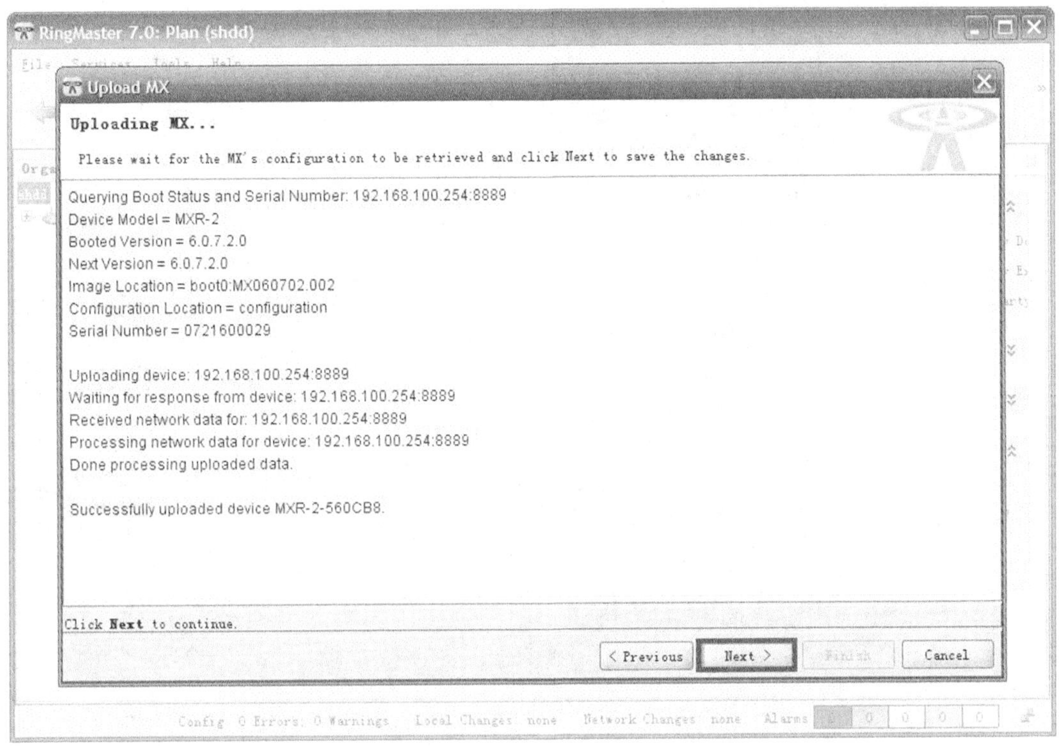

图 24-13

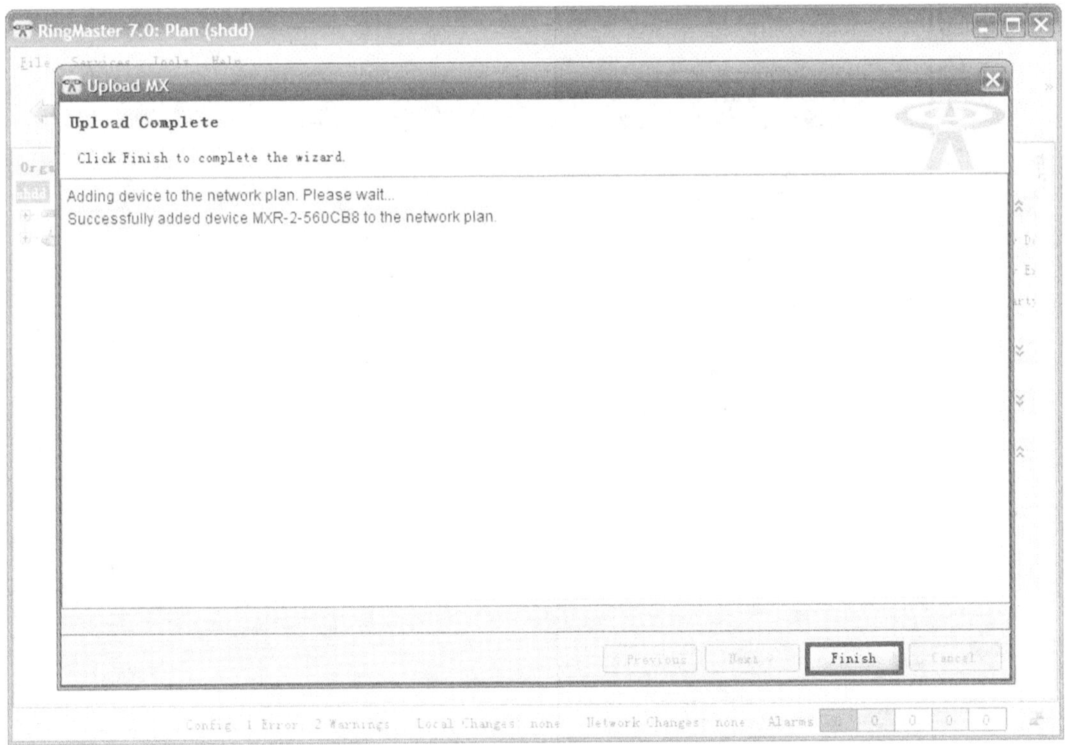

图 24-14

完成添加后，进入无线交换机的操作界面，如图 24-15 所示。

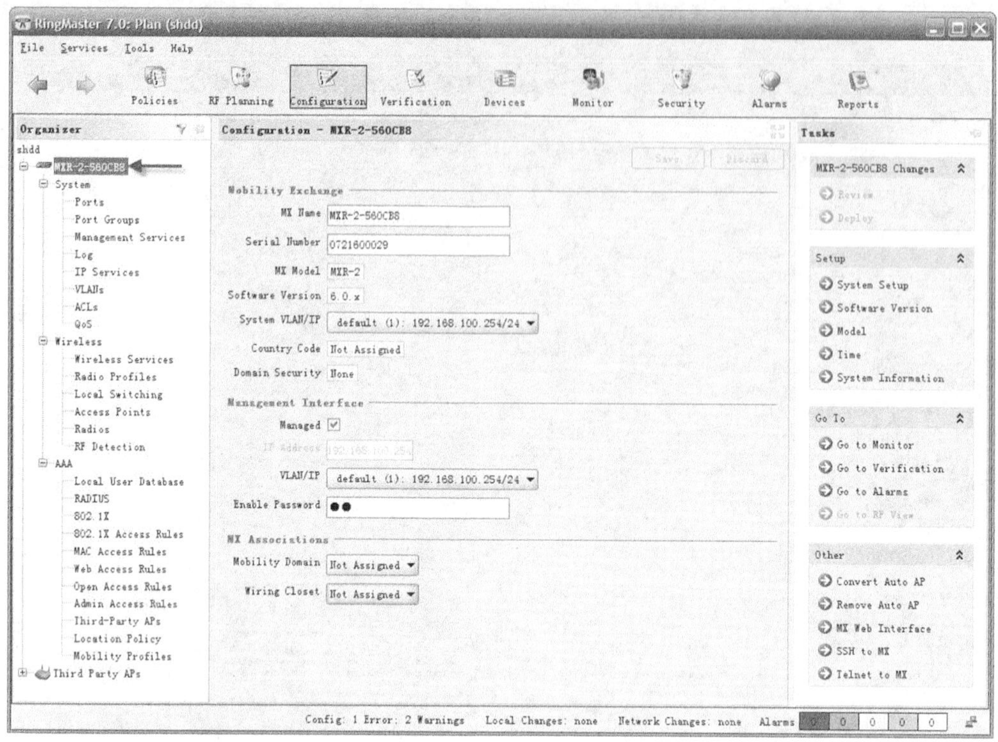

图 24-15

步骤 3　配置无线 AP。

进入“Wireless”→“Access Points”选项，添加 AP，如图 24-16 所示。

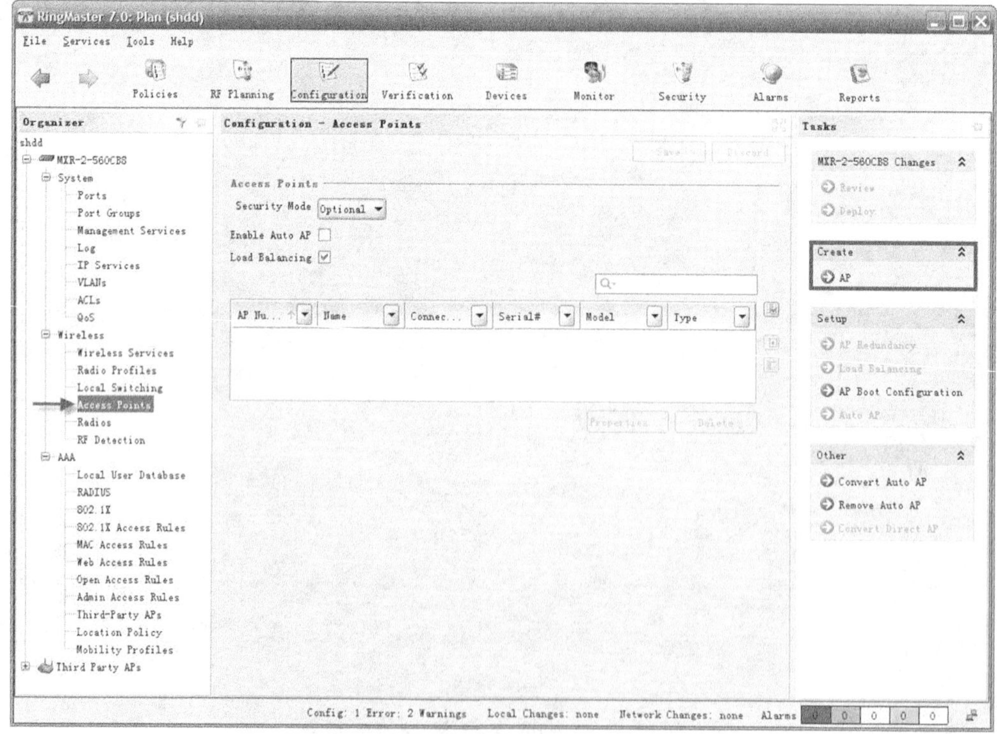

图 24-16

为添加的AP进行命名，并选择连接方式，默认使用“Distributed”模式，如图24-17所示。

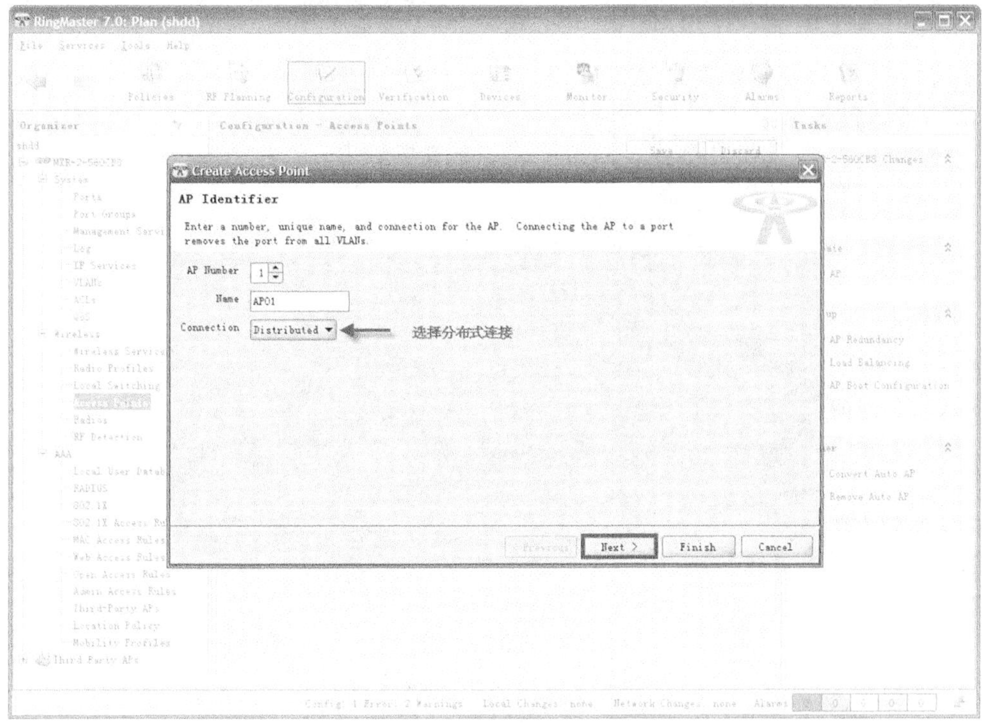

图 24-17

将需要添加的AP机身后面的SN号输入对话框，用于AP与无线交换机的注册过程，如图24-18所示。

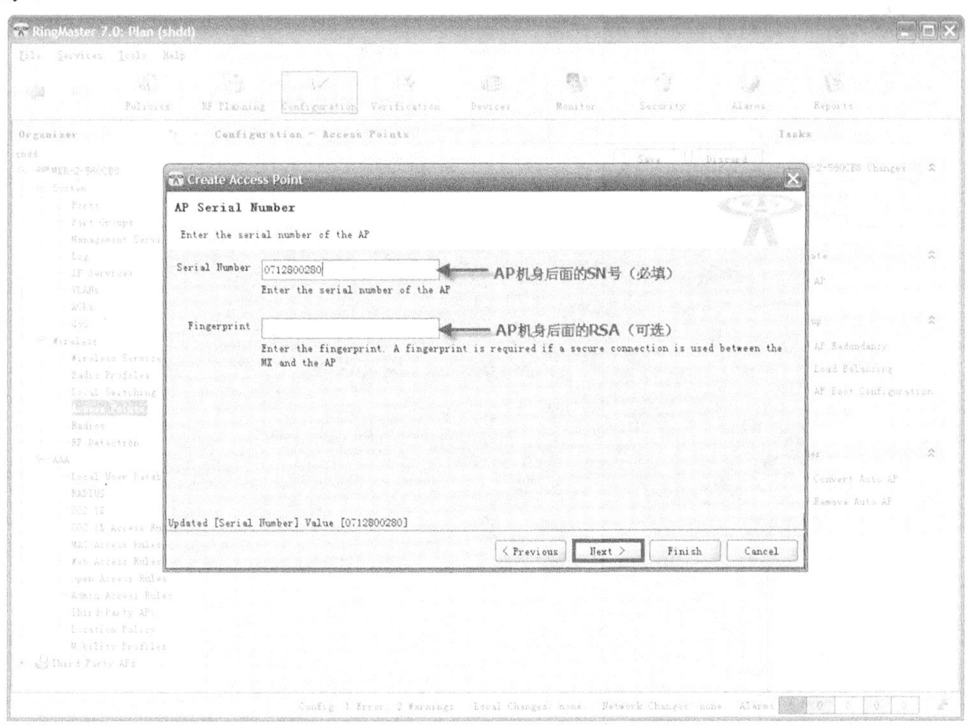

图 24-18

选择添加 AP 的具体型号和传输协议，完成 AP 的添加，如图 24-19 所示。

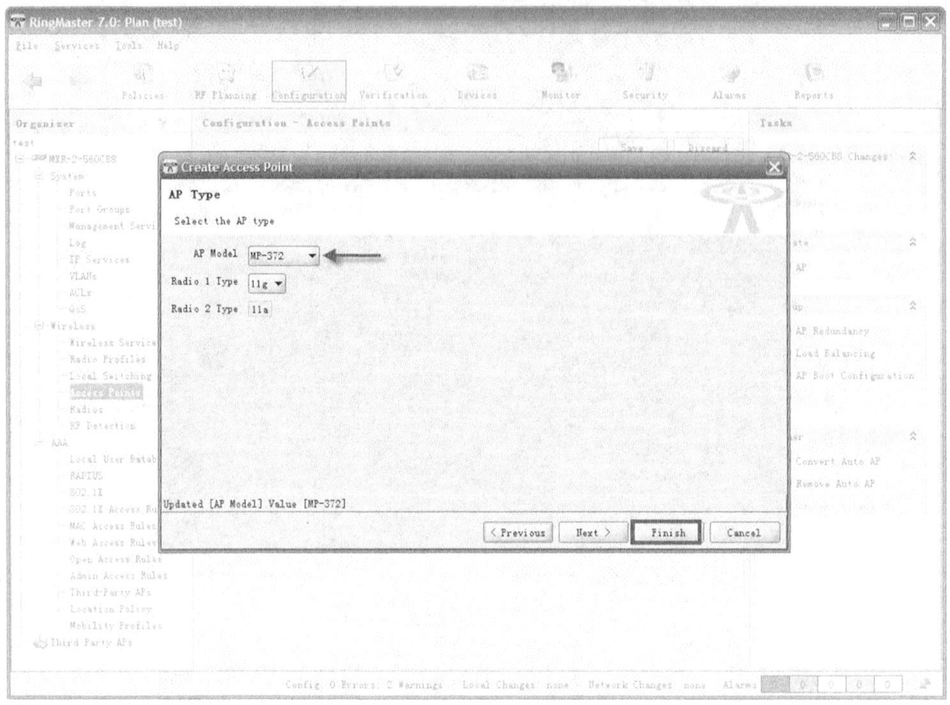

图 24-19

步骤 4　配置无线交换机的 DHCP 服务器。

进入“Syestem”→“VLANs”选项，选择“default”VLAN，进入属性配置页面，如图 24-20 所示。

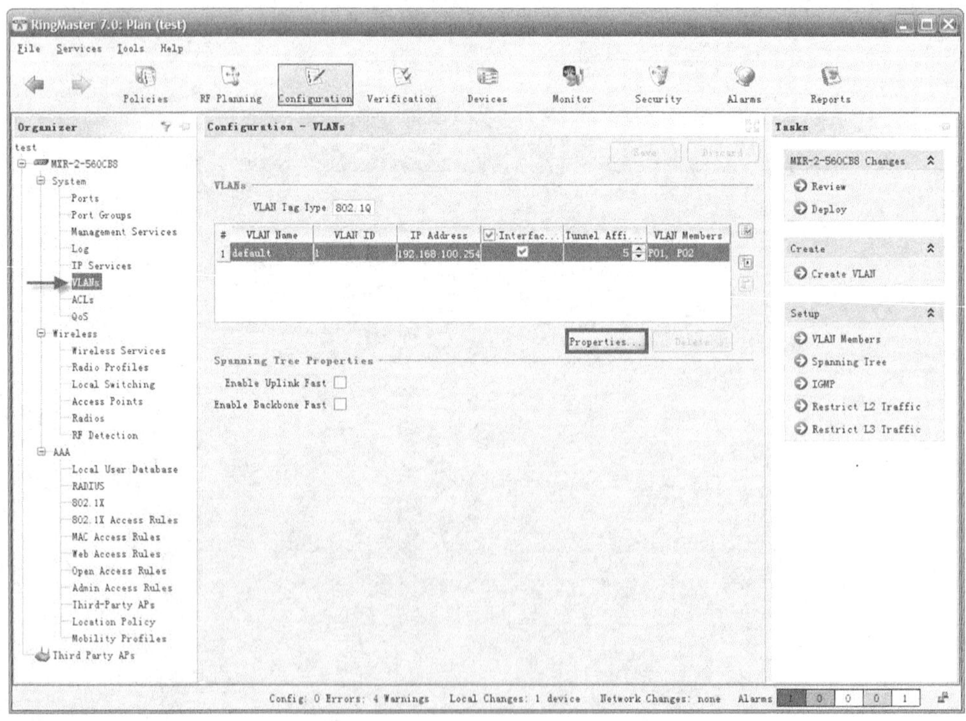

图 24-20

进入“VLAN Properties”→“DHCP Server”选项，激活 DHCP 服务器，设置地址池和 DNS，保存设置，如图 24-21 所示。

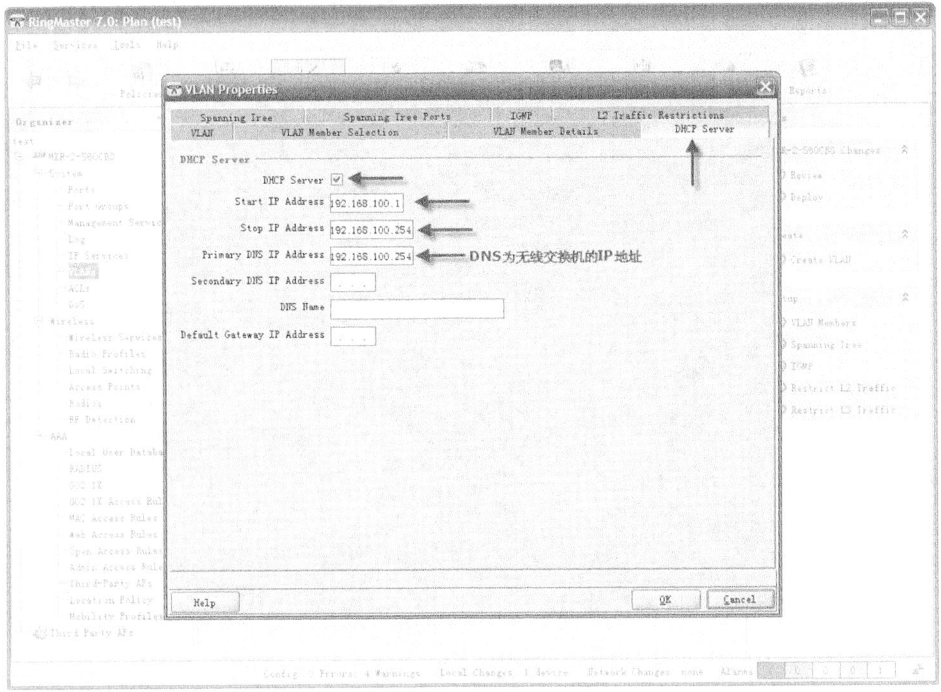

图 24-21

进入“System”→“Ports”选项，将无线交换机的端口 PoE 打开，并保存设置，如图 24-22 所示。

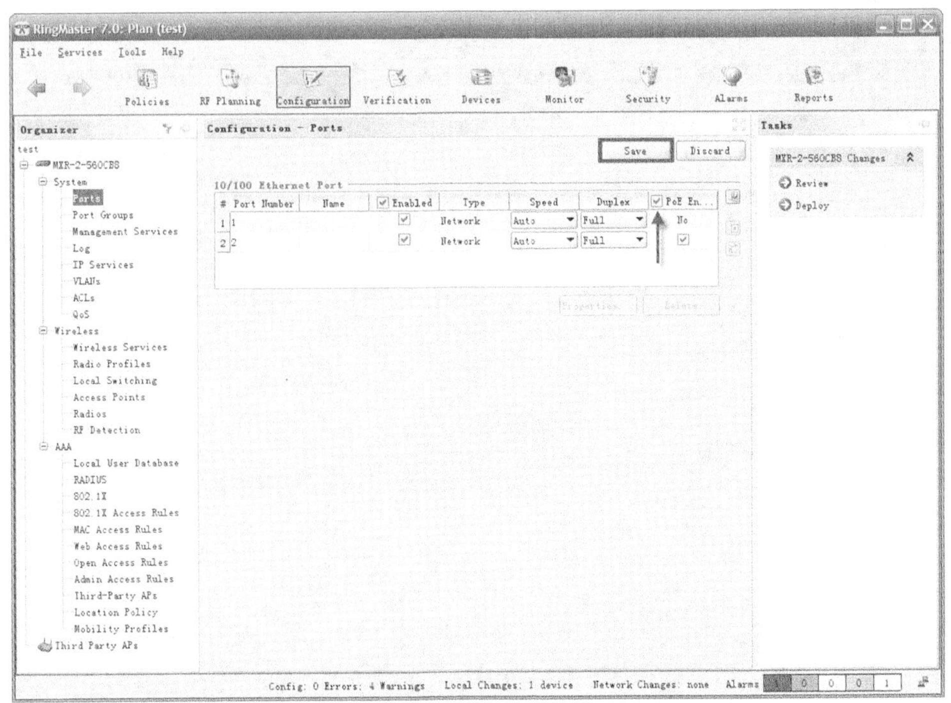

图 24-22

步骤 5　创建开放接入服务。

（1）首先，建立一个 Open Access Service Profile，如图 24-23 所示。

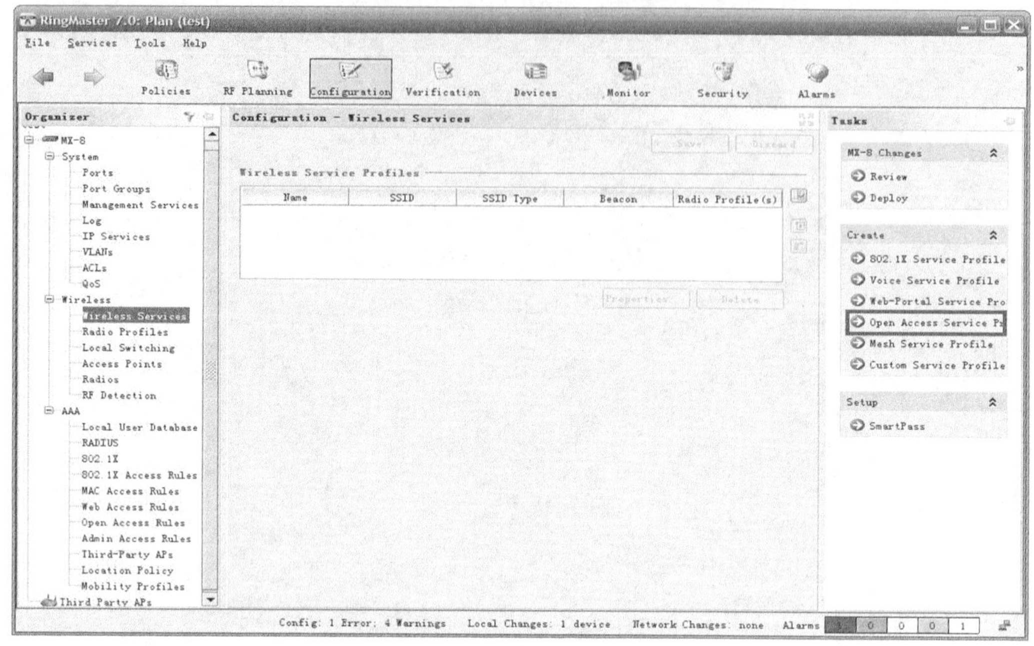

图 24-23

（2）输入 SSID 名，由于是开放式的服务，因此设置 SSID Type 为“Clear”，即不加密，如图 24-24 所示。

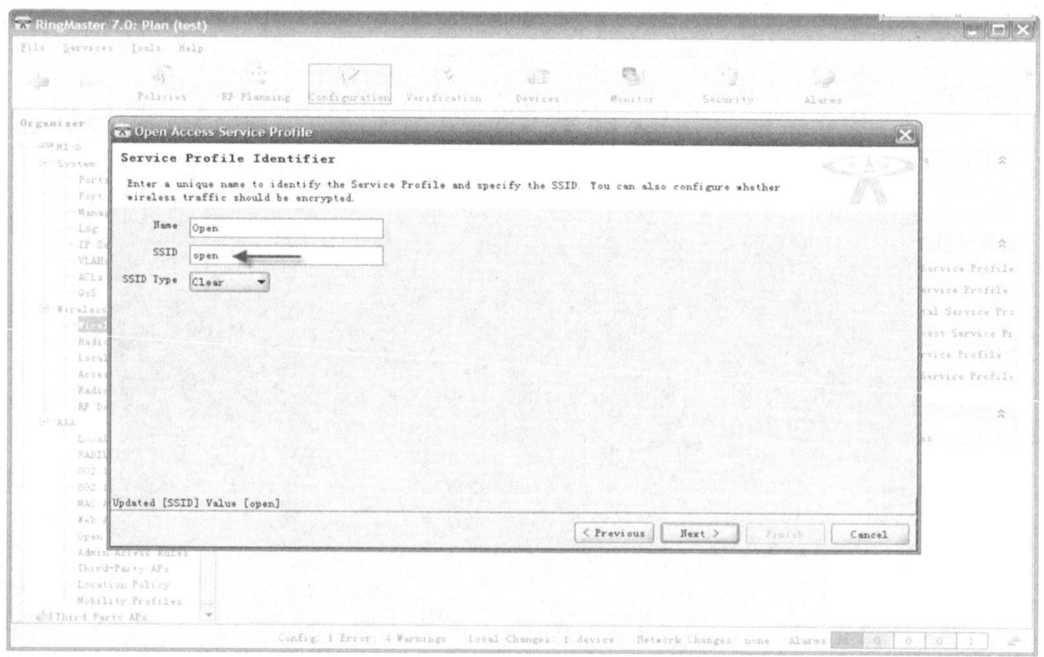

图 24-24

（3）默认将用户的 VLAN 定义为“default”VLAN，即用户联入这个 SSID 可获得默认 VLAN 的 IP 地址，如图 24-25 所示。

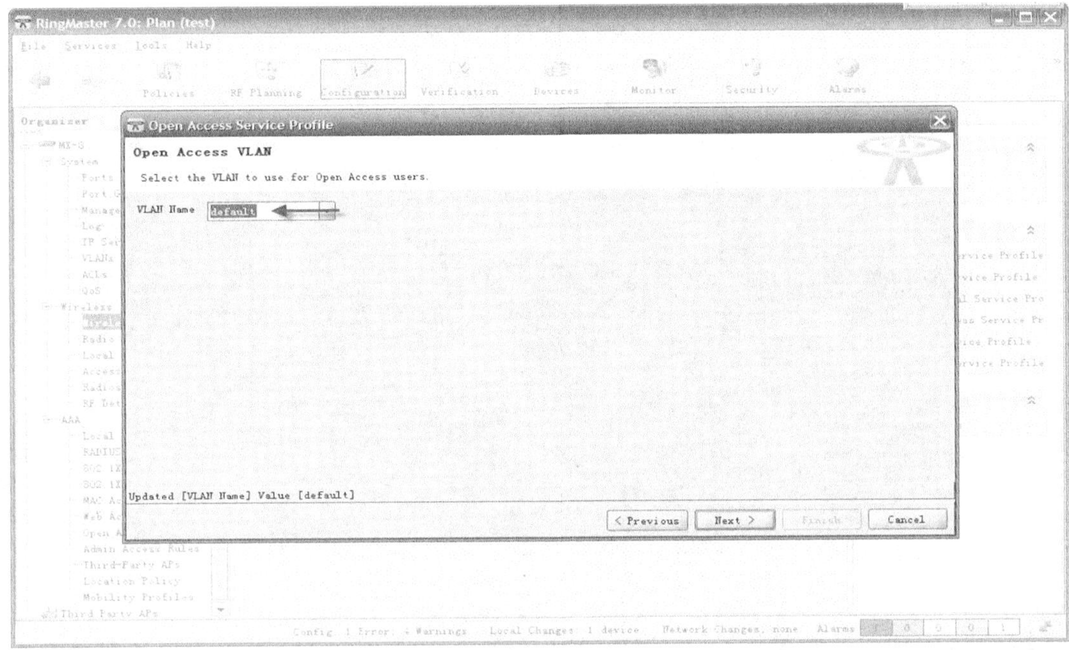

图 24-25

（4）选择默认的 Radio Profile（Radio Profile 定义了 AP 的射频规则），即该无线配置作用下的 AP 采用默认的射频规则，如图 24-26 所示。

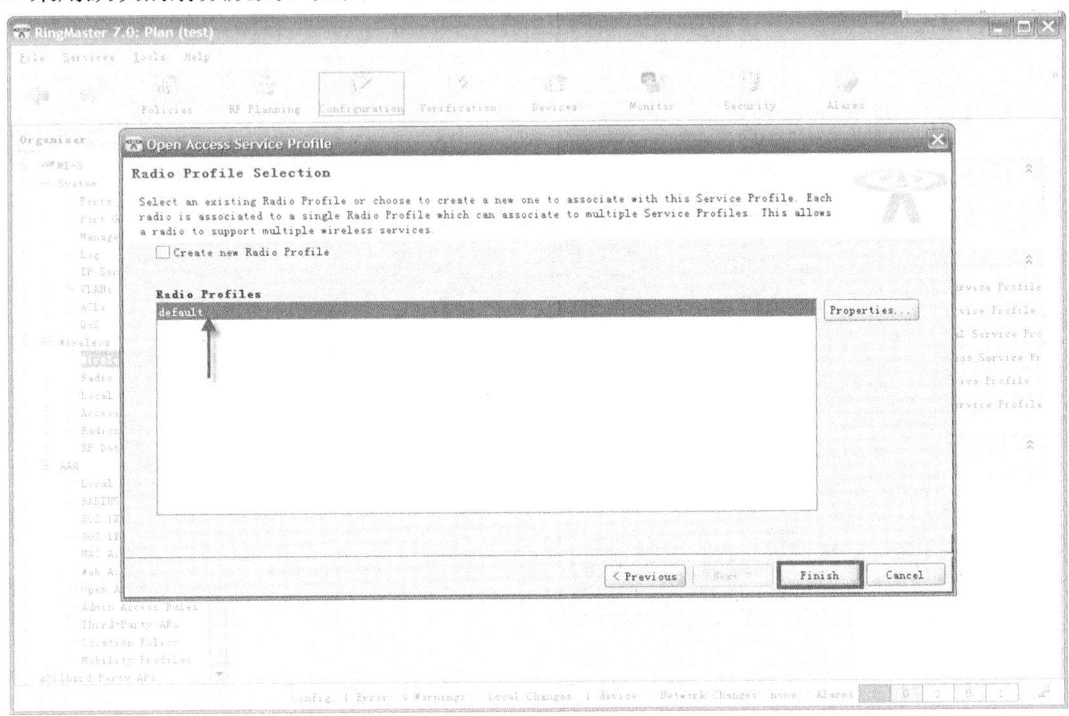

图 24-26

（5）完成“开放式无线接入服务”的配置，选择如图 24-27 所示的“Deploy”选项，下发配置到无线交换机。

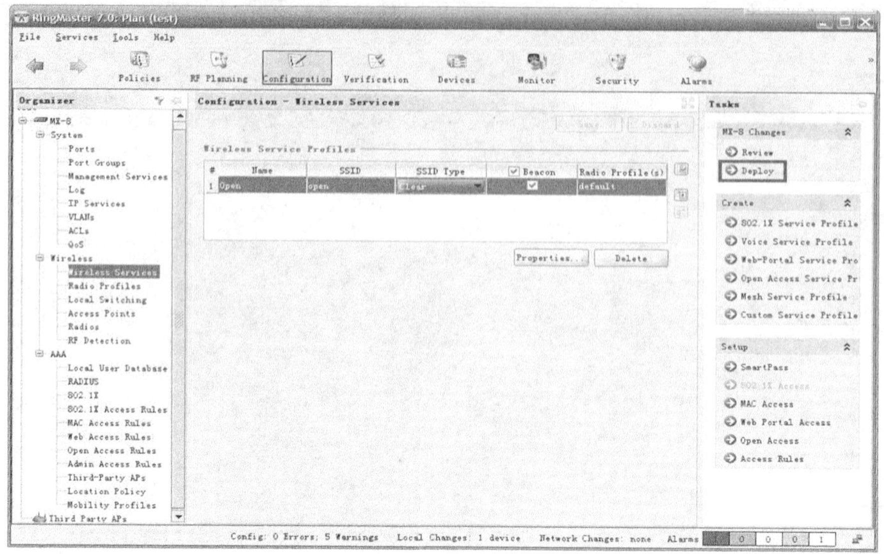

图 24-27

（6）配置完成，开放式无线接入网络建立完成。

步骤 6　测试该无线接入服务。

STA2 打开无线网卡，扫描这个 SSID，并获取 IP 地址，如图 24-28 所示。

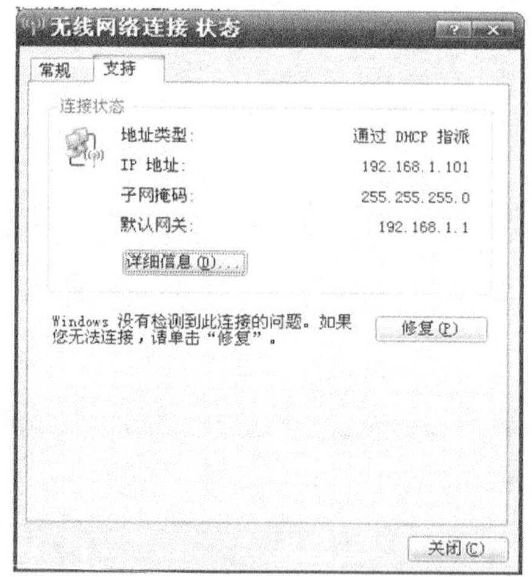

图 24-28

【注意事项】

（1）无线网卡的硬件开关必须是开启的。

（2）无线客户端的 IP 地址是自动获取的。

实验 25　配置单频多模的无线网络

【实验名称】

配置单频多模的无线网络。

【实验目的】

通过配置多模的无线网络，掌握 802.11b 和 802.11g 两种协议的实际性能表现，巩固 802.11 协议的理论知识。

【背景描述】

华新大学正在筹建无线校园网，由于学校人数众多，笔记本用户数达到全校的 50%。据调查得知，其中有不少笔记本是迅驰一代的芯片，只能支持 802.11b 协议。学校考虑到兼容性的问题，需要建设一个既能支持 802.11b，也能支持 802.11g 的无线网络，并且不能因为 802.11b 的用户存在而影响到 802.11g 用户的速率。

【需求分析】

需要建设支持 802.11b 和 802.11g 的无线网络，并且不能因为 802.11b 用户的存在而影响到 802.11g 的 54Mbps 速率。要保证 802.11b/g 的无线客户端共存，需采用支持 802.11g 保护模式的无线 AP。

【实验拓扑】

实验的拓扑图，如图 25-1 所示。

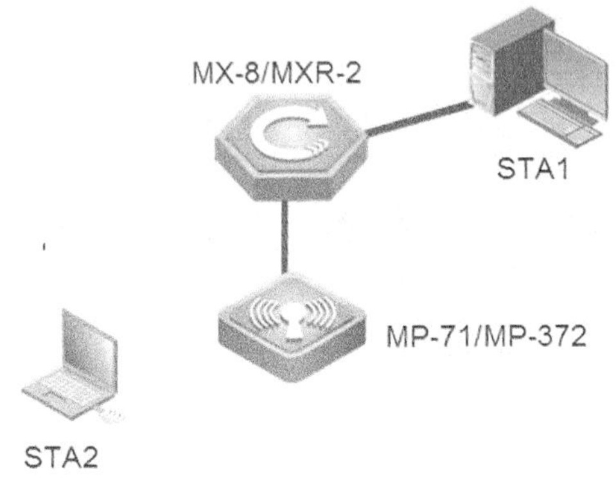

图 25-1

【实验设备】

RG-WG54U 1 块。

PC 1 台。

MP-71/MP-372 1 台。

MXR-2/MX-8 1 台。

安装有无线网管 RingMaster 的服务器 1 台。

【预备知识】

无线局域网基本知识、智能无线交换网络工作原理、智能无线交换产品基本操作。

【实验原理】

802.11b 和 802.11g 标准在理论上的最高数据传输速率可达到 11Mbps 和 54Mbps，但在实际应用中，通常有一半的“原始速度”被分组负载、校验和、帧位、错误恢复数据和其他“无用”的信息占用了。即使不考虑传播范围和障碍物对性能的削弱影响，实际吞吐率也仅能达到最高传输速率的一半，甚至更低，约为 5.5Mbps 和 20~26Mbps。

【实验步骤】

步骤 1　配置无线交换机的基本参数。

无线交换机的默认 IP 地址是 192.168.100.1/24，因此将 STA-1 的 IP 地址配置为 192.168.100.2/24，并打开浏览器，登录到 https://192.168.100.1，弹出如图 25-2 所示的界面，选择“是（Y）”按钮。

系统的默认管理用户名是 admin，密码为空，如图 25-3 所示。

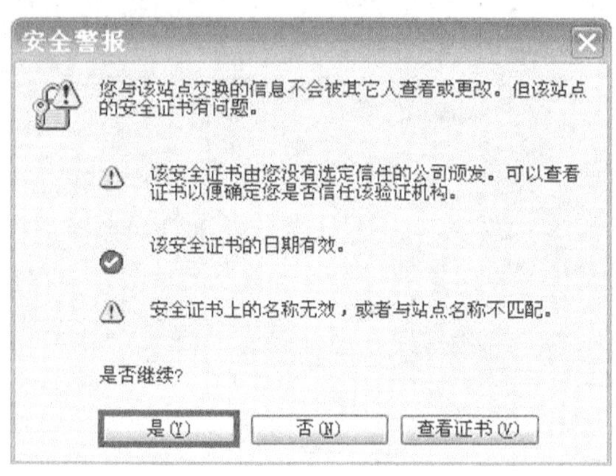

图 25-2

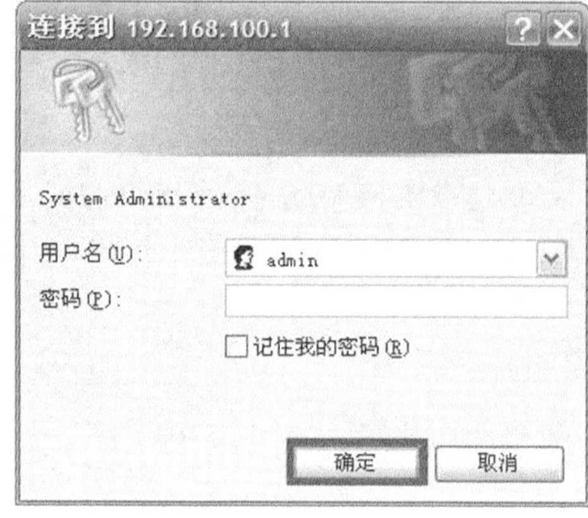

图 25-3

输入用户名和密码后，就进入了无线交换机的 Web 配置页面，单击“Start”按钮，进入快速配置指南页面，如图 25-4 所示。

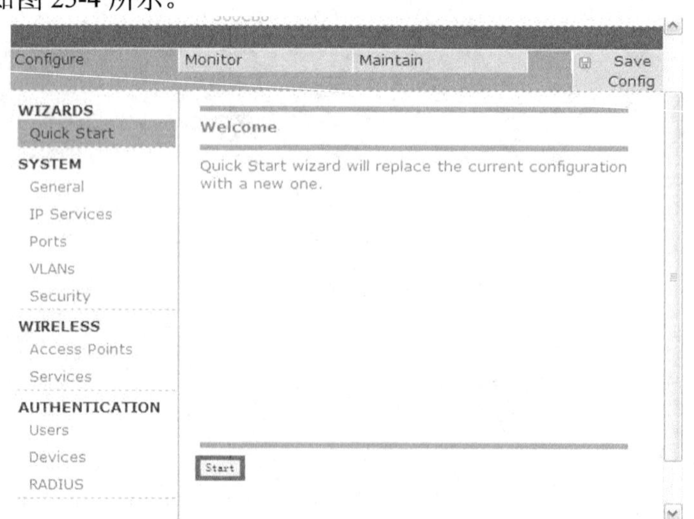

图 25-4

选择管理无线交换机的工具“RingMaster”，如图 25-5 所示。

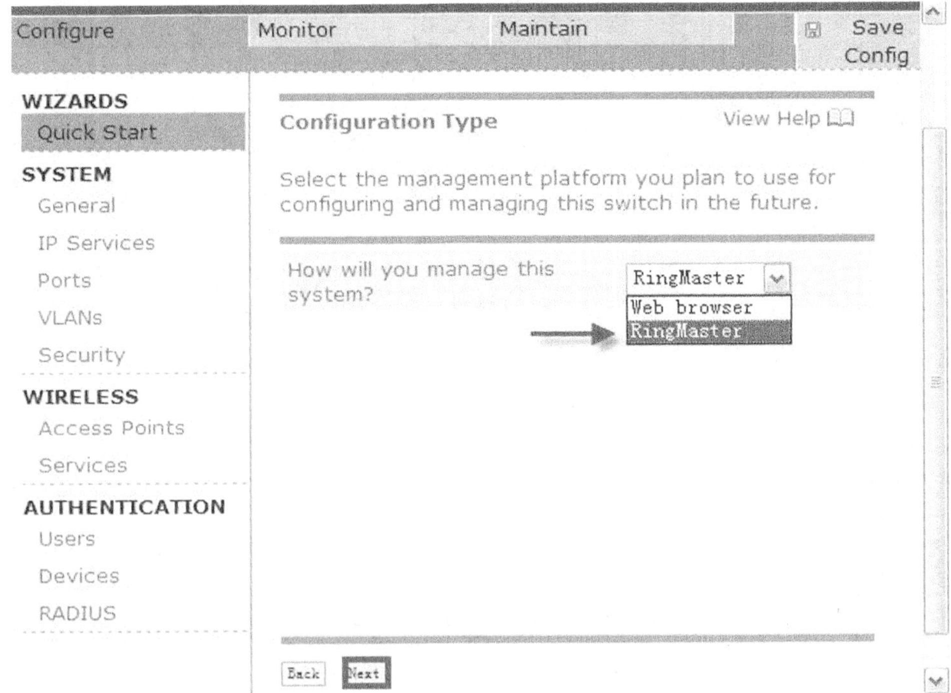

图 25-5

配置无线交换机的 IP 地址、子网掩码以及默认网关，如图 25-6 所示。

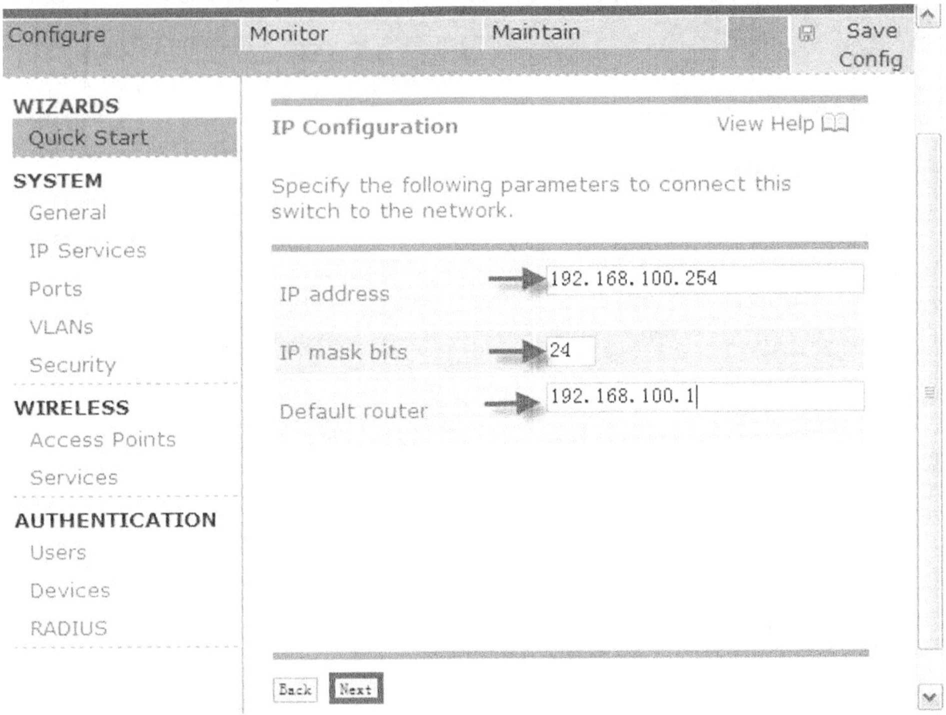

图 25-6

设置系统的管理密码，如图 25-7 所示。

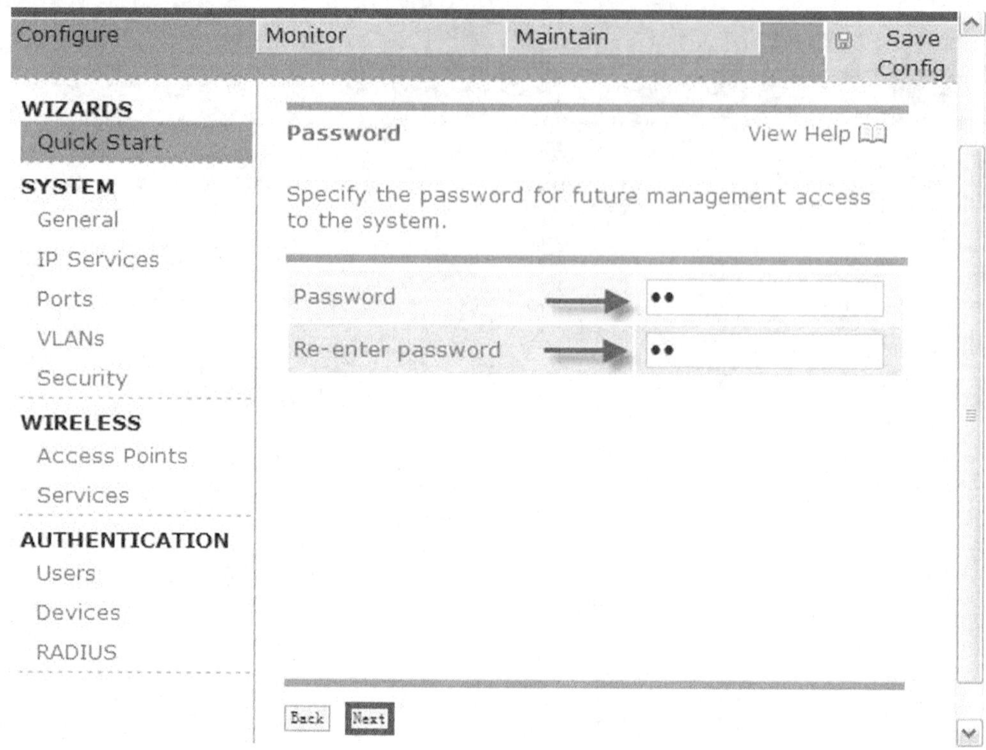

图 25-7

设置系统时间及时区，如图 25-8 所示。

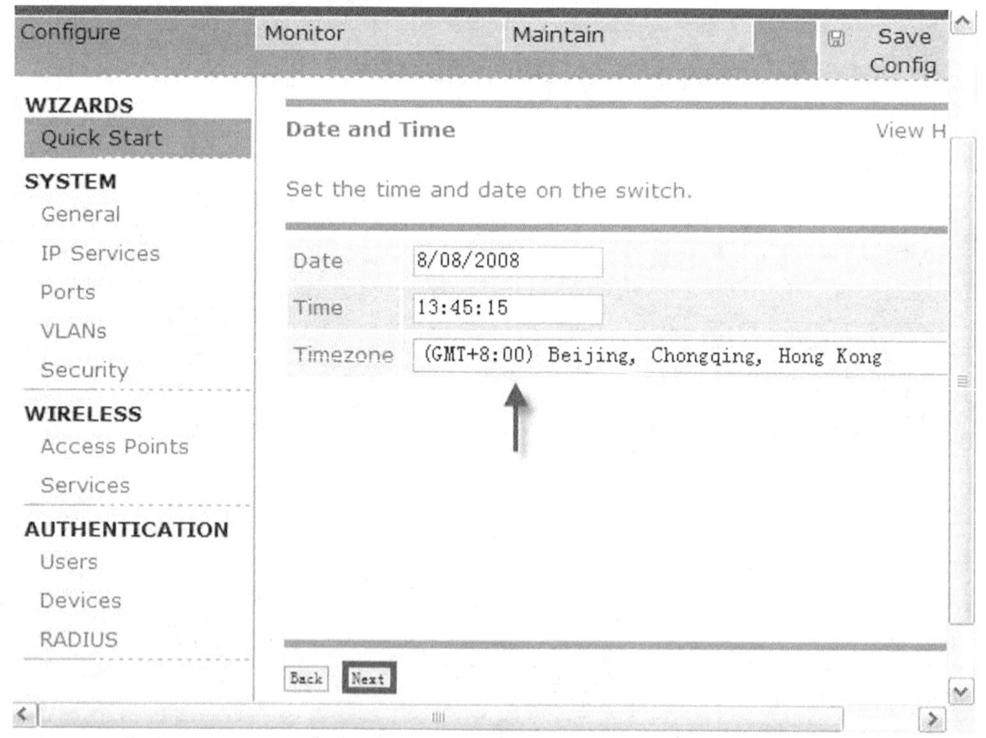

图 25-8

确认无线交换机的基本配置，如图 25-9 所示。

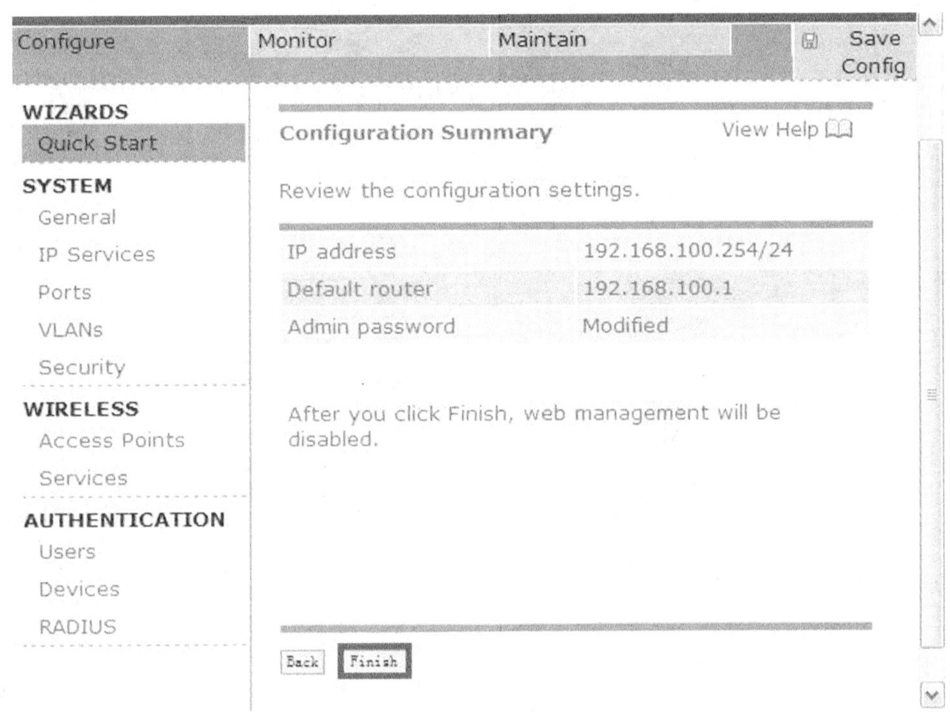

图 25-9

完成无线交换机的基本配置。

步骤 2　通过 RingMaster 网管软件来进行无线交换机的高级配置。

运行 RingMaster 软件，地址为 127.0.0.1，端口为 443，用户名和密码默认为空，如图 25-10 所示。

图 25-10

选择“Configuration”选项，进入配置界面，并添加被管理的无线交换机，如图 25-11 所示。

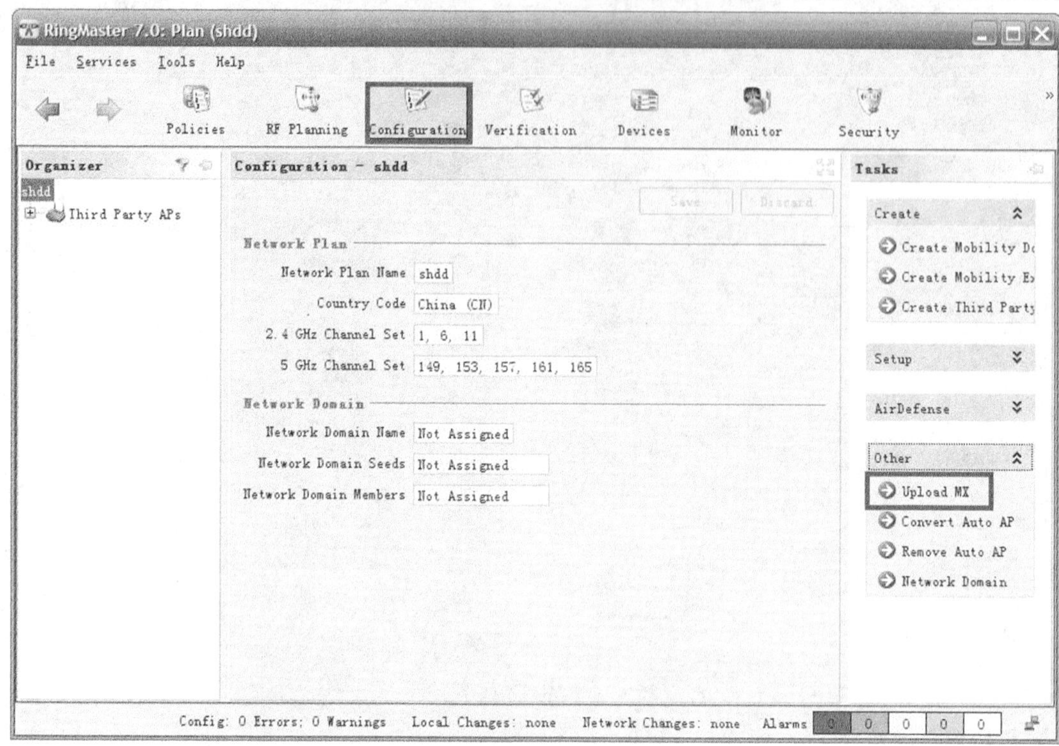

图 25-11

输入被管理的无线交换机的 IP 地址及 Enable 密码，如图 25-12 至图 25-14 所示。

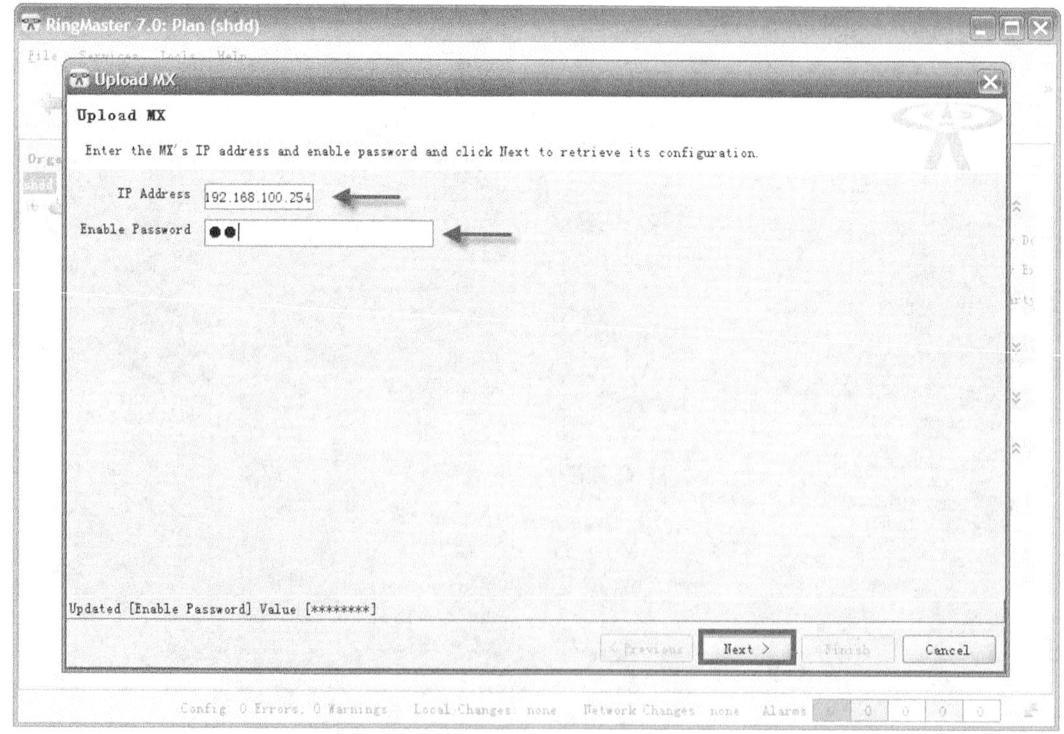

图 25-12

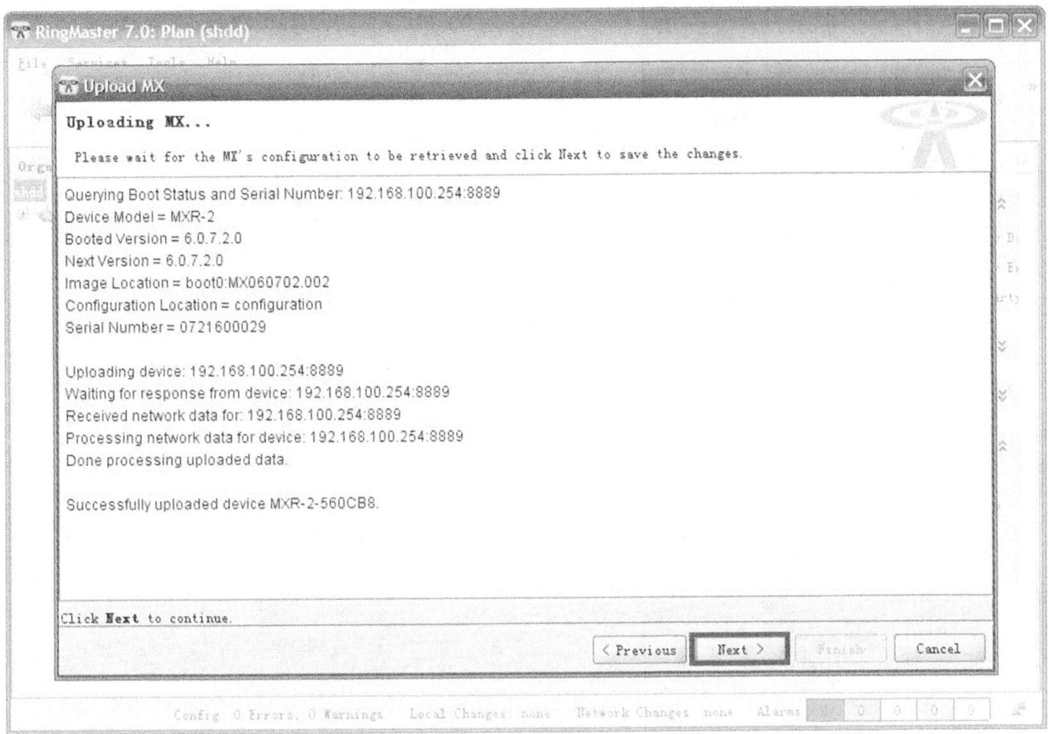

图 25-13

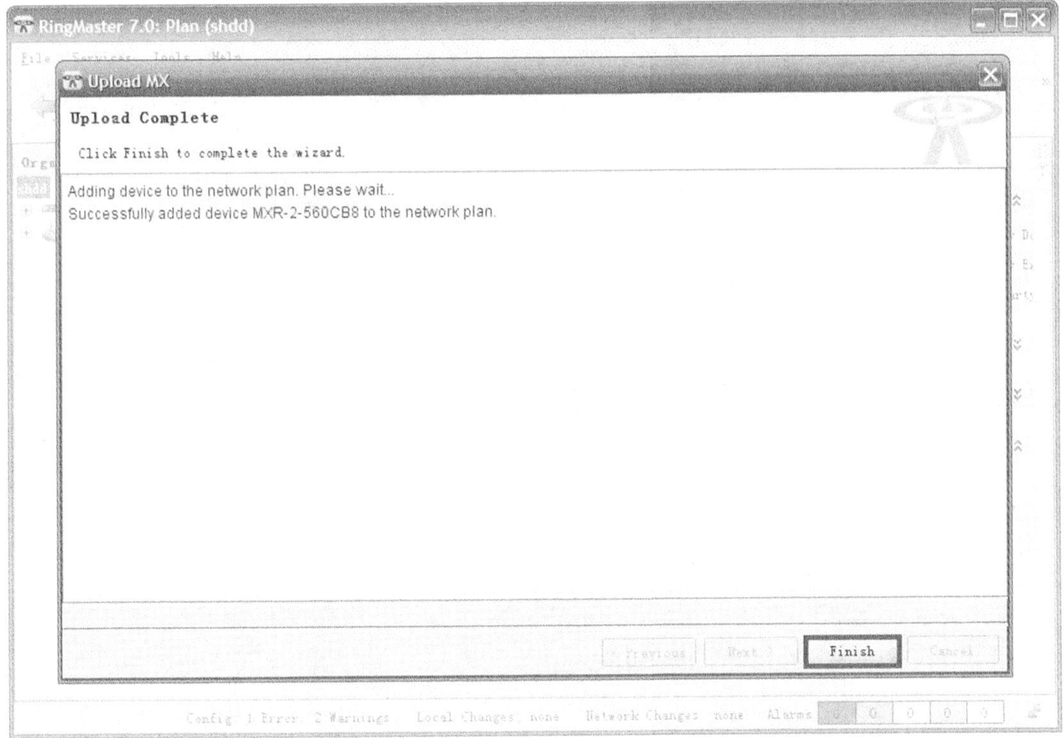

图 25-14

完成添加后，进入无线交换机的操作界面，如图 25-15 所示。

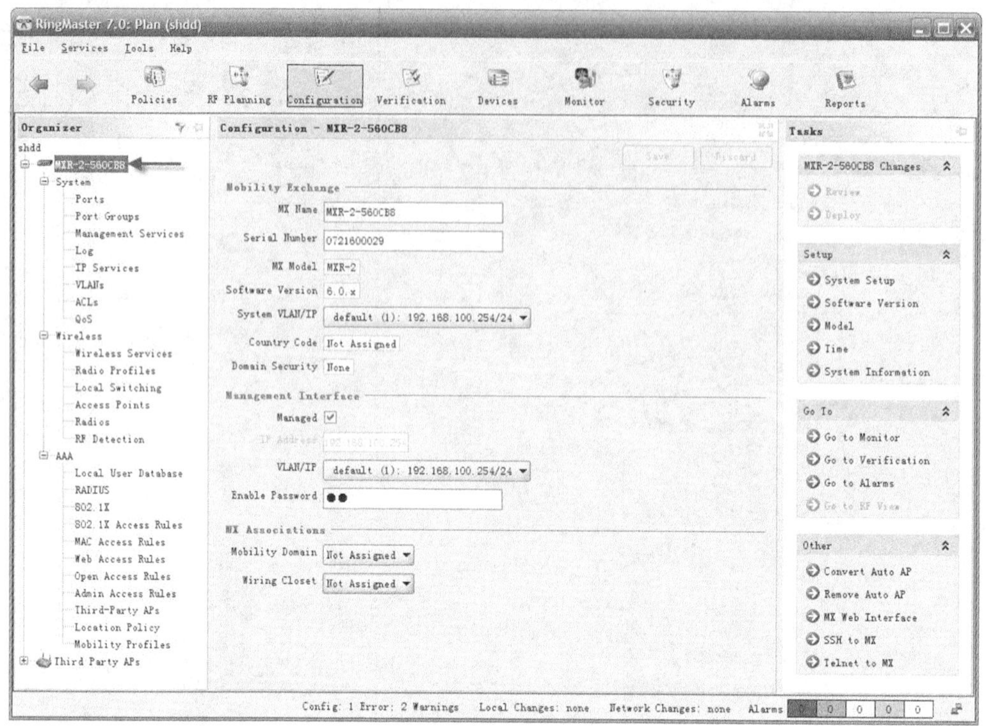

图 25-15

步骤 3　配置无线 AP。

进入“Wireless”→“Access Points”选项，添加 AP，如图 25-16 所示。

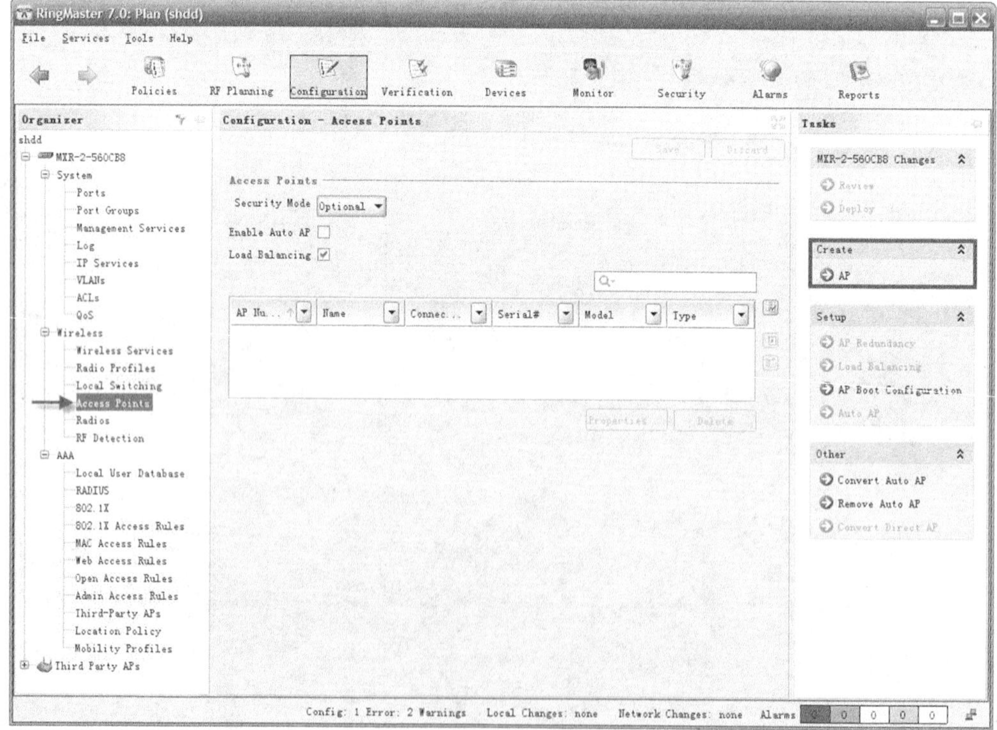

图 25-16

为添加的 AP 命名，并选择连接方式，默认使用“Distributed”模式，如图 25-17 所示。

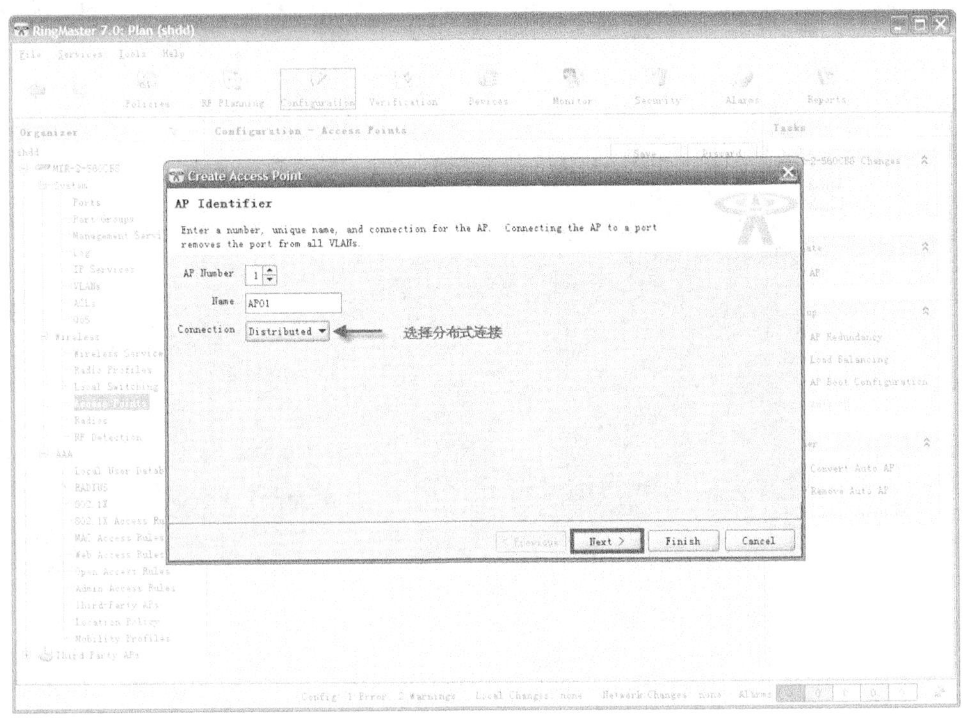

图 25-17

将需要添加的 AP 机身后面的 SN 号输入到对话框中，用于 AP 与无线交换机的注册过程，如图 25-18 所示。

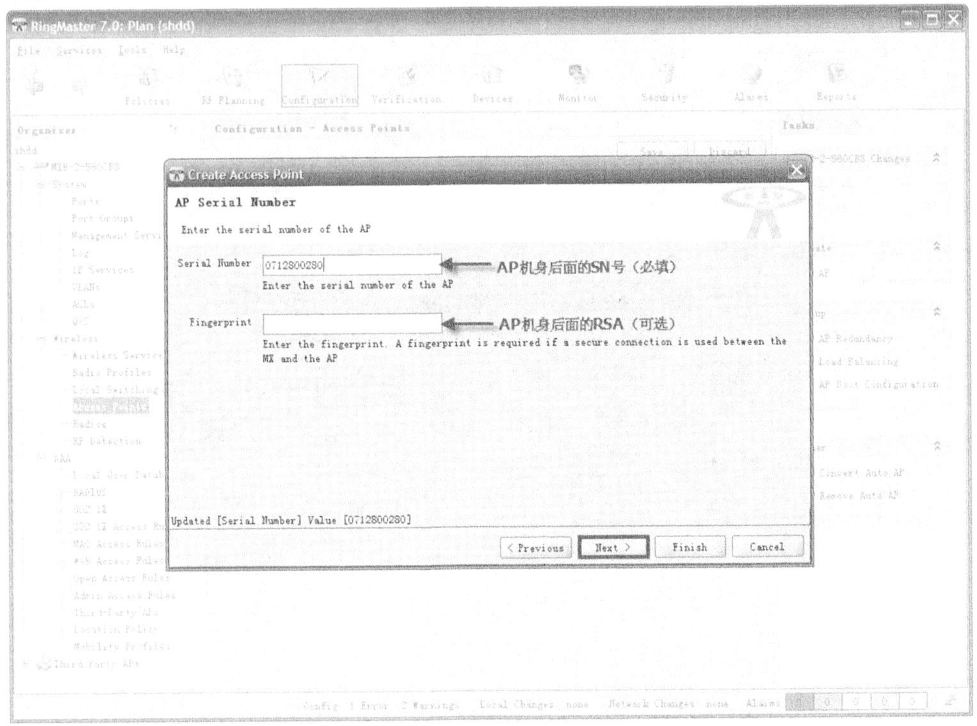

图 25-18

选择添加 AP 的具体型号和传输协议，完成 AP 的添加，如图 25-19 所示。

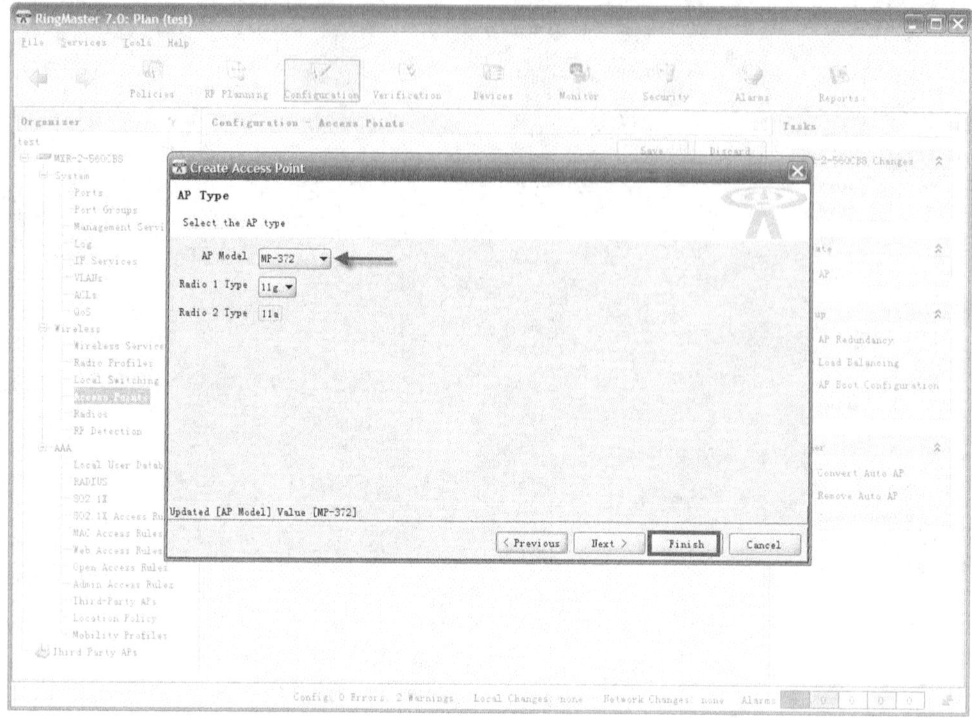

图 25-19

步骤 4　配置无线交换机的 DHCP 服务器。

进入“Syestem”→“VLANs”选项，选择“default”VLAN，进入属性配置页面，如图 25-20 所示。

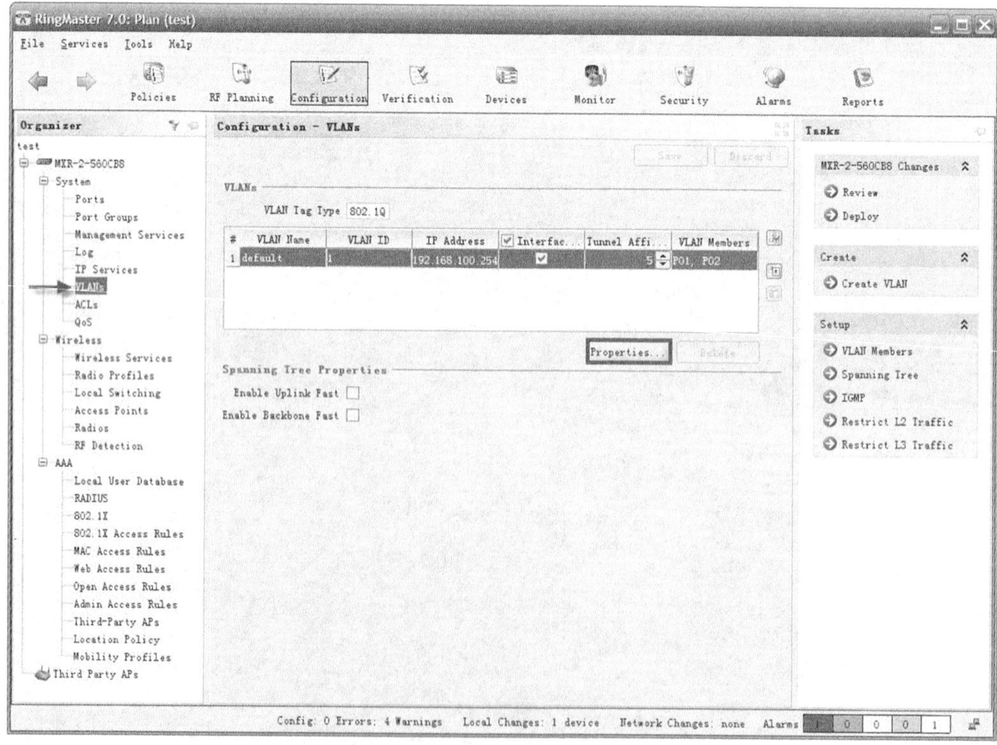

图 25-20

进入“VLAN Properties”→“DHCP Server”选项，激活DHCP服务器，设置地址池和DNS，保存设置，如图25-21所示。

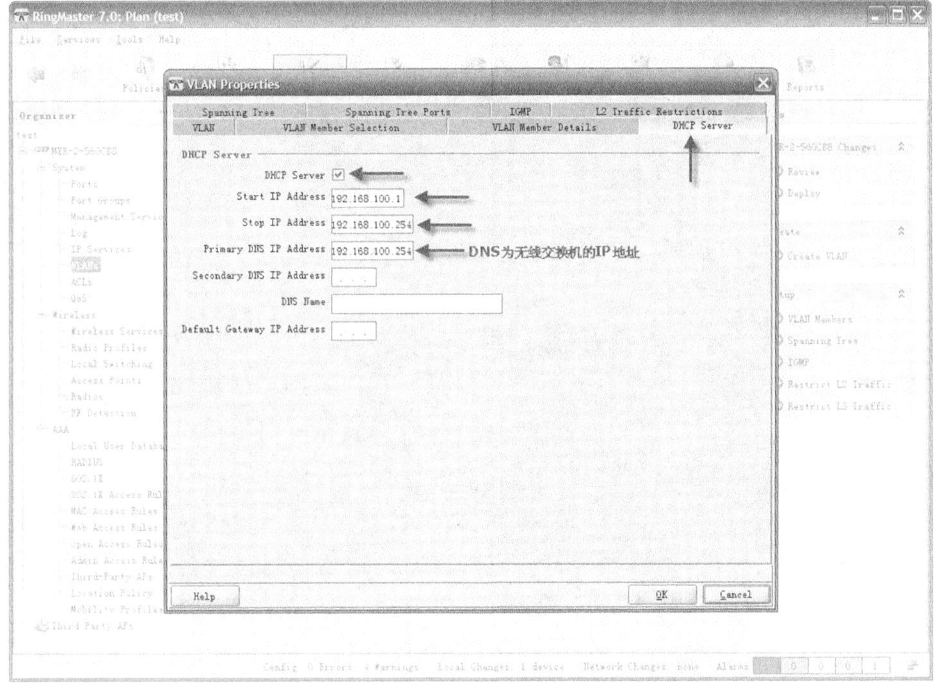

图 25-21

进入“System”→“Ports”选项，将无线交换机的端口PoE打开，并保存设置，如图25-22所示。

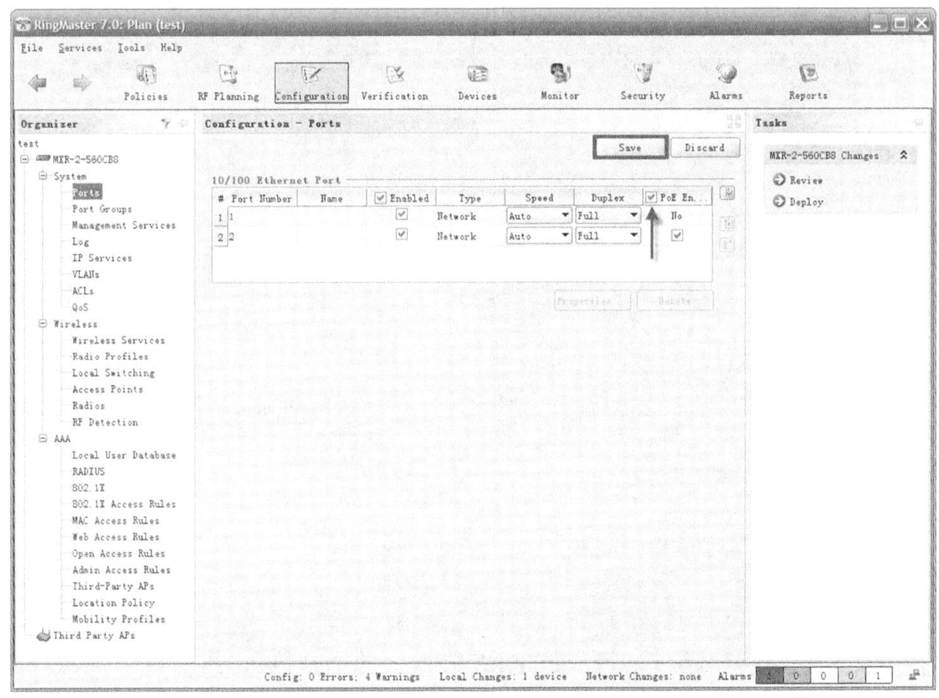

图 25-22

步骤 5　建立开放式的无线接入系统。

（1）首先，建立一个 Open Access Service Profile，如图 25-23 所示。

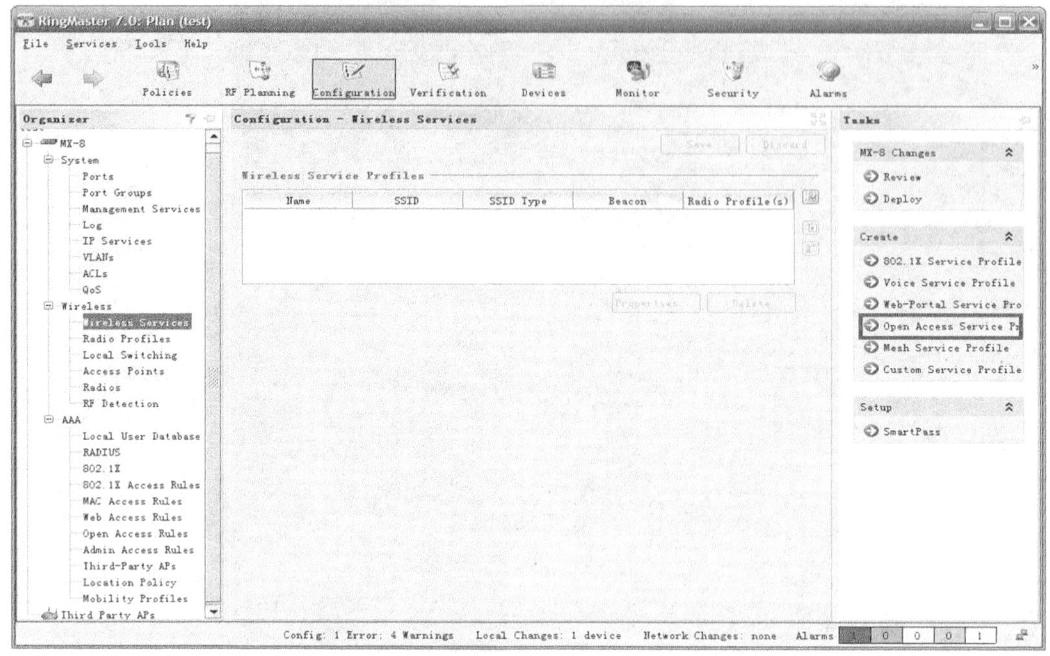

图 25-23

（2）输入 SSID 名，由于是开放式的服务，“SSID Type”为“clear”，即不加密，如图 25-24 所示。

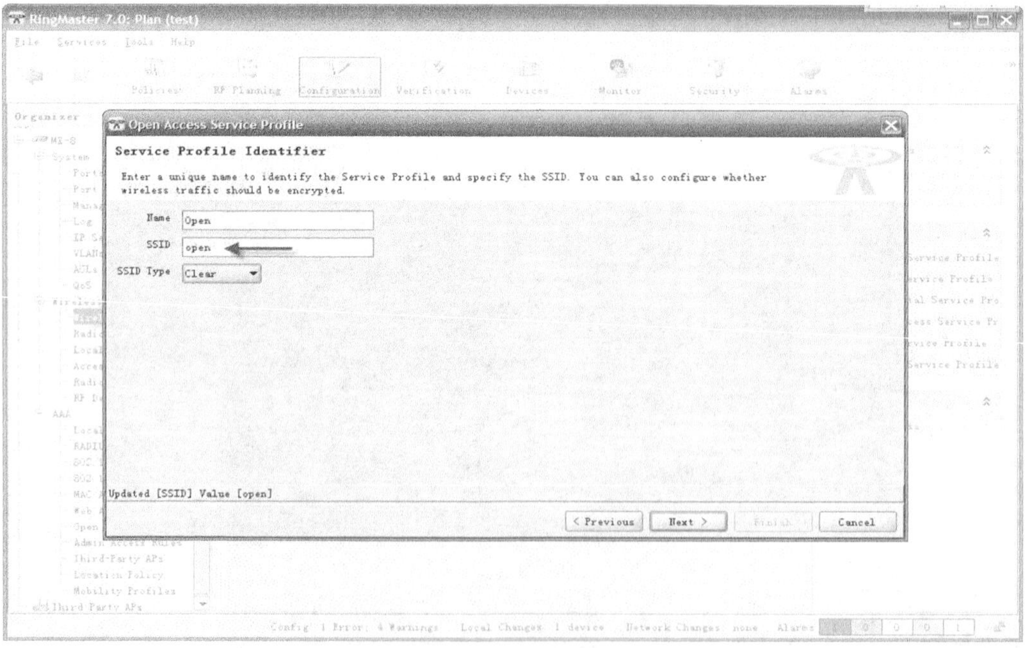

图 25-24

（3）默认将用户的 VLAN 定义为 default VLAN，即用户联入 SSID 时，可获得默认 VLAN 的 IP 地址，如图 25-25 所示。

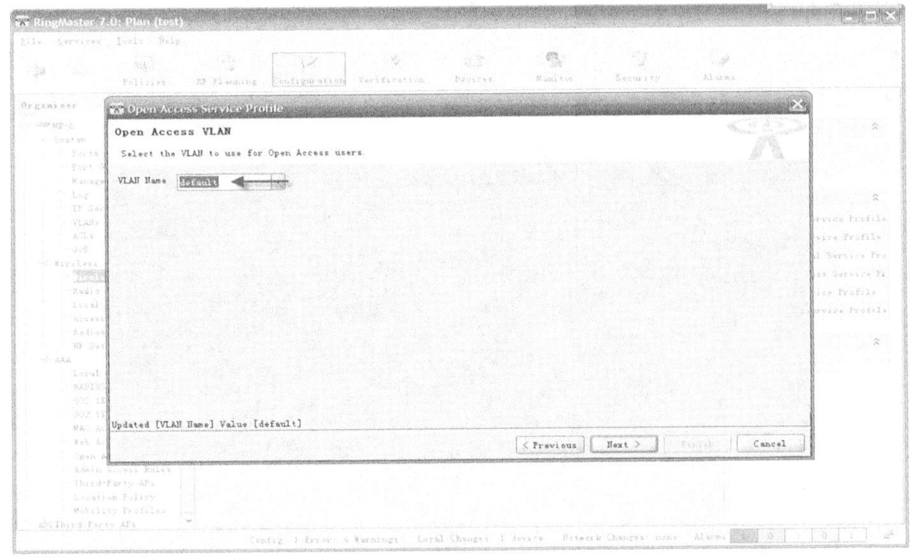

图 25-25

（4）选择默认的 Radio Profile（Radio Profile 定义了 AP 的射频规则），即该无线配置作用下的 AP 采用默认的射频规则，如图 25-26 所示。

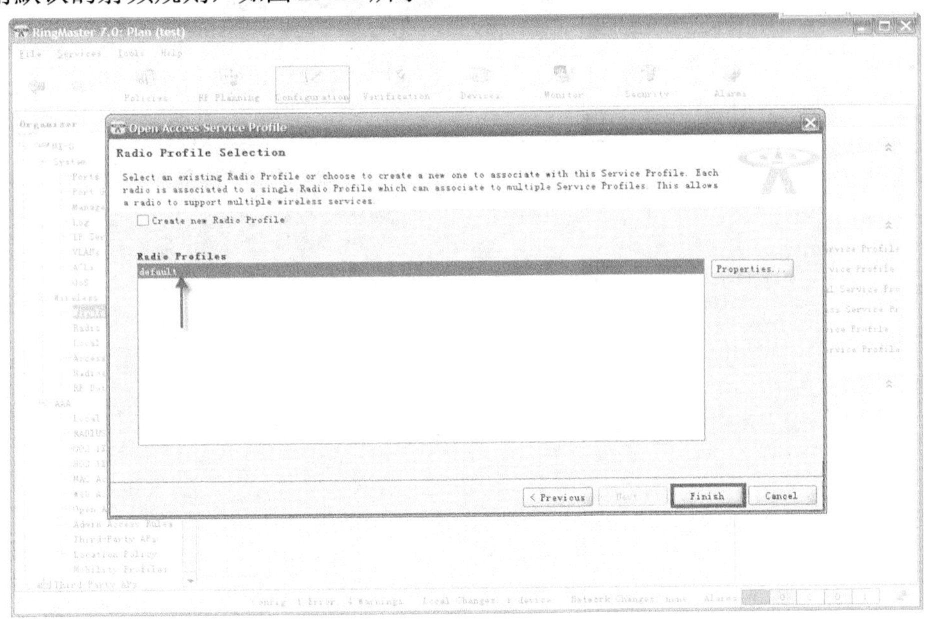

图 25-26

1）完成“开放式无线接入服务”的配置，选择如图 25-27 所示的“Deploy”，下发配置到无线交换机。

2）配置完成，开放式无线接入网络建立完成。

步骤 6　测试 802.11b 和 802.11g 无线网卡的实际吞吐量，如图 25-28 所示。

（1）将 STA2 的无线网卡配置成 802.11b 模式，关联上“OPEN”SSID，并获取 IP 地址。测试 STA2 到 STA1 的 FTP 下载速率。

（2）将 STA2 的无线网卡配置成 802.11b 模式，关联上“OPEN”SSID，并获取 IP 地址。测试 STA2 到 STA1 的 FTP 下载速率。

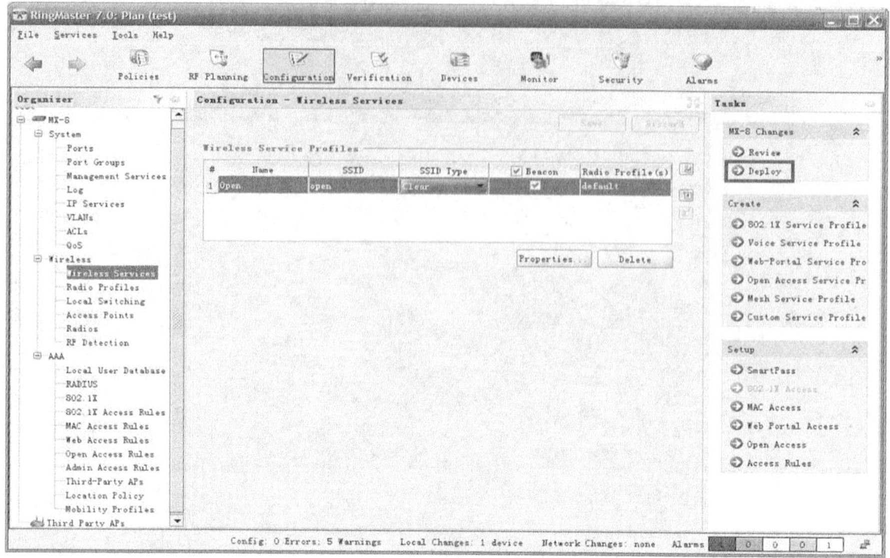

图 25-27

【预期结果】

模式	上传速率（Mbps）	下载速率（Mbps）
802.11g	21.5	22
802.11b	5.5	5.9

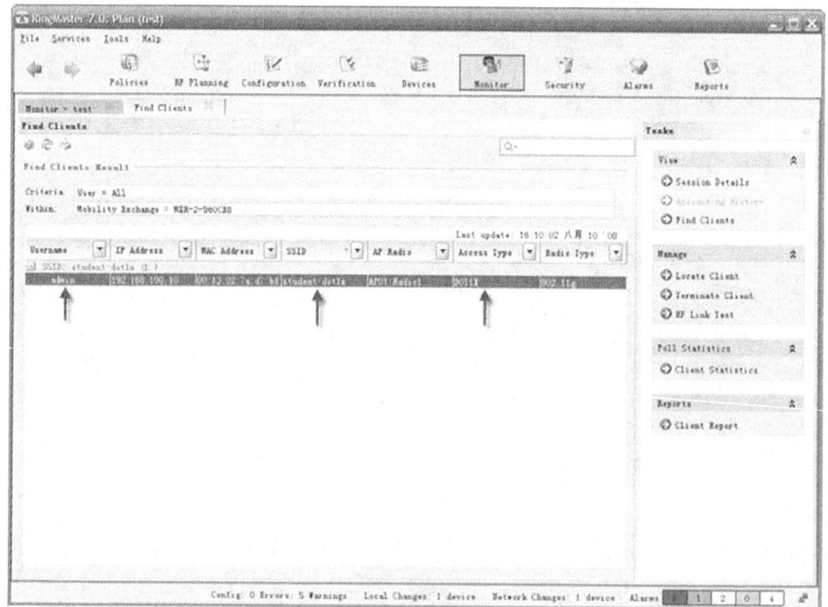

图 25-28

至此就完成了所有配置。

实验 26　搭建采用 WEP 加密方式的无线网络

【实验名称】

搭建采用 WEP 加密方式的无线网络。

【实验目的】

掌握 WEP 加密方式，无线网络的概念及搭建方法。

【背景描述】

小张从学校毕业后直接进入一家企业担任网络管理员，发现公司内搜到的 SSID，直接就可以接入无线网络，没有任何认证加密手段，由于无线网络不像有线网，有严格的物理范围，例如，要接入网络必须要有一根网线插上才能上，而无线网不同，无线信号可能会广播到公司办公室以外的地方，或者大楼外，或者别的公司，都可以搜到，这样收到信号的人就可以随意地接入到网络里来，很不安全。于是小张建议，采用 WEP 加密的方式来对无线网进行加密及接入控制，只有输入正确密钥的用户才可以接入到无线网络里来，并且无线网中的数据传输也是加密的。

【需求分析】

防止非法用户连接进来、防止无线信号被窃听。共享密钥的接入认证。数据加密，防止非法窃听。

【实验拓扑】

实验的拓扑图，如图 26-1 所示。

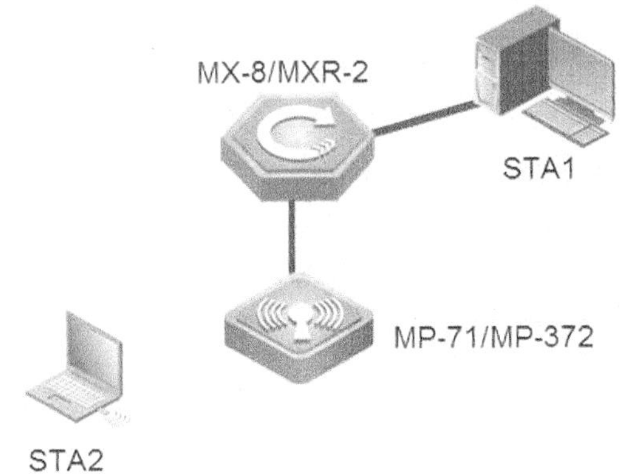

图 26-1

【实验设备】

MXR-2/MX-8 1 台。

MP-71/MP-372 1 台。

无线网卡 1 块。

安装有 RingMaster 网管的服务器 1 台。

【预备知识】

无线局域网基本知识、智能无线产品的基本原理、Ringmaster 的基本操作能力。

【实验原理】

WEP 加密方式的无线网络是采用共享密钥形式的接入、加密方式，即，在 AP 上设置了相应的 WEP 密钥，在客户端也需要输入和 AP 端一样的密钥才可以正常接入，并且 AP 与无线客户端的通信也通过了 WEP 加密。即使空中有人抓取到无线数据包，也看不到里面的内容。

但是，WEP 加密方式存在漏洞，现在有些软件可以对此密钥进行破解，所以，这种加密方式不是最安全的加密方式。但是由于大部分的客户端都支持 WEP，所以现在此方式的应用场合还是很多的。

【实验步骤】

步骤 1　配置无线交换机的基本参数。

无线交换机的默认 IP 地址是 192.168.100.1/24，因此将 STA-1 的 IP 地址配置为 192.168.100.2/24，并打开浏览器，登录到 https://192.168.100.1，弹出如图 26-2 所示的界面，选择“是(Y)”按钮。

系统的默认管理用户名是 admin，密码为空，如图 26-3 所示。

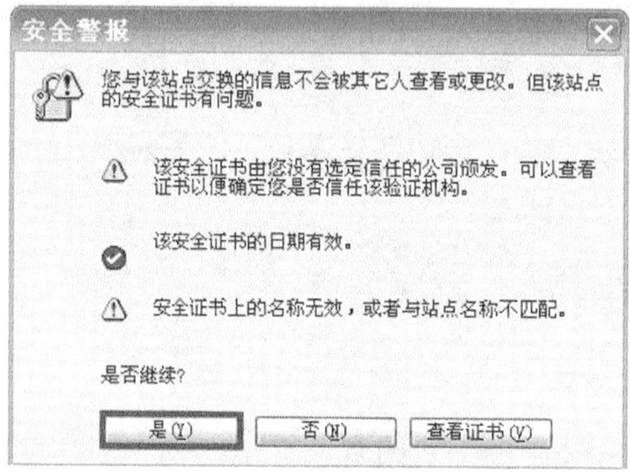

图 26-2

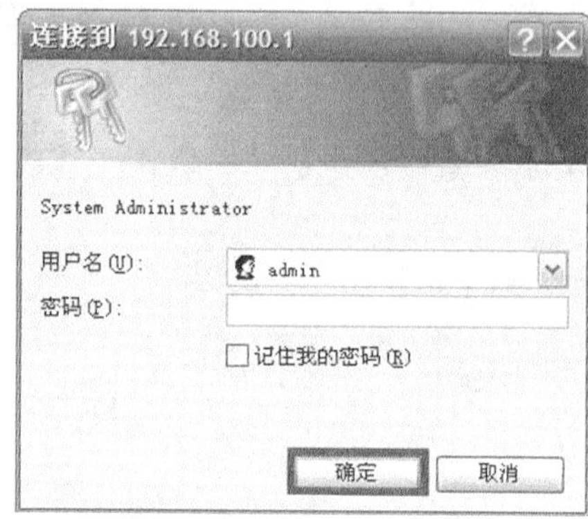

图 26-3

输入用户名和密码后，就进入了无线交换机的 Web 配置页面，单击“Start”按钮，进入快速配置指南页面，如图 26-4 所示。

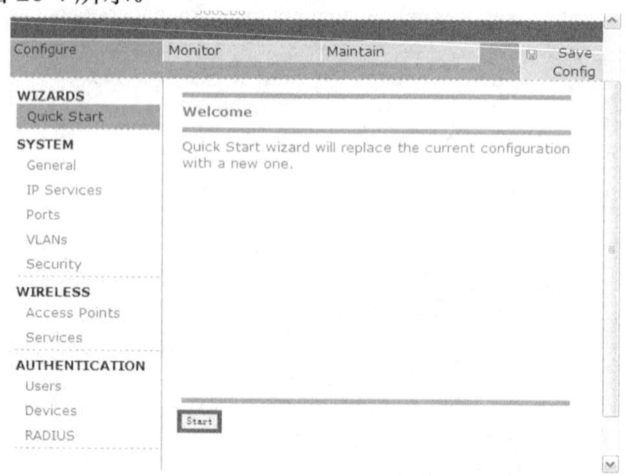

图 26-4

选择管理无线交换机的工具“RingMaster”，如图 26-5 所示。

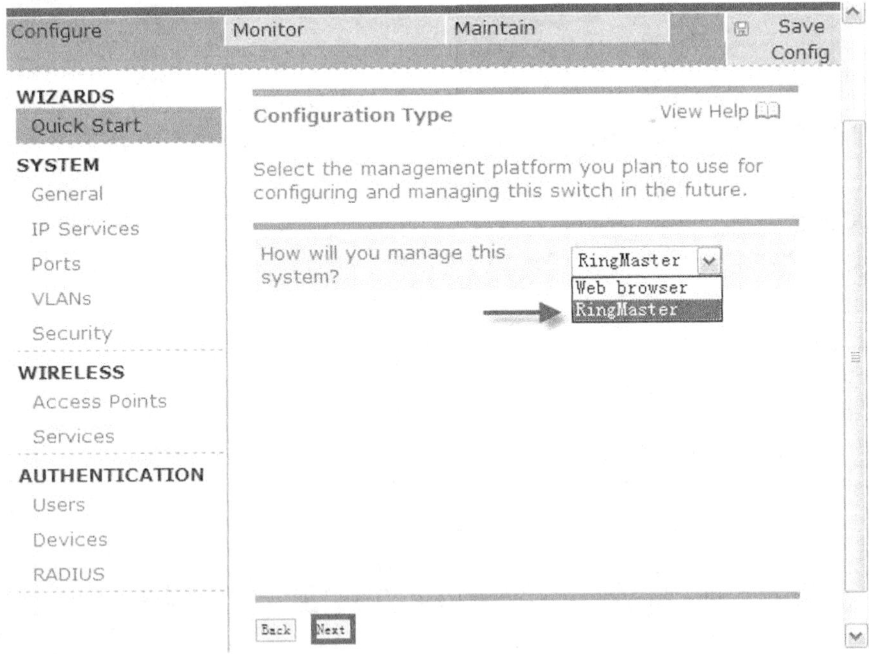

图 26-5

配置无线交换机的 IP 地址、子网掩码以及默认网关，如图 26-6 所示。

Configure
Monitor
Maintain
Save Config
WIZARDS
Quick Start
SYSTEM
General
IP Services
Ports
VLANs
Security
WIRELESS
Access Points
Services
AUTHENTICATION
Users
Devices
RADIUS
IP Configuration
View Help
Specify the following parameters to connect this switch to the network.
IP address 192.168.100.254
IP mask bits 24
Default router 192.168.100.1
Back
Next

图 26-6

设置系统的管理密码，如图 26-7 所示。

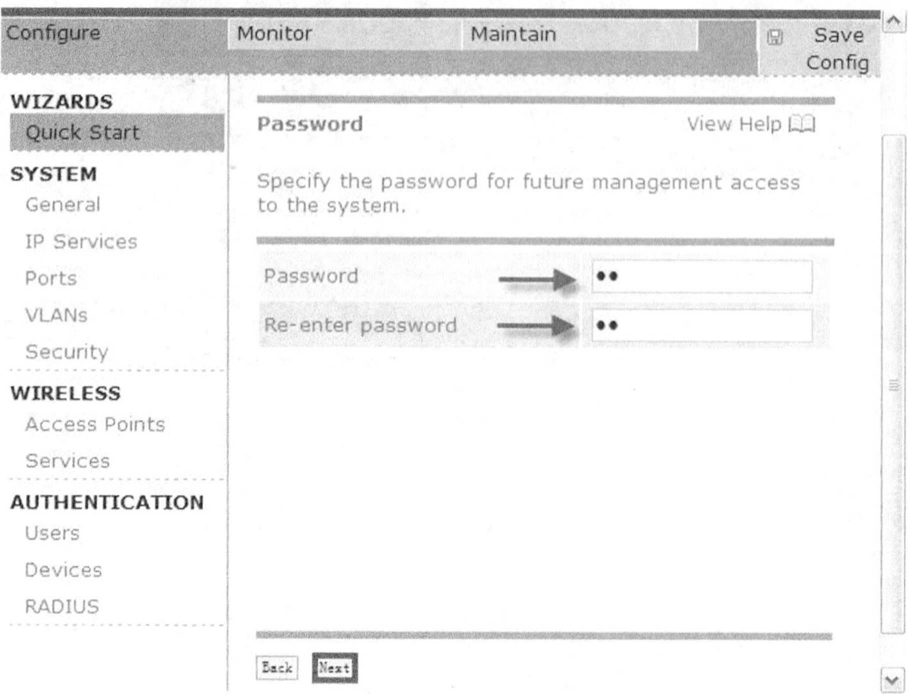

图 26-7

设置系统时间及时区，如图 26-8 所示。

Configure　Monitor　Maintain　Save Config

WIZARDS
Quick Start
SYSTEM
General
IP Services
Ports
VLANs
Security
WIRELESS
Access Points
Services
AUTHENTICATION
Users
Devices
RADIUS

Date and Time　View H

Set the time and date on the switch.

Date　8/08/2008
Time　13:45:15
Timezone　(GMT+8:00) Beijing, Chongqing, Hong Kong

Back　Next

图 26-8

确认无线交换机的基本配置，如图 26-9 所示。

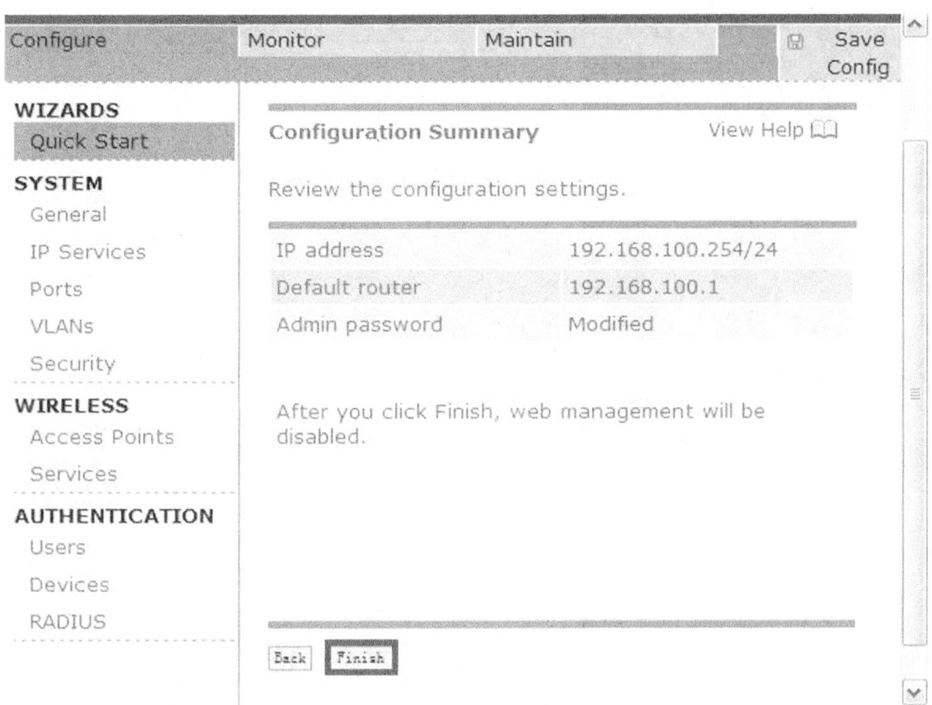

图 26-9

完成无线交换机的基本配置。

步骤 2　通过 RingMaster 网管软件来进行无线交换机的高级配置。

运行 RingMaster 软件，地址为 127.0.0.1，端口为 443，用户名和密码默认为空，如图 26-10 所示。

图 26-10

选择“Configuration”选项，进入配置界面，并添加被管理的无线交换机，如图 26-11 所示。

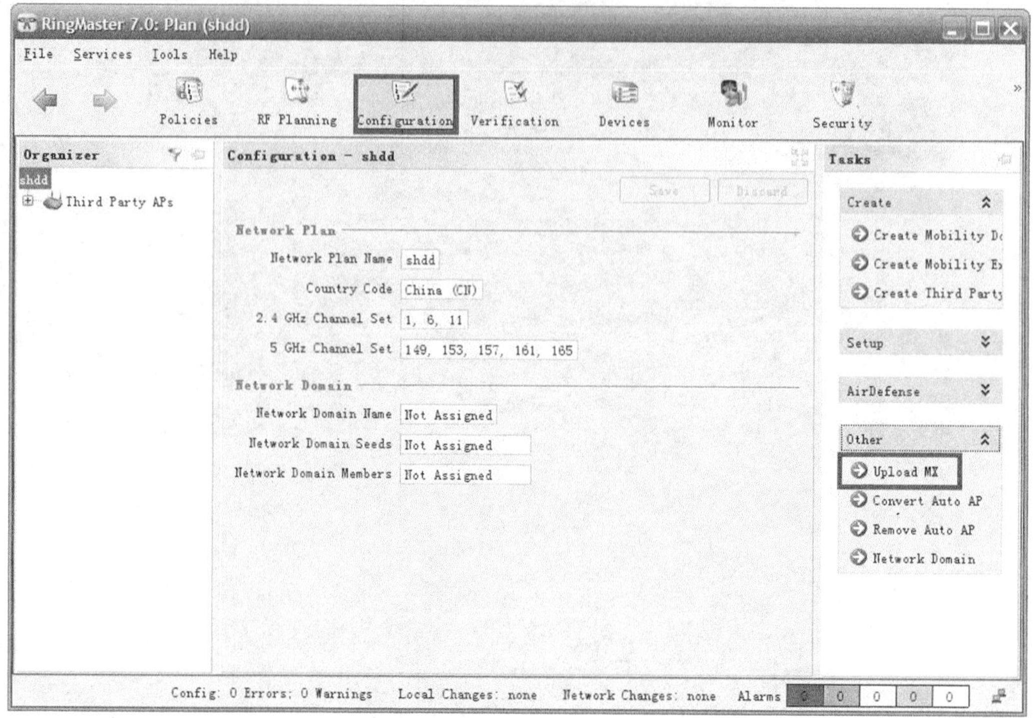

图 26-11

输入被管理的无线交换机的 IP 地址及 Enable 密码，如图 26-12 至图 26-14 所示。

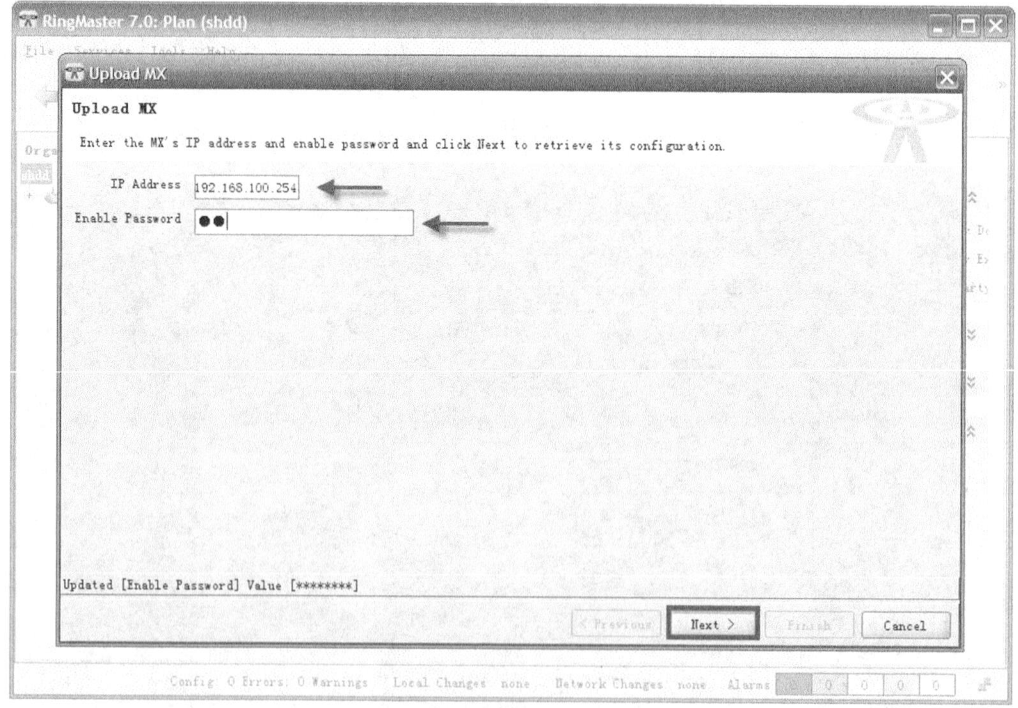

图 26-12

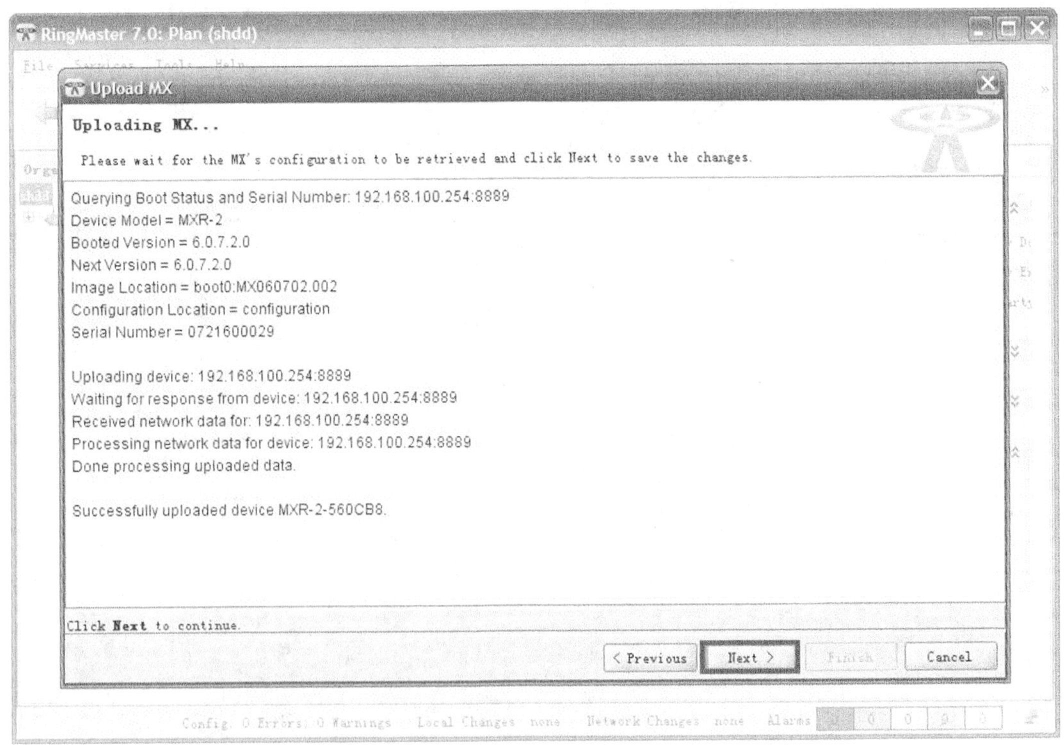

图 26-13

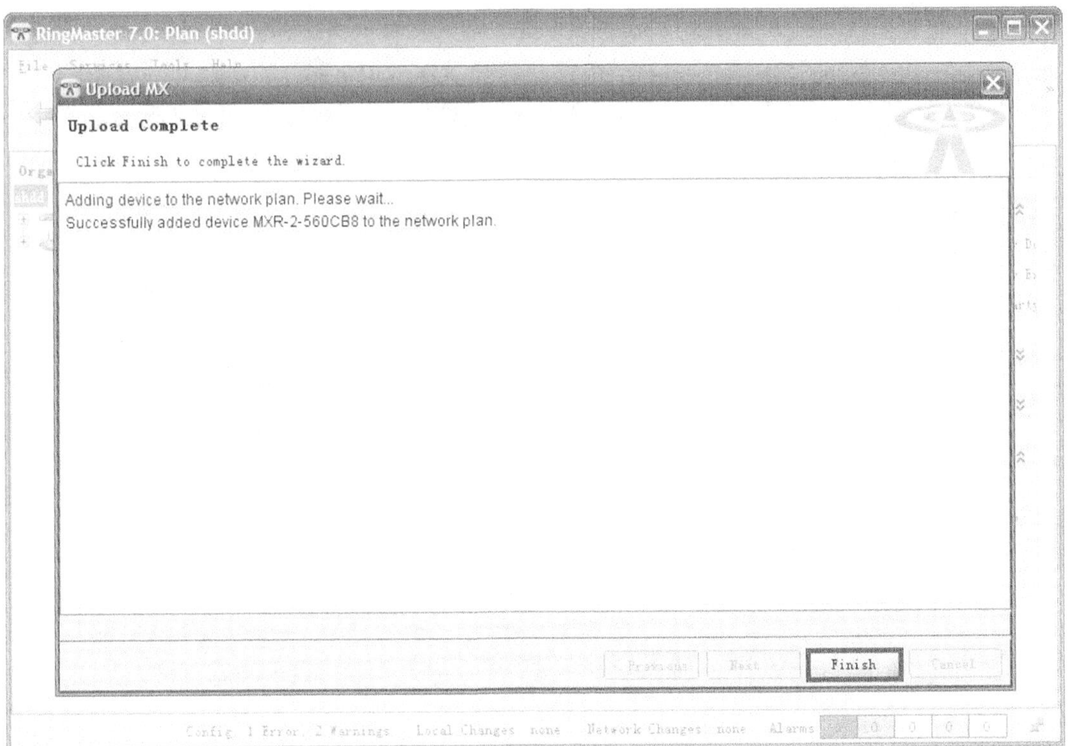

图 26-14

完成添加后，进入无线交换机的操作界面，如图 26-15 所示。

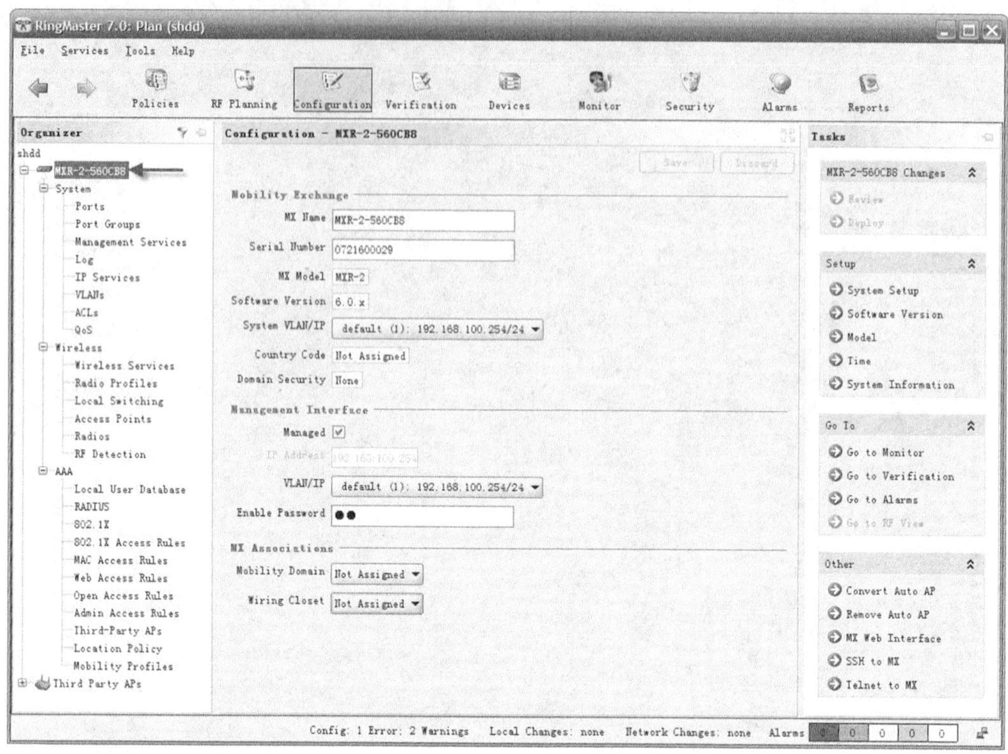

图 26-15

步骤 3　配置无线 AP。

进入“Wireless”→“Access Points”选项，添加 AP，如图 26-16 所示。

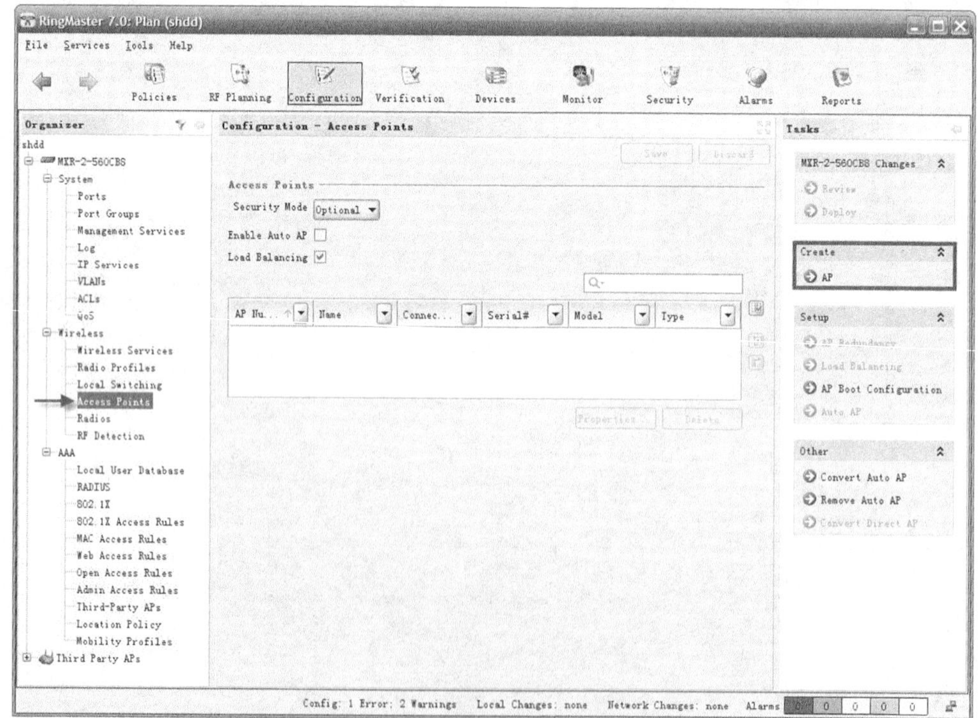

图 26-16

为添加的 AP 进行命名，并选择连接方式，默认使用“Distributed”模式，如图 26-17 所示。

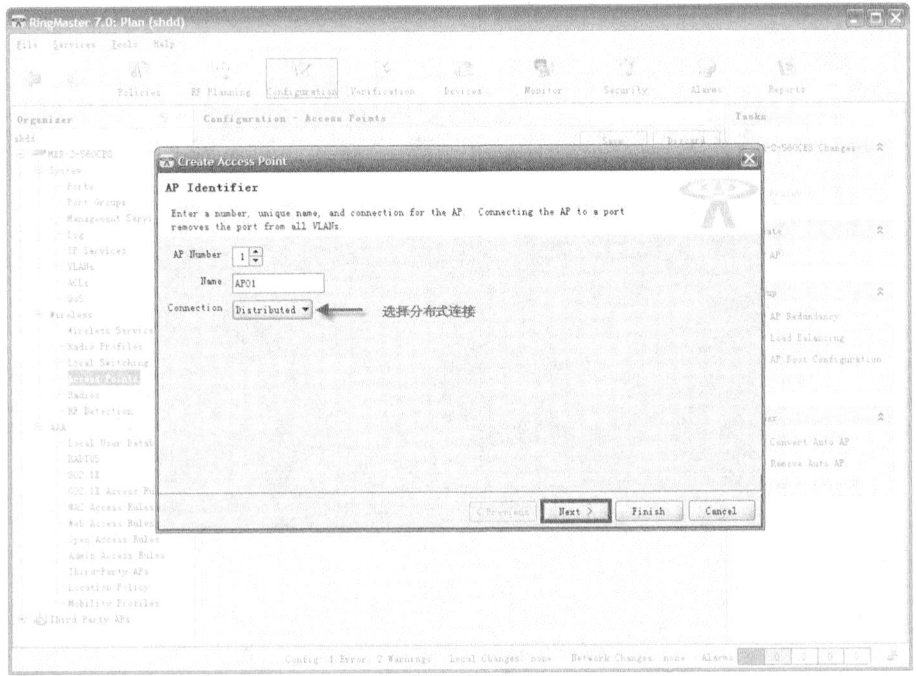

图 26-17

将需要添加的 AP 机身后面的 SN 号输入到对话框中，用于 AP 与无线交换机的注册过程，如图 26-18 所示。

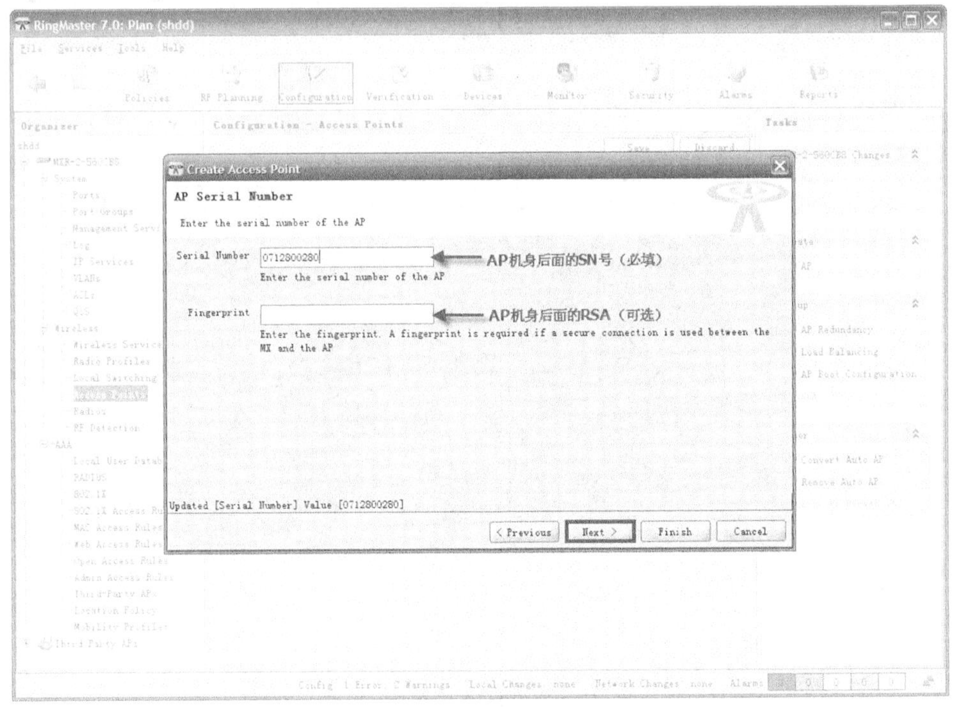

图 26-18

选择添加 AP 后的具体型号和传输协议，完成 AP 的添加，如图 26-19 所示。

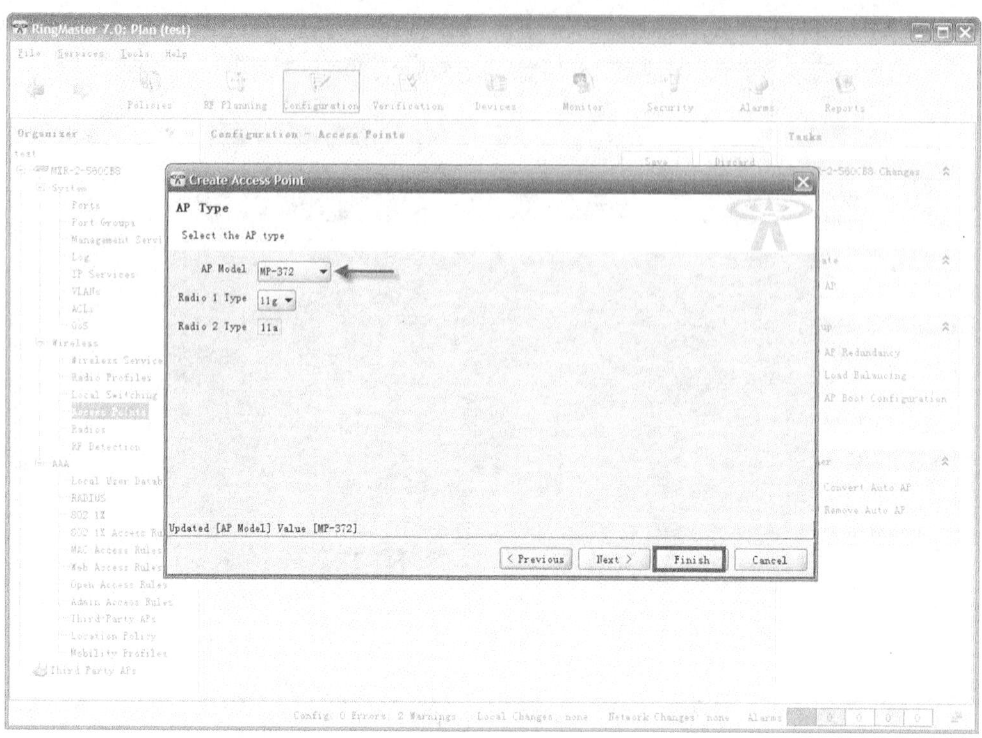

图 26-19

步骤 4　配置无线交换机的 DHCP 服务器。

进入“Syestem”→“VLANs”选项，选择“default”VLAN，进入属性配置，如图 26-20 所示。

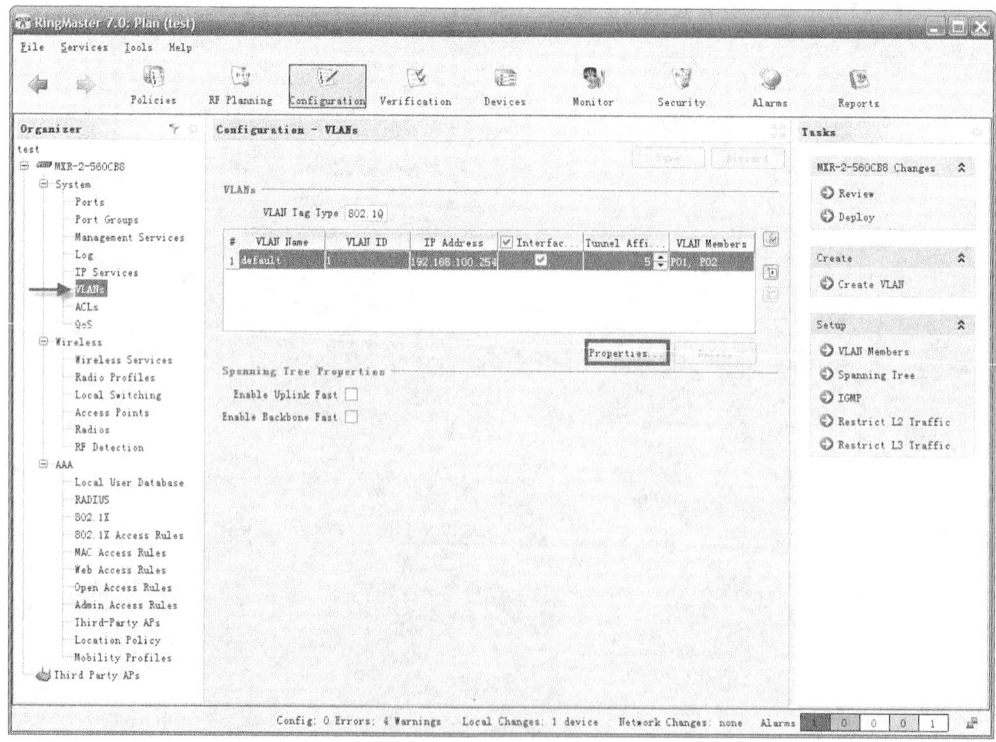

图 26-20

进入“VLAN Properties”→“DHCP Server”选项，激活 DHCP 服务器，设置地址池和 DNS，保存设置内容，如图 26-21 所示。

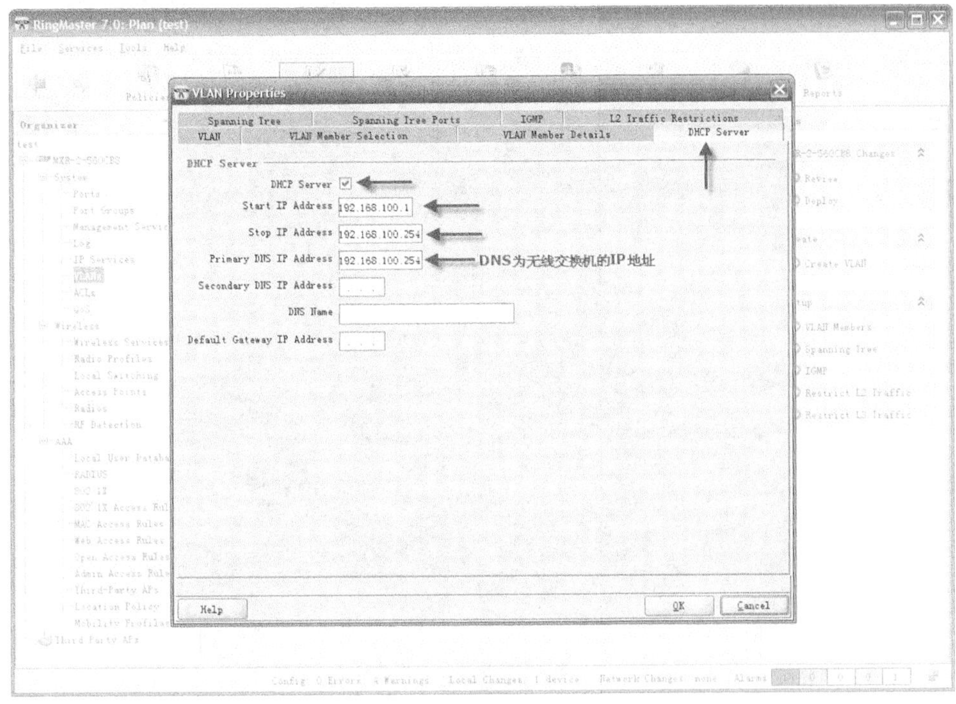

图 26-21

进入“System”→“Ports”选项，将无线交换机的端口 PoE 打开，并保存设置，如图 26-22 所示。

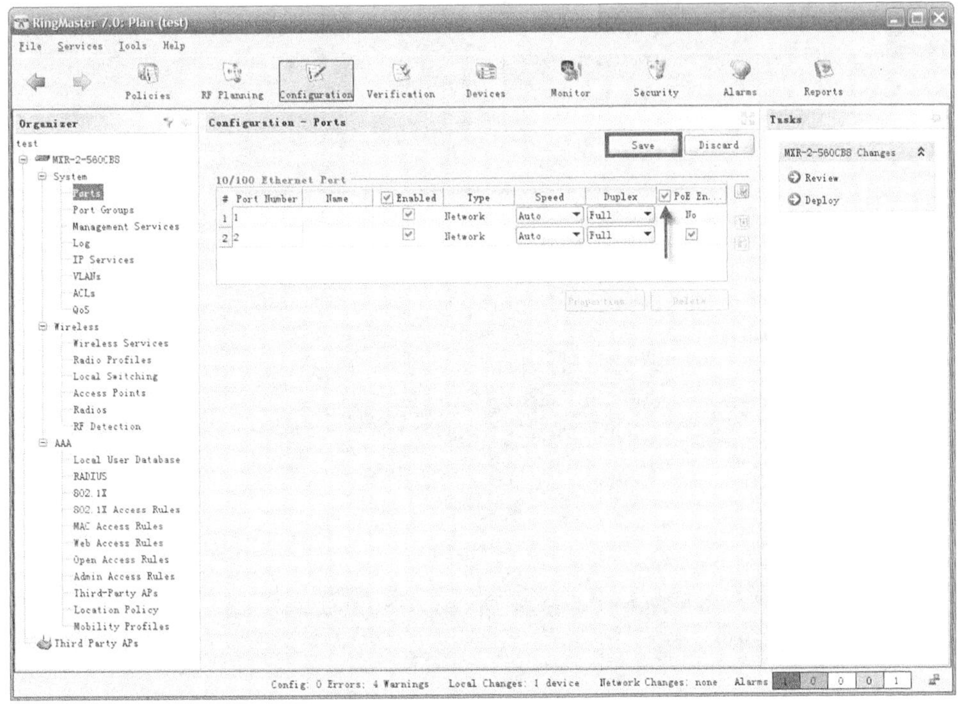

图 26-22

步骤 5　配置 Wireless Services。

在菜单“Configuration”下，选择“Wireless”→“Wireless Services”选项，如图 26-23 所示。

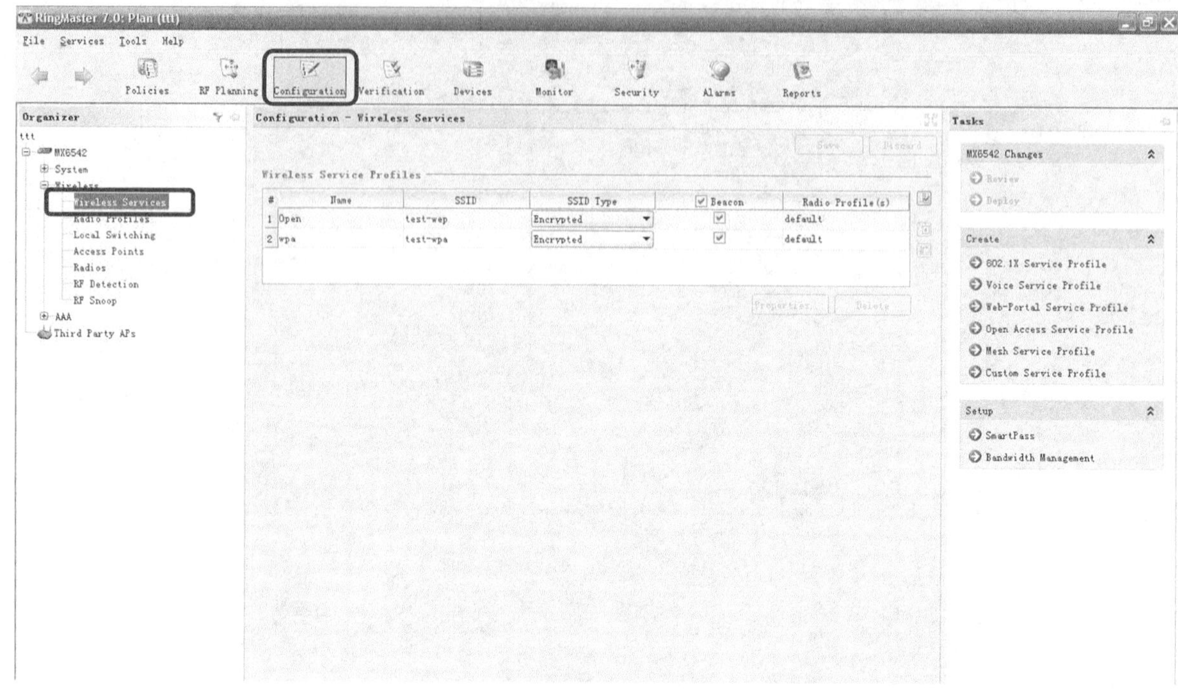

图 26-23

创建一个“Service Profile”：在管理页面的右边，“Create”的下面选择“Open Access Service Profile”选项，如图 26-24 所示。

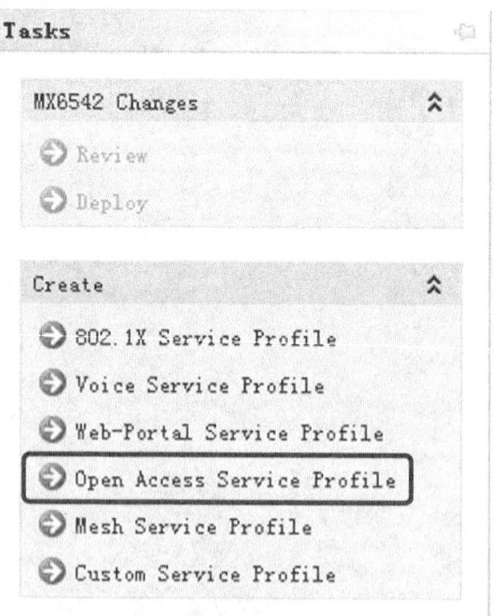

图 26-24

然后输入实验使用的 Service-profile 名为 Open，SSID 为 test-wep，SSID 类型为 Encrypted，即加密的，如图 26-25 所示。

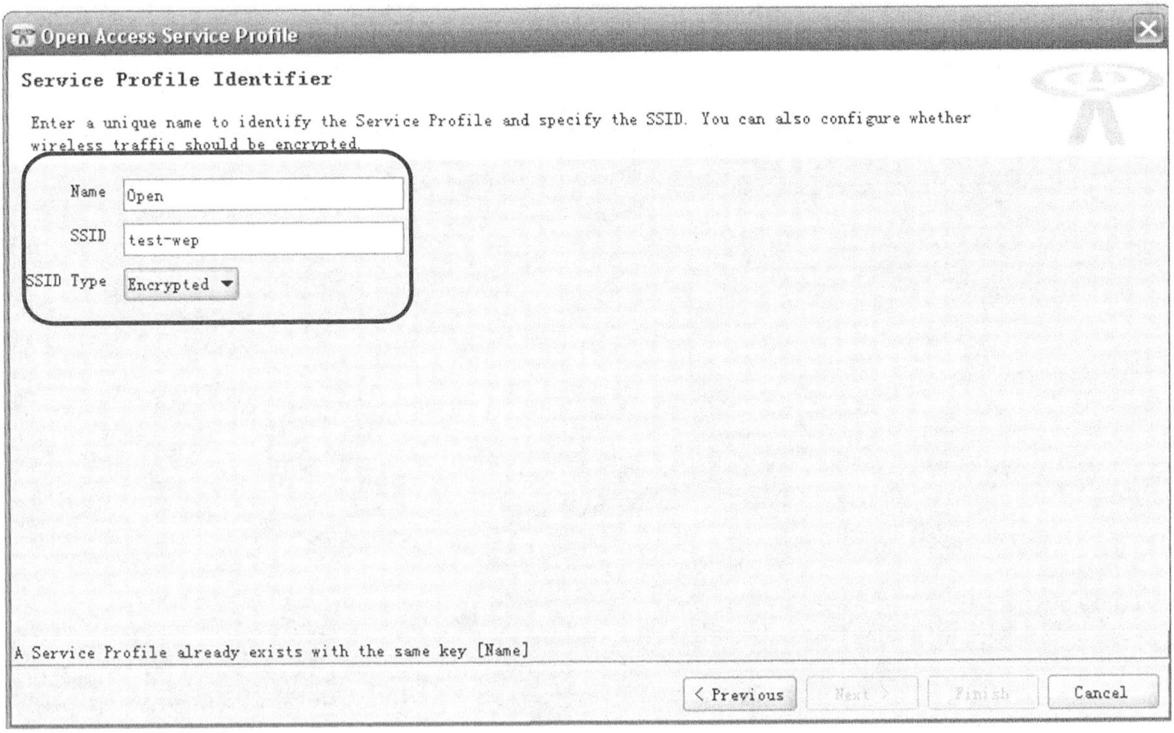

图 26-25

选择使用静态的 WEP 加密方式，如图 26-26 所示。

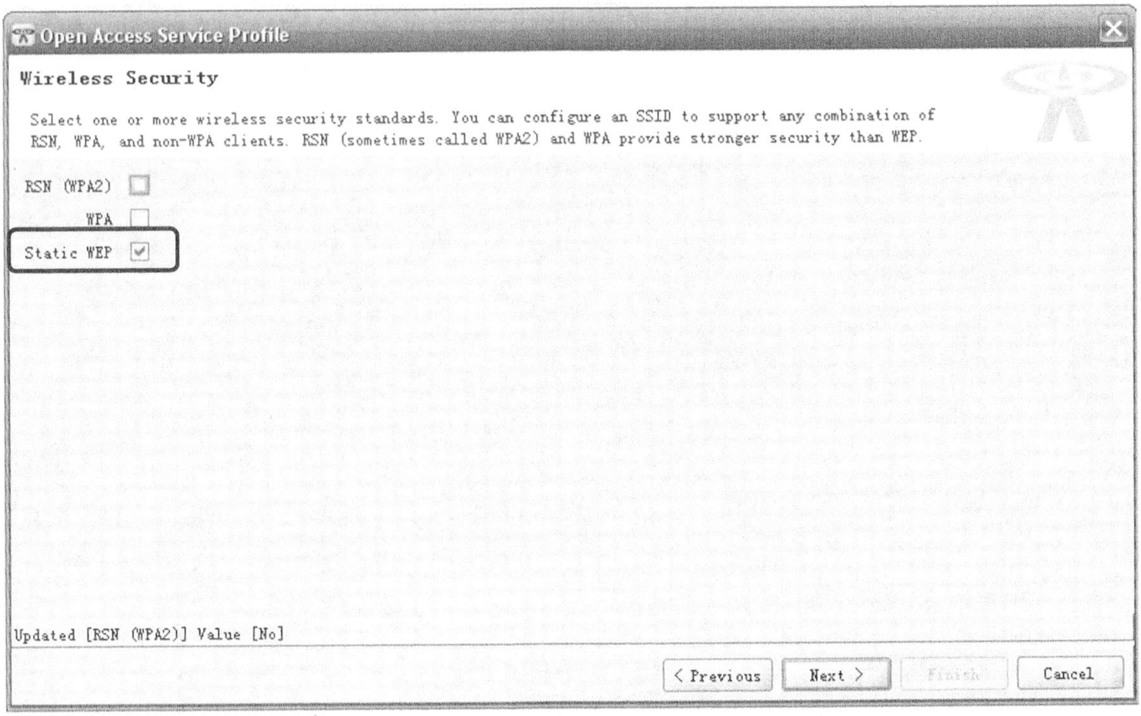

图 26-26

输入密钥“1234567890”，此后，无线客户端都需要输入正确的密钥才能接入进来，如图 26-27 所示。

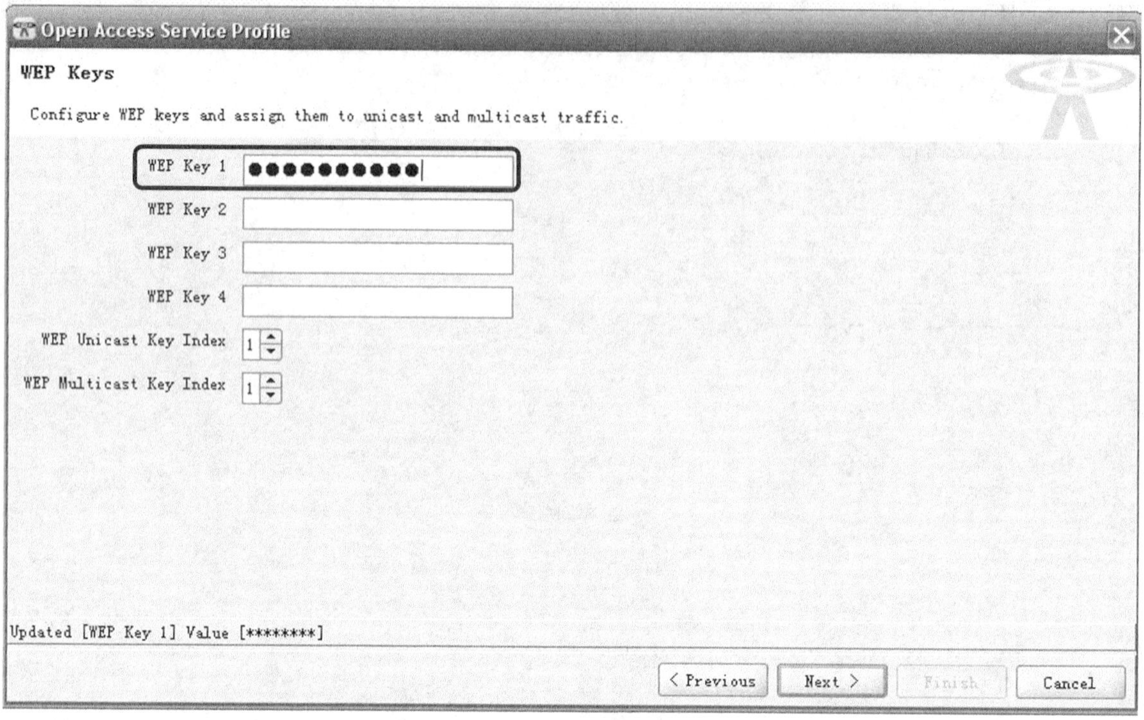

图 26-27

VLAN Name 为 default，如图 26-28 所示。

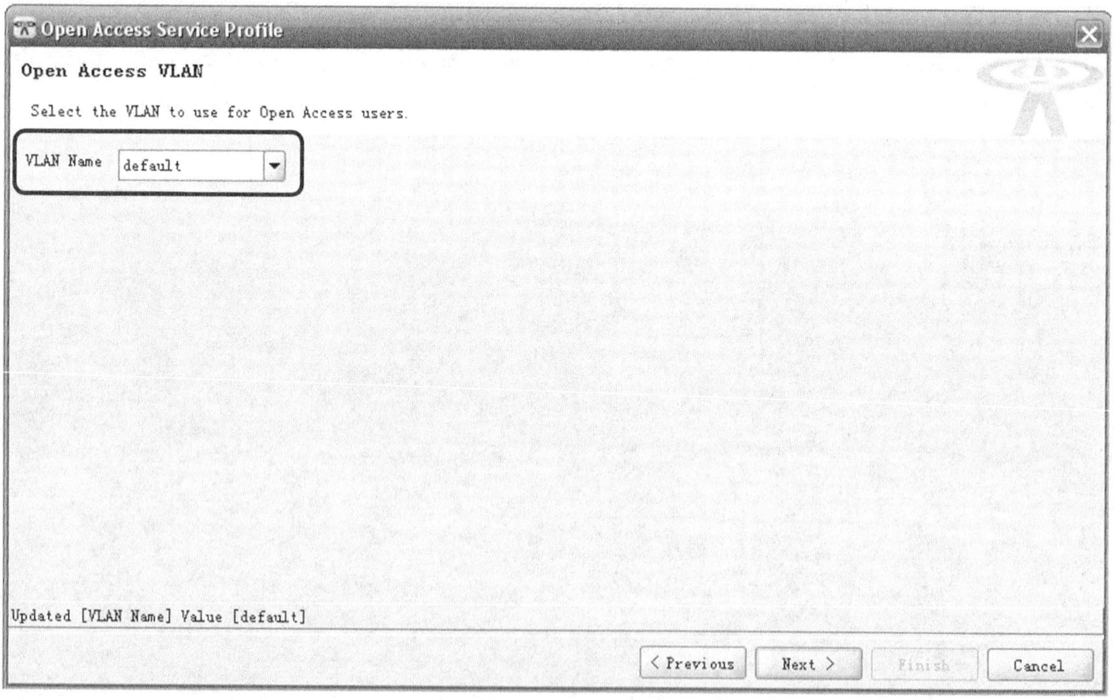

图 26-28

Radio Profiles 使用 default，然后单击“Finish”按钮，如图 26-29 所示。

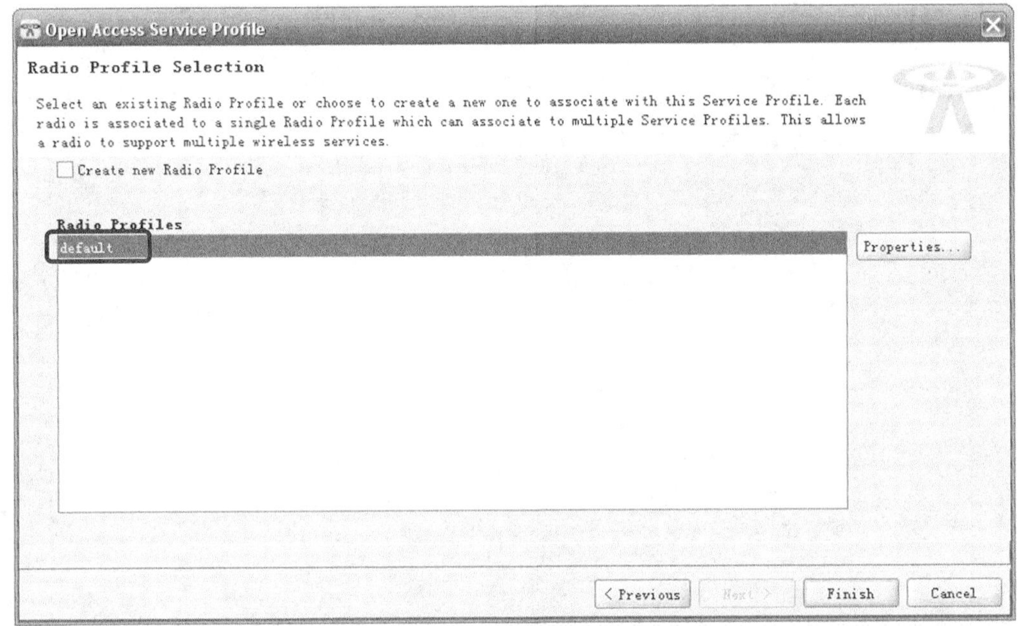

图 26-29

至此就成功地创建了一个名为 Open 的 Service Profile，如图 26-30 所示。

Wireless Service Profiles

#	Name	SSID	SSID Type	Beacon	Radio Profile(s)
1	Open	test-wep	Encrypted	✓	default

图 26-30

然后单击窗口右边的“Deploy”选项，将刚才所做的配置下发到无线交换机中，如图 26-31 所示。

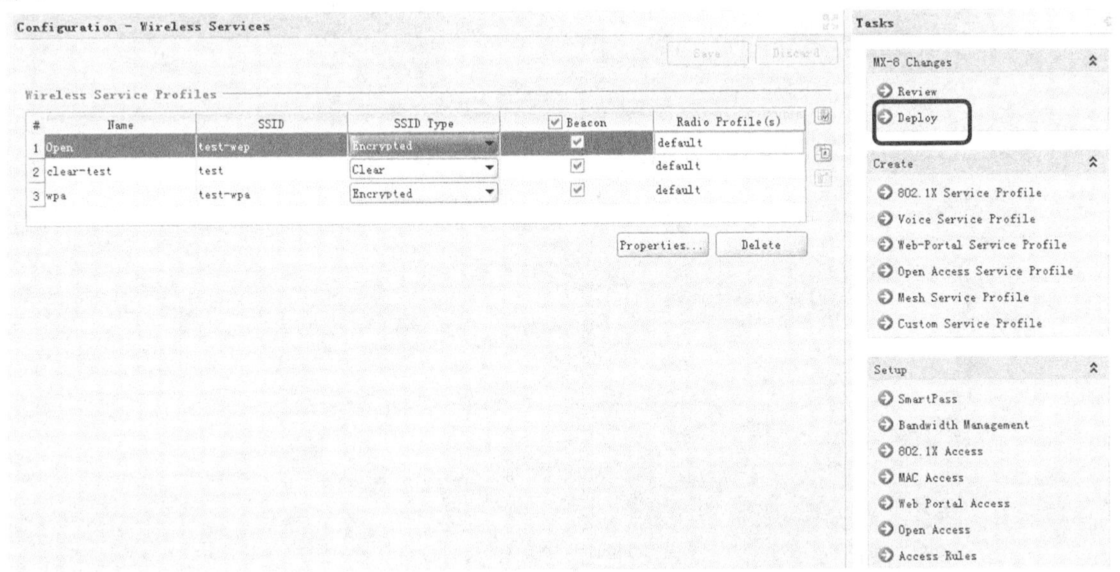

图 26-31

当弹出窗口中出现如图 26-32 所示的“Deploy completed”时，表示配置下发完成。

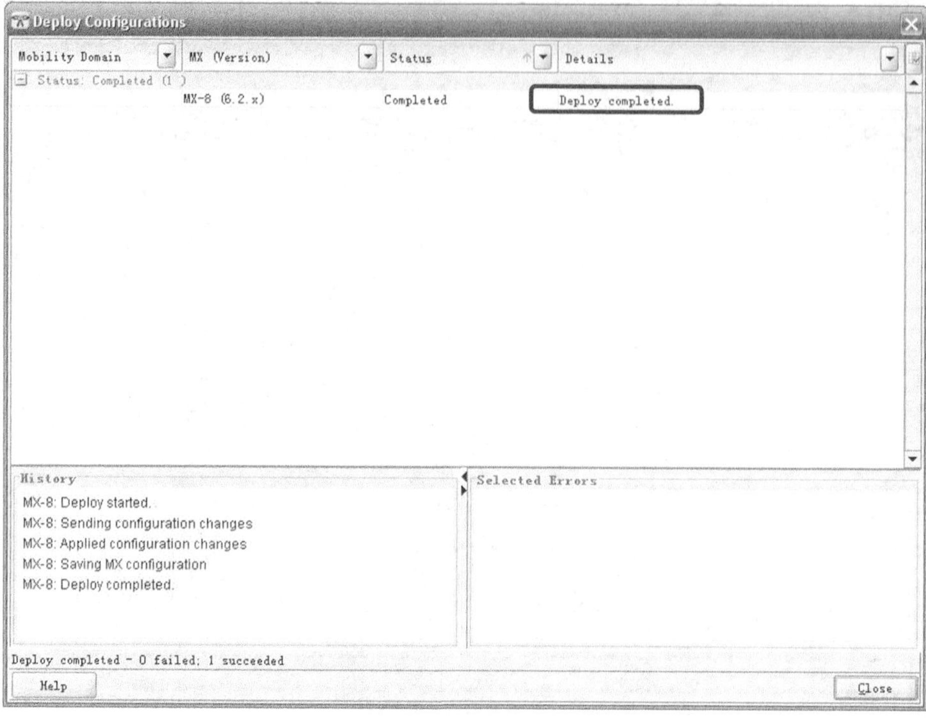

图 26-32

此时配置全部完成，无线网络便会广播出采用 WEP 加密方式的 SSID“test-wep”。

步骤 6　测试无线客户端的连接情况。

打开无线网卡，搜寻无线网络，会发现名为“test-wep”的 SSID，联入该 SSID，如图 26-33 所示。

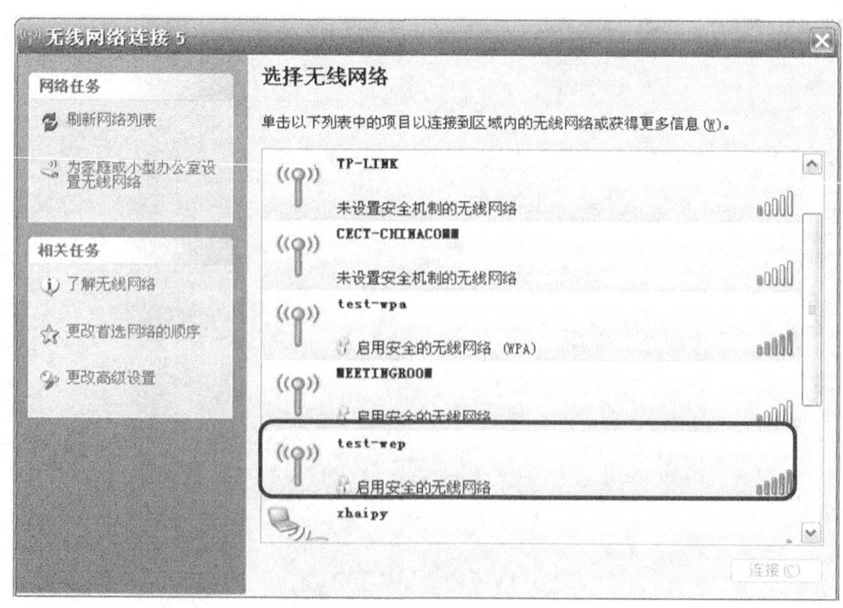

图 26-33

选中该SSID，单击“连接”按钮，此时会提示输入WEP密钥，输入密钥“1234567890”，如图26-34所示。

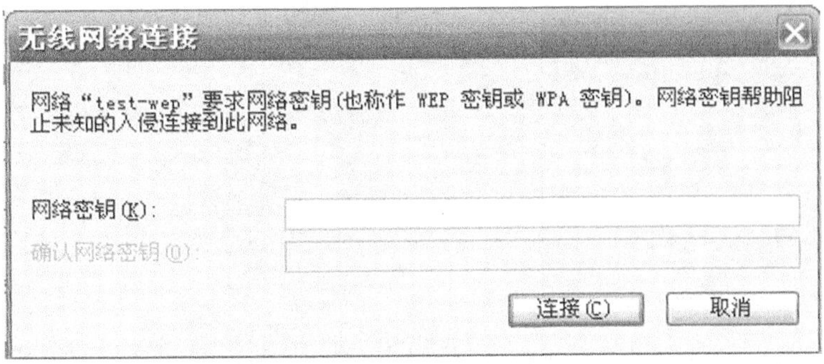

无线网络连接

网络“test-wep”要求网络密钥(也称作 WEP 密钥或 WPA 密钥)。网络密钥帮助阻止未知的入侵连接到此网络。

网络密钥(K):

确认网络密钥(O):

连接(C) 取消

图26-34

单击“连接”按钮后，无线客户端便可以正确地连接到无线网络上了。

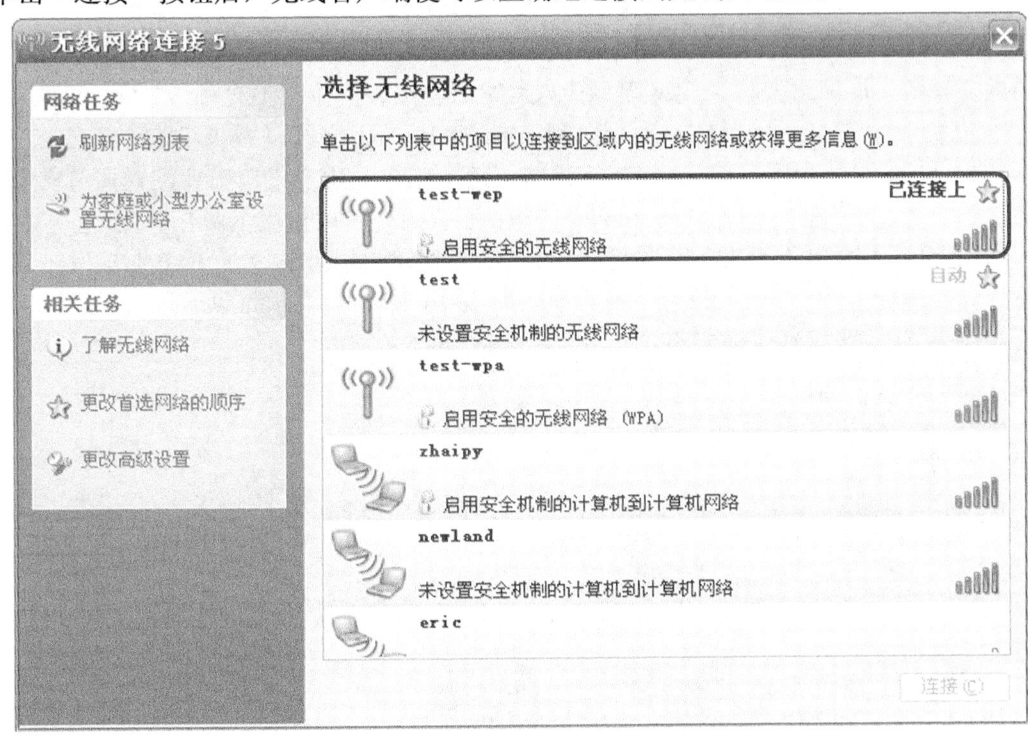

图26-35

无线客户端可以Ping通无线交换机地址。

综合实验 1　交换篇

【实验名称】

综合实验 1　交换篇。

【实验目的】

构建交换网络的三级结构。

【背景描述】

你是某系统集成公司的技术工程师，公司现在承接一个企业网的搭建项目，经过现场勘测及充分与客户沟通，你建议该网络采用经典的三级网络构架，现项目方案已经得到客户的认可，并且请你负责整个网络的实施。

【需求分析】

通过合理的三层网络架构，实现用户接入网络的安全、快捷，不允许 VLAN10 的用户去访问 VLAN30 的 FTP 服务，VLAN20 不受限制；VLAN10 的用户接口需要配置端口安全，设置最大连接数为 3，如果违规则采取 shutdown 措施，VLAN20 的用户接口需要配置端口安全，设置最大连接数为 2，如果违规则采取 shutdown 措施；配置静态路由使用全网互通；配置 NAT 功能，使用内网用户使用 200.1.1.3 -200.1.1.6 这段地址去访问互联网；将内网的 FTP 服务发布到互联网上，使用内网地址为 192.168.13.254，公网地址为 200.1.1.7，并要求可以通过内网地址访问 FTP 服务器；使用 ACL 防止冲击波病毒。

【网络拓扑原型】

网络拓扑原型图，如图 27-1 所示。

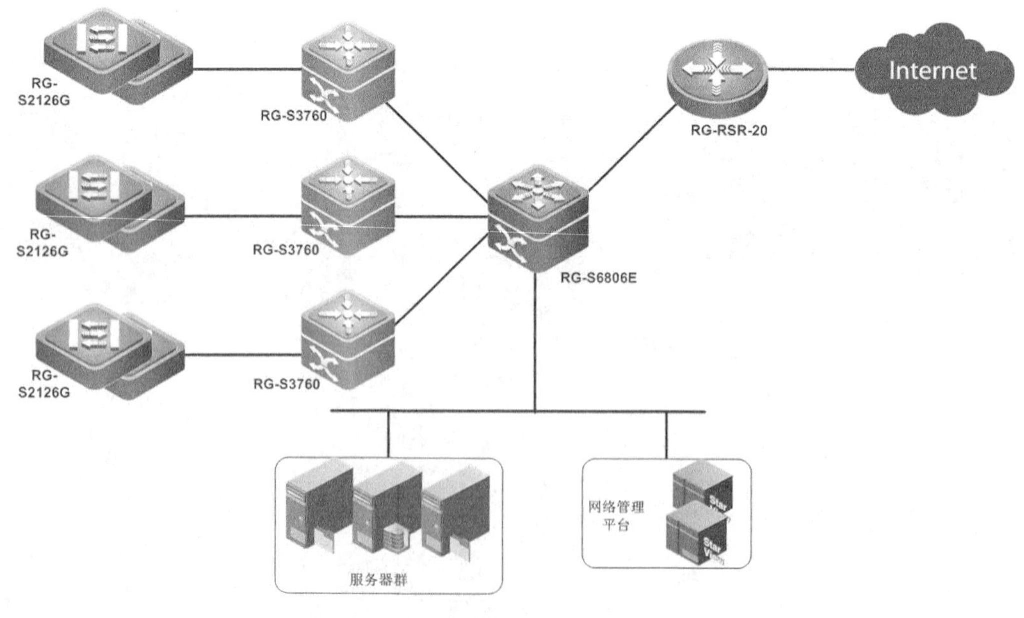

图 27-1

【实验拓扑】

实验拓扑图，如图 27-2 所示。

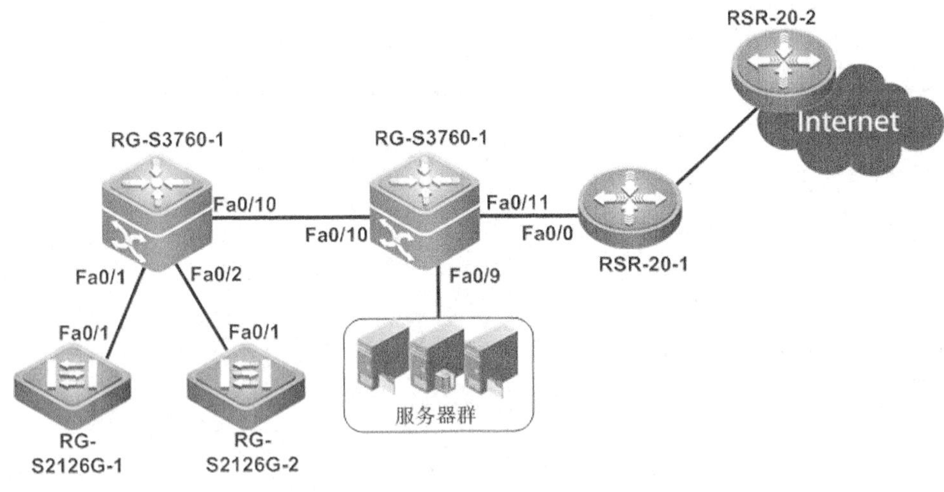

图 27-2

【实验设备】

二层交换机 RG-S2126G 2 台。
三层交换机 RG-S3760 2 台。
路由器 RG-RSR-20 2 台。
计算机 1 台。

【客户需求】

整个网络采用核心—汇聚—接入三级结构，通过出口路由器做 NAT 供内网用户访问外网，同时要求 VLAN10 内网用户不能访问 VLAN30 的 FTP 服务，VLAN20 不作限制。接入层交换机要实现防冲击波的功能。

【实验拓扑说明】

整个实验用 RG-S2126G1 模拟 VLAN10 用户接入交换机，RG-S2126G2 模拟 VLAN20 用户接入交换机，VLAN10 与 VLAN20 的用户通过 RG-S3760-1 实现 VLAN 间路由。

RG-S3760-1 与 RG-S3760-2 之间通过静态路由，实现内网用户的对外数据包转发及对内网服务器的访问。

RG-S3760-2 与 R1 连接，在 R1 上启用 NAT 功能，保证内网用户可以访问外网，实验拓扑中以 R2 模拟 Internet。

【IP 地址规划】

设备名称	端口名称	IP 地址
S3760-1	VLAN10	192.168.11.1/24
	VLAN20	192.168.12.1/24
	VLAN30	192.168.13.1/24
S3760-2	VLAN10	192.168.11.2/24
	VLAN20	192.168.12.2/24
	VLAN30	192.168.13.2/24
	Fa0/11	172.16.1.1/30

（续表）

设备名称	端口名称	IP 地址
RSR20-1	Fa0/0	172.16.1.2/30
	Fa0/1	200.1.1.1/28
RSR20-2	Fa0/1	200.1.1.2/28

【实验步骤】

步骤 1　配置接入层交换机。

```
switch(config)#hostanme S2126G-1
S2126G-1 (config)#vlan 10
S2126G-1 (config-vlan)#exit
S2126G-1 (config)# interface range fa 0/2-24
S2126G-1 (config-if-range)#switchport access vlan 10
S2126G-1 (config-if-range)#exit
S2126G-1 (config)# interface range fa 0/1
S2126G-1 (config-if)#switchport mode trunk
S2126G-1 (config-if)#exit
S2126G-1 (config)# interface range fa 0/2-24
S2126G-1 (config-if-range)#switchport port-security
S2126G-1 (config-if-range)# switchport port-security maximum 3
S2126G-1 (config-if-range)#switchport port-security violation shutdown
S2126G-1 (config-if-range)#switchport port-security
S2126G-1 (config-if-range)#exit
switch(config)#hostname S2126G-2
S2126G-2(config)#vlan 20
S2126G-2 (config-vlan)#exit
S2126G-2(config)#interface range fa 0/2-24
S2126G-2 (config-if-range)#switchport access vlan 20
S2126G-2 (config-if-range)#exit
S2126G-2 (config)# interface range fa 0/1
S2126G-2 (config-if)#switchport mode trunk
S2126G-2 (config-if)#exit
S2126G-2 (config)# interface range fa 0/2-24
S2126G-2 (config-if-range)#switchport port-security
S2126G-2 (config-if-range)# switchport port-security maximum 2
S2126G-2 (config-if-range)#switchport port-security violation shutdown
S2126G-2 (config-if-range)#switchport port-security
S2126G-2 (config-if-range)#exit
```

步骤 2　配置汇聚层交换机。

```
switch(config)#host RG-S3760-1
RG-S3760-1 (config)# vlan 10
RG-S3760-1 (config-vlan)#exit
```

```
RG-S3760-1 (config)# vlan 20
RG-S3760-1 (config-vlan)#exit
RG-S3760-1 (config)# vlan 30
RG-S3760-1 (config-vlan)#exit
RG-S3760-1 (config)#interface vlan 10
RG-S3760-1 (config-if)# ip address 192.168.11.1 255.255.255.0
RG-S3760-1 (config-if)#no shutdown
RG-S3760-1 (config-if)#exit
RG-S3760-1 (config)#interface vlan 20
RG-S3760-1 (config-if)# ip address 192.168.12.1 255.255.255.0
RG-S3760-1 (config-if)#no shutdown
RG-S3760-1 (config-if)#exit
RG-S3760-1 (config)#interface vlan 30
RG-S3760-1 (config-if)# ip address 192.168.13.1 255.255.255.0
RG-S3760-1 (config-if)#no shutdown
RG-S3760-1 (config-if)#exit
RG-S3760-1 (config)#interface FastEthernet 0/10
RG-S3760-1 (config-if)#switchport mode trunk
RG-S3760-1 (config-if)#exit
RG-S3760-1 (config)#interface range FastEthernet 0/1-2
RG-S3760-1 (config-if-range)#switchport mode trunk
RG-S3760-1 (config-if)#exit
RG-S3760-1 (config)#access-list 110 deny tcp 192.168.11.0 0.0.0.255 192.168.13.0
0.0.0.255 eq ftp
RG-S3760-1 (config)#access-list 110 deny tcp 192.168.11.0 0.0.0.255 192.168.13.0
0.0.0.255 eq ftp-data
RG-S3760-1 (config)#access-list 110 permit ip any any
RG-S3760-1 (config)#interface VLAN 10
RG-S3760-1 (config-if)#ip access-group 110 in
RG-S3760-1 (config-if)#exit
RG-S3760-1 (config)#ip route  0.0.0.0 0.0.0.0  VLAN 10
    switch(config)#hostname RG-S3760-2
RG-S3760-2 (config)# vlan 10
RG-S3760-2 (config-vlan)#exit
RG-S3760-2 (config)# vlan 20
RG-S3760-2 (config-vlan)#exit
RG-S3760-2 (config)# vlan 30
RG-S3760-2 (config-vlan)#exit
RG-S3760-2 (config)#interface vlan 10
RG-S3760-2 (config-if)# ip address 192.168.11.2 255.255.255.0
RG-S3760-2 (config-if)#no shutdown
```

```
RG-S3760-2 (config-if)#exit
RG-S3760-2 (config)#interface vlan 20
RG-S3760-2 (config-if)# ip address 192.168.12.2 255.255.255.0
RG-S3760-2 (config-if)#no shutdown
RG-S3760-2 (config-if)#exit
RG-S3760-2 (config)#interface vlan 30
RG-S3760-2 (config-if)# ip address 192.168.13.2 255.255.255.0
RG-S3760-2 (config-if)#no shutdown
RG-S3760-2 (config-if)#exit
RG-S3760-2 (config)#interface FastEthernet 0/9
RG-S3760-2 (config-if)#switchport mode access
RG-S3760-2 (config-if)#switchport access vlan 30
RG-S3760-2 (config-if)#exit
RG-S3760-2 (config)#interface FastEthernet 0/10
RG-S3760-2 (config-if)#switchport mode trunk
RG-S3760-2 (config-if)#exit
RG-S3760-2 (config)#interface FastEthernet 0/11
RG-S3760-2 (config-if)#no switchport
RG-S3760-2 (config-if)#ip address 172.16.1.1 255.255.255.252
RG-S3760-2 (config)#ip route  0.0.0.0 0.0.0.0  172.16.1.2
```

步骤 3　配置路由器。

```
router(config)#hostname RG-RSR20-1
RG-RSR20-1(config)# interface FastEthernet 0/0
RG-RSR20-1(config-if)#ip address 172.16.1.2 255.255.255.252
RG-RSR20-1(config)#interface FastEthernet 0/1
RG-RSR20-1(config-if)#ip address 200.1.1.1 255.255.255.240
RG-RSR20-1(config-if)#exit
RG-RSR20-1(config)# access-list 10 permit 192.168.11.0 0.0.0.255
RG-RSR20-1(config)# access-list 10 permit 192.168.12.0 0.0.0.255
RG-RSR20-1(config)# access-list 10 permit 192.168.13.0 0.0.0.255
RG-RSR20-1(config)# interface FastEthernet 0/0
RG-RSR20-1(config-if)# ip nat inside
RG-RSR20-1(config-if)#exit
RG-RSR20-1(config)#interface FastEthernet 0/1
RG-RSR20-1(config-if)# ip nat outside
RG-RSR20-1(config-if)#exit
RG-RSR20-1(config)#  ip  nat  pool  internet  200.1.1.3  200.1.1.6  netmask
255.255.255.240
RG-RSR20-1(config)#ip nat inside source list 10 pool internet overload
RG-RSR20-1(config)#ip nat inside source static tcp 192.168.13.254 21 200.1.1.7
```

```
21 permit-inside

RG-RSR20-1(config)#ip nat inside source static tcp 192.168.13.254 20 200.1.1.7
20 permit-inside
RG-RSR20-1(config)#access-list 115 deny udp any any eq tftp
RG-RSR20-1(config)#access-list 115 deny tcp any any eq 135
RG-RSR20-1(config)#access-list 115 deny udp any any eq 135
RG-RSR20-1(config)#access-list 115 deny udp any any eq netbios-ns
RG-RSR20-1(config)#access-list 115 deny udp any any eq netbios-dgm
RG-RSR20-1(config)#access-list 115 deny tcp any any eq 139
RG-RSR20-1(config)#access-list 115 deny udp any any eq netbios-ss
RG-RSR20-1(config)#access-list 115 deny tcp any any eq 445
RG-RSR20-1(config)#access-list 115 deny tcp any any eq 593
RG-RSR20-1(config)#access-list 115 deny tcp any any eq 4444
RG-RSR20-1(config)#access-list 115 permit ip any any
RG-RSR20-1(config)#interface FastEthernet 0/1
RG-RSR20-1(config-if)#ip access-group 115 in
RG-RSR20-1(config-if)#ip access-group 115 out
RG-RSR20-1(config-if)#exit
RG-RSR20-1(config)#ip route  0.0.0.0 0.0.0.0  200.1.1.2
RG-RSR20-1(config)#ip route  192.168.11.0 255.255.255.0  172.16.1.1
RG-RSR20-1(config)#ip route  192.168.12.0 255.255.255.0  172.16.1.1
RG-RSR20-1(config)#ip route  192.168.13.0 255.255.255.0  172.16.1.1
router(config)#hostname RG-RSR20-2
RG-RSR20-2(config)# interface FastEthernet 0/0
RG-RSR20-2(config-if)#ip address 200.1.1.2 255.255.255.240
RG-RSR20-2(config-if)#exit
RG-RSR20-2(config)#interface Loopback 0
RG-RSR20-2(config-if)#ip address 210.1.1.1 255.255.255.0
RG-RSR20-2(config-if)#exit
RG-RSR20-2(config)#ip route  0.0.0.0 0.0.0.0  200.1.1.1
RG-RSR20-2(config)#enable password 123
RG-RSR20-2(config)#line vty 0 4
RG-RSR20-2(config-line)#password 123
RG-RSR20-2(config-line)#login
RG-RSR20-2(config-line)#exit
```

步骤 4　配置 FTP 服务器。

配置内网的 FTP 服务器步骤如下。

（1）选择“开始”→“控制面板”→“添加或删除程序”，如图 27-3 所示。

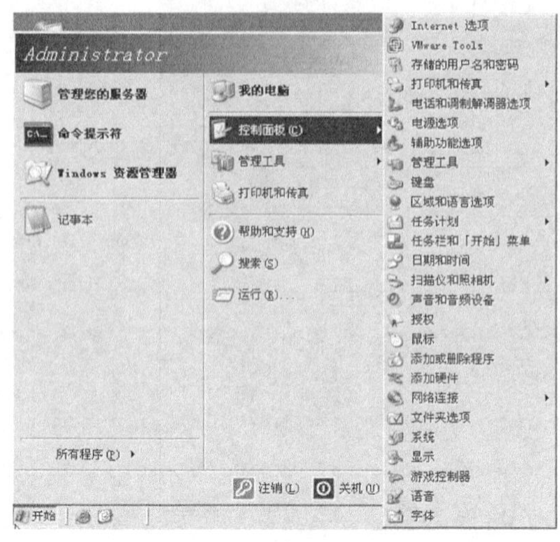

图 27-3

（2）单击“添加或删除程序”按扭，如图 27-4 所示。

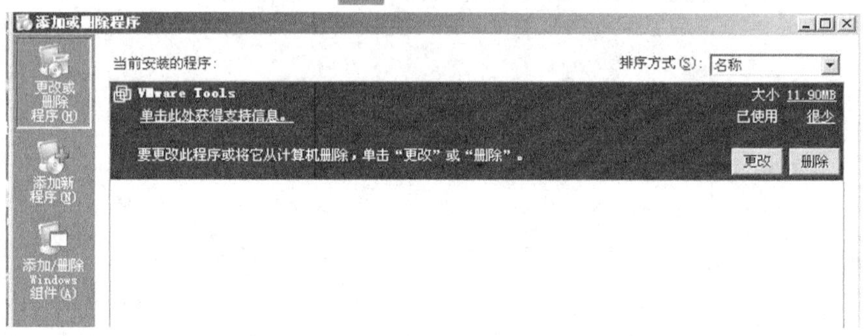

图 27-4

（3）选择“应用程序服务器”选项，单击“详细信息”按钮，如图 27-5 所示。

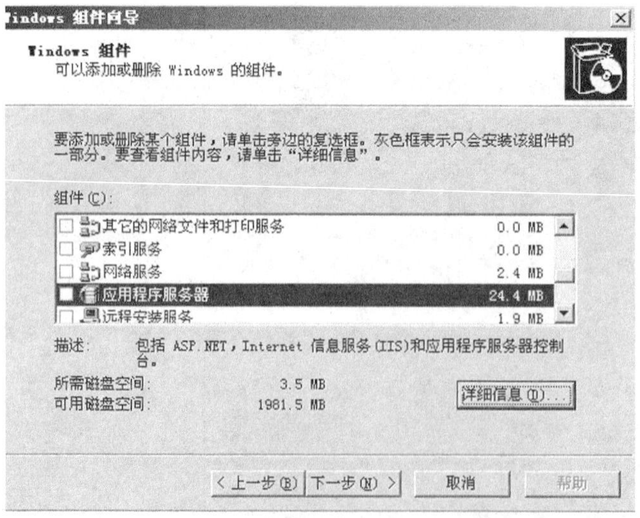

图 27-5

（4）打开“应用程序服务器”对话框，如图 27-6 所示。

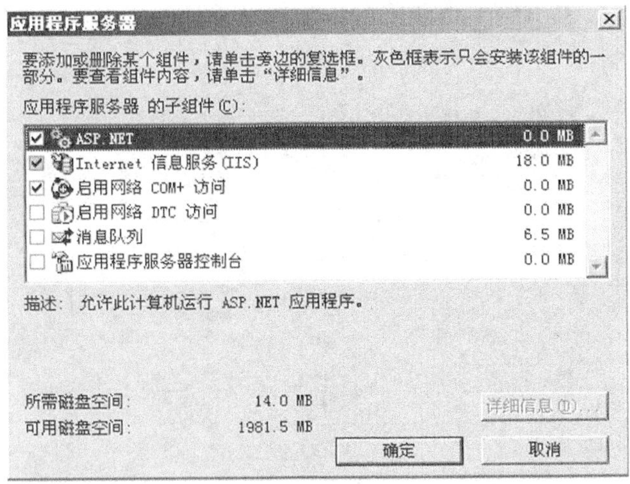

图 27-6

（5）选择“Internet 信息服务（IIS）”，单击“详细信息”按钮，如图 27-7 所示。。

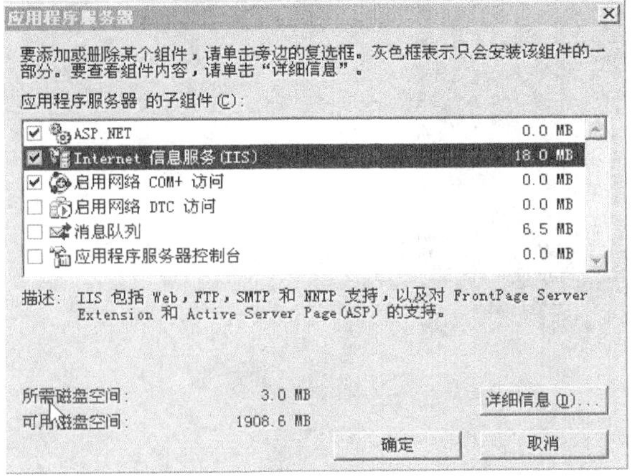

图 27-7

（6）选择“文件传输协议（FTP）服务”，如图 27-8 所示。

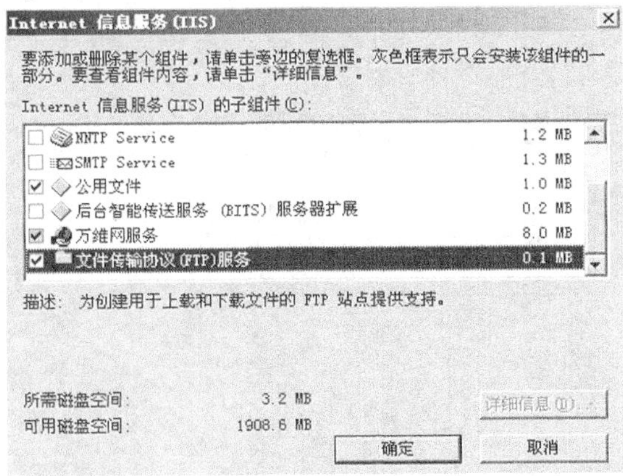

图 27-8

（7）单击“确定”按钮，开始安装，如图 27-9 所示。注意这时需要将 Windows server 2003 盘放入光驱中。

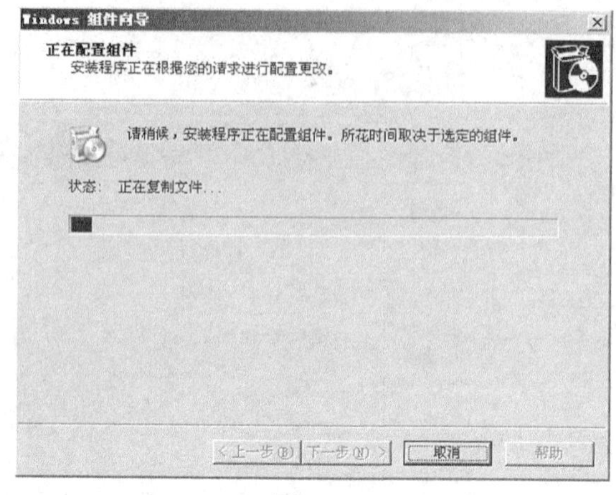

图 27-9

（8）单击“完成”，安装完成，如图 27-10 所示。

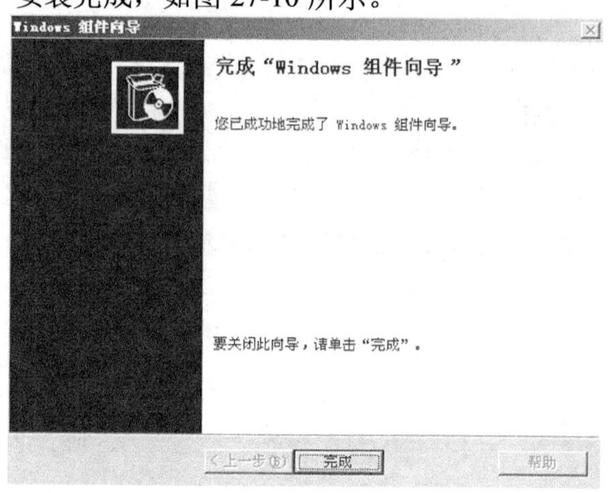

图 27-10

（9）选择“开始”→“管理工具”→“Internet 信息服务（IIS）管理器”，如图 27-11 所示。

（10）打开管理器，选择“FTP 站点”，右击在弹出的快捷菜单中选择“属性”，如图 27-12 所示。

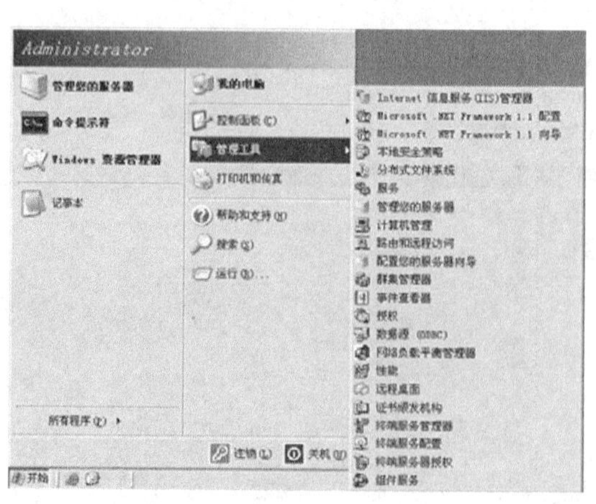

图 27-11

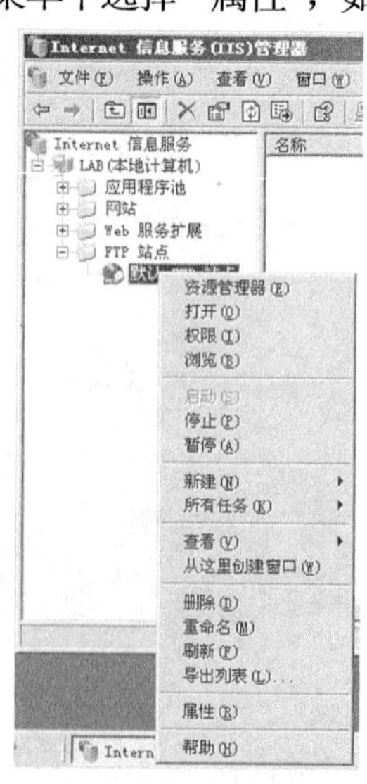

图 27-12

（11）选择“主目录”→“FTP 站点目录”→“浏览”，选择文件夹，如图 27-13 所示。

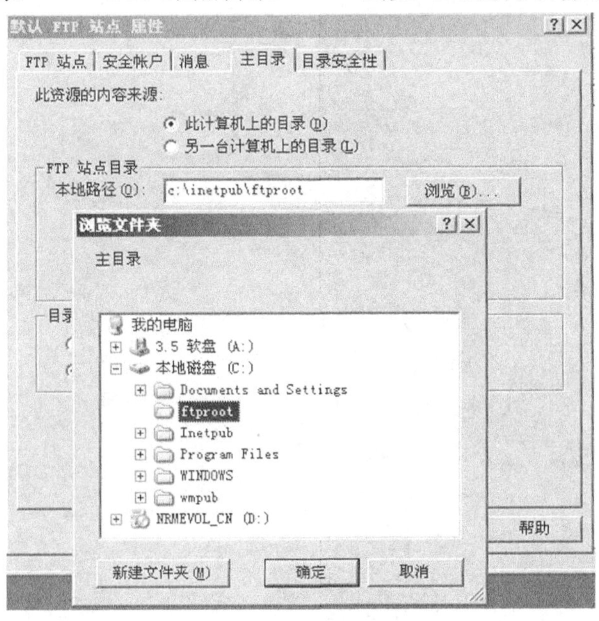

图 27-13

（12）单击“确定”按钮，配置完成。

步骤 5　验证测试。

```
S2126G-1#show vlan
VLAN Name          Status    Ports
------------------------------------------------------------------------
  1 VLAN0001        STATIC
 10 VLAN0010        STATIC  Fa0/1, Fa0/2, Fa0/3, Fa0/4
                             Fa0/5, Fa0/6, Fa0/7, Fa0/8
                             Fa0/9, Fa0/10, Fa0/11, Fa0/12
                             Fa0/13, Fa0/14, Fa0/15, Fa0/16
                             Fa0/17, Fa0/18, Fa0/19, Fa0/20
                             Fa0/21, Fa0/22, Fa0/23, Fa0/24
S2126G-1#show port-security
Secure Port      MaxSecureAddr(count) CurrentAddr(count) Security Action
------------------------------------------------------------------------
FastEthernet 0/2                 3                0          Shutdown
FastEthernet 0/3                 3                0          Shutdown
FastEthernet 0/4                 3                0          Shutdown
FastEthernet 0/5                 3                0          Shutdown
FastEthernet 0/6                 3                0          Shutdown
FastEthernet 0/7                 3                0          Shutdown
FastEthernet 0/8                 3                0          Shutdown
FastEthernet 0/9                 3                0          Shutdown
FastEthernet 0/10                3                0          Shutdown
```

```
FastEthernet 0/11              3             0          Shutdown
FastEthernet 0/12              3             0          Shutdown
FastEthernet 0/13              3             0          Shutdown
FastEthernet 0/14              3             0          Shutdown
FastEthernet 0/15              3             0          Shutdown
FastEthernet 0/16              3             0          Shutdown
FastEthernet 0/17              3             0          Shutdown
FastEthernet 0/18              3             0          Shutdown
FastEthernet 0/19              3             0          Shutdown
FastEthernet 0/20              3             0          Shutdown
FastEthernet 0/21              3             0          Shutdown
FastEthernet 0/22              3             0          Shutdown
FastEthernet 0/23              3             0          Shutdown
FastEthernet 0/24              3             0          Shutdown
S2126G-2#show vlan
VLAN Name          Status    Ports
--------------------------------------------------------------------------
1 VLAN0001         STATIC
20 VLAN0020        STATIC    Fa0/1, Fa0/2, Fa0/3, Fa0/4
                                    Fa0/5, Fa0/6, Fa0/7, Fa0/8
                                    Fa0/9, Fa0/10, Fa0/11, Fa0/12
                                    Fa0/13, Fa0/14, Fa0/15, Fa0/16
                                    Fa0/17, Fa0/18, Fa0/19, Fa0/20
                                    Fa0/21, Fa0/22, Fa0/23, Fa0/24
S2126G-2#show port-security
Secure Port     MaxSecureAddr(count) CurrentAddr(count) Security Action
--------------------------------------------------------------------------
FastEthernet 0/2               2             0          Shutdown
FastEthernet 0/3               2             0          Shutdown
FastEthernet 0/4               2             0          Shutdown
FastEthernet 0/5               2             0          Shutdown
FastEthernet 0/6               2             0          Shutdown
FastEthernet 0/7               2             0          Shutdown
FastEthernet 0/8               2             0          Shutdown
FastEthernet 0/9               2             0          Shutdown
FastEthernet 0/10              2             0          Shutdown
FastEthernet 0/11              2             0          Shutdown
FastEthernet 0/12              2             0          Shutdown
FastEthernet 0/13              2             0          Shutdown
FastEthernet 0/14              2             0          Shutdown
FastEthernet 0/15              2             0          Shutdown
FastEthernet 0/16              2             0          Shutdown
```

```
FastEthernet 0/17              2          0         Shutdown
FastEthernet 0/18              2          0         Shutdown
FastEthernet 0/19              2          0         Shutdown
FastEthernet 0/20              2          0         Shutdown
FastEthernet 0/21              2          0         Shutdown
FastEthernet 0/22              2          0         Shutdown
FastEthernet 0/23              2          0         Shutdown
FastEthernet 0/24              2          0         Shutdown
```

RG-S3760-1#show vlan

```
VLAN Name         Status    Ports
------------------------------------------------------------------------
   1 VLAN0001     STATIC    Fa0/1, Fa0/2, Fa0/3, Fa0/4
                            Fa0/5, Fa0/6, Fa0/7, Fa0/8
                            Fa0/9, Fa0/10, Fa0/11, Fa0/12
                            Fa0/13, Fa0/14, Fa0/15, Fa0/16
                            Fa0/17, Fa0/18, Fa0/19, Fa0/20
                            Fa0/21, Fa0/22, Fa0/23, Fa0/24
                            Gi0/25, Gi0/26, Gi0/27, Gi0/28
  10 VLAN0010     STATIC    Fa0/1, Fa0/2, Fa0/10
  20 VLAN0020     STATIC    Fa0/1, Fa0/2, Fa0/10
  30 VLAN0030     STATIC    Fa0/1, Fa0/2, Fa0/10
```

RG-S3760-1#show access-lists

```
ip access-list extended 110
 10 deny tcp 192.168.11.0 0.0.0.255 192.168.13.0 0.0.0.255 eq ftp
 20 deny tcp 192.168.11.0 0.0.0.255 192.168.13.0 0.0.0.255 eq ftp-data
 30 permit ip any any
```

RG-S3760-1#show ip access-group

```
ip access-group 110 in
Applied On interface VLAN 10.
```

RG-S3760-2#show vlan

```
VLAN Name           Status    Ports
------------------------------------------------------------------------
   1 VLAN0001       STATIC    Fa0/1, Fa0/2, Fa0/3, Fa0/4
                              Fa0/5, Fa0/6, Fa0/7, Fa0/8
                              Fa0/10, Fa0/12, Fa0/13, Fa0/14
                              Fa0/15, Fa0/16, Fa0/17, Fa0/18
                              Fa0/19, Fa0/20, Fa0/21, Fa0/22
                              Fa0/23, Fa0/24, Gi0/25, Gi0/26
                              Gi0/27, Gi0/28
  10 VLAN0010       STATIC    Fa0/10
  20 VLAN0020       STATIC    Fa0/10
  30 VLAN0030       STATIC    Fa0/9, Fa0/10
```

```
RG-S3760-2#show ip route
Codes:  C - connected, S - static,  R - RIP B - BGP
        O - OSPF, IA - OSPF inter area
        N1 - OSPF NSSA external type 1, N2 - OSPF NSSA external type 2
        E1 - OSPF external type 1, E2 - OSPF external type 2
        i - IS-IS, L1 - IS-IS level-1, L2 - IS-IS level-2, ia - IS-IS inter area
        * - candidate default
Gateway of last resort is 172.16.1.2 to network 0.0.0.0
S*   0.0.0.0/0 [1/0] via 172.16.1.2
C    172.16.1.0/30 is directly connected, FastEthernet 0/11
C    172.16.1.1/32 is local host.
C    192.168.11.0/24 is directly connected, VLAN 10
C    192.168.11.2/32 is local host.
C    192.168.12.0/24 is directly connected, VLAN 20
C    192.168.12.2/32 is local host.
C    192.168.13.0/24 is directly connected, VLAN 30
C    192.168.13.2/32 is local host.
```

RG-RSR20-1#show ip int brief

```
Interface                        IP-Address(Pri)      OK?       Status
FastEthernet 0/0                  172.16.1.2/30        YES        UP
FastEthernet 0/1                  200.1.1.1/28         YES        UP
```

RG-RSR20-1#show ip route

```
Codes:  C - connected, S - static,  R - RIP B - BGP
        O - OSPF, IA - OSPF inter area
        N1 - OSPF NSSA external type 1, N2 - OSPF NSSA external type 2
        E1 - OSPF external type 1, E2 - OSPF external type 2
        i - IS-IS, L1 - IS-IS level-1, L2 - IS-IS level-2, ia - IS-IS inter area
        * - candidate default
Gateway of last resort is 200.1.1.2 to network 0.0.0.0
S*   0.0.0.0/0 [1/0] via 200.1.1.2
C    172.16.1.0/30 is directly connected, FastEthernet 0/0
C    172.16.1.2/32 is local host.
S    192.168.11.0/24 [1/0] via 172.16.1.1
S    192.168.12.0/24 [1/0] via 172.16.1.1
S    192.168.13.0/24 [1/0] via 172.16.1.1
C    200.1.1.0/28 is directly connected, FastEthernet 0/1
C    200.1.1.1/32 is local host.
```

RG-RSR20-1#sh ip nat statistics

```
Total translations: 0, max entries permitted: 30000
 Peak translations: 0 @ 01:09:56 ago
Outside interfaces: FastEthernet 0/1
Inside interfaces: FastEthernet 0/0
```

```
Rule statistics:
[ID: 3] inside source static
 hit: 0
 match (before routing):
  tcp packet with destination-ip 200.1.1.7 destination-port 21
 action :
  translate ip packet's destination-ip use ip 192.168.13.254 with port set to 21
 ***sticky rule ***
 inside source static
 hit: 0
 match (after routing):
  tcp packet from inside interface with destination-ip 192.168.13.254
 action :
  translate ip packet's source-ip use ip 200.1.1.7
[ID: 2] inside source static
 hit: 0
 match (before routing):
  tcp packet with destination-ip 200.1.1.7 destination-port 20
 action :
  translate ip packet's destination-ip use ip 192.168.13.254 with port set to 20
 ***sticky rule ***
 inside source static
 hit: 0
 match (after routing):
  tcp packet from inside interface with destination-ip 192.168.13.254
 action :
  translate ip packet's source-ip use ip 200.1.1.7
[ID: 1] inside source dynamic
 hit: 0
 match (after routing):
  ip packet with source-ip match access-list 10
 action :
  translate ip packet's source-ip use pool internet
```

RG-RSR20-1#show access-lists

```
ip access-list standard 10
 10 permit 192.168.11.0 0.0.0.255
 20 permit 192.168.12.0 0.0.0.255
 30 permit 192.168.13.0 0.0.0.255
ip access-list extended 115
 10 deny udp any any eq tftp
 20 deny tcp any any eq 135
 30 deny udp any any eq 135
```

```
 40 deny udp any any eq netbios-ns
 50 deny udp any any eq netbios-dgm
 60 deny tcp any any eq 139
 70 deny udp any any eq netbios-ss
 80 deny tcp any any eq 445
 90 deny tcp any any eq 593
 100 deny tcp any any eq 4444
 110 permit ip any any
RG-RSR20-1#show ip access-group
ip access-group 115 in
ip access-group 115 out
Applied On interface FastEthernet 0/1.
```

使用主机 Telnet 登录路由器 RG-RSR20-2，可以查看到转换条目，如图 27-14 所示。

图 27-14

```
RG-RSR20-1#show ip nat translations
Pro Inside global      Inside local        Outside local      Outside global
tcp 200.1.1.3:1052     192.168.11.200:1052 210.1.1.1:23       210.1.1.1:23
```

使用外网的一台主机访问 FTP 服务器，如图 27-15 所示。

在 IE 浏览器中输入 ftp://200.1.1.7。

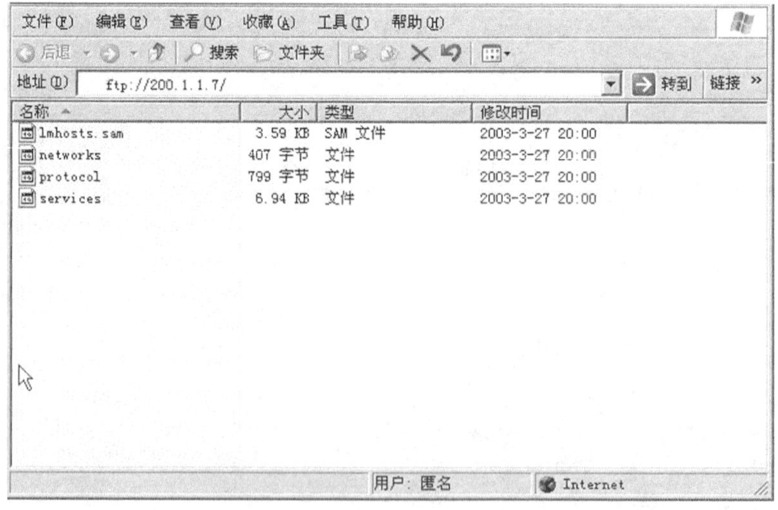

图 27-15

【参考配置】

```
S2126G-1#show running-config
```

```
Building configuration...
Current configuration : 4337 bytes
!
version  RGNOS  10.1.00(4),  Release(18443)(Tue  Jul  17  19:51:54  CST  2007
-ubu6server)
hostname S2126G-1
!
vlan 1
!
vlan 10
!
interface FastEthernet 0/1
 switchport mode trunk
!
interface FastEthernet 0/2
 switchport access vlan 10
 switchport port-security maximum 3
 switchport port-security violation shutdown
 switchport port-security
!
interface FastEthernet 0/3
 switchport access vlan 10
 switchport port-security maximum 3
 switchport port-security violation shutdown
 switchport port-security
!
interface FastEthernet 0/4
 switchport access vlan 10
 switchport port-security maximum 3
 switchport port-security violation shutdown
 switchport port-security
!
interface FastEthernet 0/5
 switchport access vlan 10
 switchport port-security maximum 3
 switchport port-security violation shutdown
 switchport port-security
!
interface FastEthernet 0/6
 switchport access vlan 10
 switchport port-security maximum 3
 switchport port-security violation shutdown
```

```
 switchport port-security
!
interface FastEthernet 0/7
 switchport access vlan 10
 switchport port-security maximum 3
 switchport port-security violation shutdown
 switchport port-security
!
interface FastEthernet 0/8
 switchport access vlan 10
 switchport port-security maximum 3
 switchport port-security violation shutdown
 switchport port-security
!
interface FastEthernet 0/9
 switchport access vlan 10
 switchport port-security maximum 3
 switchport port-security violation shutdown
 switchport port-security
!
interface FastEthernet 0/10
 switchport access vlan 10
 switchport port-security maximum 3
 switchport port-security violation shutdown
 switchport port-security
!
interface FastEthernet 0/11
 switchport access vlan 10
 switchport port-security maximum 3
 switchport port-security violation shutdown
 switchport port-security
!
interface FastEthernet 0/12
 switchport access vlan 10
 switchport port-security maximum 3
 switchport port-security violation shutdown
 switchport port-security
!
interface FastEthernet 0/13
 switchport access vlan 10
 switchport port-security maximum 3
 switchport port-security violation shutdown
```

```
 switchport port-security
!
interface FastEthernet 0/14
 switchport access vlan 10
 switchport port-security maximum 3
 switchport port-security violation shutdown
 switchport port-security
!
interface FastEthernet 0/15
 switchport access vlan 10
 switchport port-security maximum 3
 switchport port-security violation shutdown
 switchport port-security
!
interface FastEthernet 0/16
 switchport access vlan 10
 switchport port-security maximum 3
 switchport port-security violation shutdown
 switchport port-security
!
interface FastEthernet 0/17
 switchport access vlan 10
 switchport port-security maximum 3
 switchport port-security violation shutdown
 switchport port-security
!
interface FastEthernet 0/18
 switchport access vlan 10
 switchport port-security maximum 3
 switchport port-security violation shutdown
 switchport port-security
!
interface FastEthernet 0/19
 switchport access vlan 10
 switchport port-security maximum 3
 switchport port-security violation shutdown
 switchport port-security
!
interface FastEthernet 0/20
 switchport access vlan 10
 switchport port-security maximum 3
 switchport port-security violation shutdown
 switchport port-security
!
interface FastEthernet 0/21
```

```
 switchport access vlan 10
 switchport port-security maximum 3
 switchport port-security violation shutdown
 switchport port-security
!
interface FastEthernet 0/22
 switchport access vlan 10
 switchport port-security maximum 3
 switchport port-security violation shutdown
 switchport port-security
!
interface FastEthernet 0/23
 switchport access vlan 10
 switchport port-security maximum 3
 switchport port-security violation shutdown
 switchport port-security
!
interface FastEthernet 0/24
 switchport access vlan 10
 switchport port-security maximum 3
 switchport port-security violation shutdown
 switchport port-security
!
interface GigabitEthernet 0/25
!
interface GigabitEthernet 0/26
!
interface GigabitEthernet 0/27
!
interface GigabitEthernet 0/28
!
line con 0
line vty 0 4
 login
!
!
end
S2126G-2#show running-config
Building configuration...
Current configuration : 4337 bytes
!
version RGNOS 10.1.00(4), Release(18443)(Tue Jul 17 19:51:54 CST 2007
-ubu6server)
hostname S2126G-2
!
```

```
vlan 1
!
vlan 20
!
interface FastEthernet 0/1
 switchport mode trunk
!
interface FastEthernet 0/2
 switchport access vlan 20
 switchport port-security maximum 2
 switchport port-security violation shutdown
 switchport port-security
!
interface FastEthernet 0/3
 switchport access vlan 20
 switchport port-security maximum 2
 switchport port-security violation shutdown
 switchport port-security
!
interface FastEthernet 0/4
 switchport access vlan 20
 switchport port-security maximum 2
 switchport port-security violation shutdown
 switchport port-security
!
interface FastEthernet 0/5
 switchport access vlan 20
 switchport port-security maximum 2
 switchport port-security violation shutdown
 switchport port-security
!
interface FastEthernet 0/6
 switchport access vlan 20
 switchport port-security maximum 2
 switchport port-security violation shutdown
 switchport port-security
!
interface FastEthernet 0/7
 switchport access vlan 20
 switchport port-security maximum 2
 switchport port-security violation shutdown
 switchport port-security
!
interface FastEthernet 0/8
 switchport access vlan 20
```

```
 switchport port-security maximum 2
 switchport port-security violation shutdown
 switchport port-security
!
interface FastEthernet 0/9
 switchport access vlan 20
 switchport port-security maximum 2
 switchport port-security violation shutdown
 switchport port-security
!
interface FastEthernet 0/10
 switchport access vlan 20
 switchport port-security maximum 2
 switchport port-security violation shutdown
 switchport port-security
!
interface FastEthernet 0/11
 switchport access vlan 20
 switchport port-security maximum 2
 switchport port-security violation shutdown
 switchport port-security
!
interface FastEthernet 0/12
 switchport access vlan 20
 switchport port-security maximum 2
 switchport port-security violation shutdown
 switchport port-security
!
interface FastEthernet 0/13
 switchport access vlan 20
 switchport port-security maximum 2
 switchport port-security violation shutdown
 switchport port-security
!
interface FastEthernet 0/14
 switchport access vlan 20
 switchport port-security maximum 2
 switchport port-security violation shutdown
 switchport port-security
!
interface FastEthernet 0/15
 switchport access vlan 20
 switchport port-security maximum 2
 switchport port-security violation shutdown
 switchport port-security
```

```
!
interface FastEthernet 0/16
 switchport access vlan 20
 switchport port-security maximum 2
 switchport port-security violation shutdown
 switchport port-security
!
interface FastEthernet 0/17
 switchport access vlan 20
 switchport port-security maximum 2
 switchport port-security violation shutdown
 switchport port-security
!
interface FastEthernet 0/18
 switchport access vlan 20
 switchport port-security maximum 2
 switchport port-security violation shutdown
 switchport port-security
!
interface FastEthernet 0/19
 switchport access vlan 20
 switchport port-security maximum 2
 switchport port-security violation shutdown
 switchport port-security
!
interface FastEthernet 0/20
 switchport access vlan 20
 switchport port-security maximum 2
 switchport port-security violation shutdown
 switchport port-security
!
interface FastEthernet 0/21
 switchport access vlan 20
 switchport port-security maximum 2
 switchport port-security violation shutdown
 switchport port-security
!
interface FastEthernet 0/22
 switchport access vlan 20
 switchport port-security maximum 2
 switchport port-security violation shutdown
 switchport port-security
!
interface FastEthernet 0/23
 switchport access vlan 20
```

```
 switchport port-security maximum 2
 switchport port-security violation shutdown
 switchport port-security
!
interface FastEthernet 0/24
 switchport access vlan 20
 switchport port-security maximum 2
 switchport port-security violation shutdown
 switchport port-security
!
interface GigabitEthernet 0/25
!
interface GigabitEthernet 0/26
!
interface GigabitEthernet 0/27
!
interface GigabitEthernet 0/28
!
line con 0
line vty 0 4
 login
!
end
```

RG-S3760-1#show running-config

```
Building configuration...
Current configuration : 1776 bytes
!
version RGNOS 10.1.00(4), Release(18437)(Tue Jul 17 19:13:23 CST 2007
-ubu1server)
hostname RG-S3760-1
!
vlan 1
!
vlan 10
!
vlan 20
!
vlan 30
!
ip access-list extended 110
 10 deny tcp 192.168.11.0 0.0.0.255 192.168.13.0 0.0.0.255 eq ftp
 20 deny tcp 192.168.11.0 0.0.0.255 192.168.13.0 0.0.0.255 eq ftp-data
 30 permit ip any any
!
interface FastEthernet 0/1
```

```
 switchport mode trunk
!
interface FastEthernet 0/2
 switchport mode trunk
!
interface FastEthernet 0/3
!
interface FastEthernet 0/4
!
interface FastEthernet 0/5
!
interface FastEthernet 0/6
!
interface FastEthernet 0/7
!
interface FastEthernet 0/8
!
interface FastEthernet 0/9
!
interface FastEthernet 0/10
 switchport mode trunk
!
interface FastEthernet 0/11
!
interface FastEthernet 0/12
!
interface FastEthernet 0/13
!
interface FastEthernet 0/14
!
interface FastEthernet 0/15
!
interface FastEthernet 0/16
!
interface FastEthernet 0/17
!
interface FastEthernet 0/18
!
interface FastEthernet 0/19
!
interface FastEthernet 0/20
!
interface FastEthernet 0/21
!
interface FastEthernet 0/22
```

```
!
interface FastEthernet 0/23
!
interface FastEthernet 0/24
!
interface GigabitEthernet 0/25
!
interface GigabitEthernet 0/26
!
interface GigabitEthernet 0/27
!
interface GigabitEthernet 0/28
!
interface VLAN 10
 ip access-group 110 in
 ip address 192.168.11.1 255.255.255.0
!
interface VLAN 20
 ip address 192.168.12.1 255.255.255.0
!
interface VLAN 30
 ip address 192.168.13.1 255.255.255.0
!
ip route  0.0.0.0 0.0.0.0  VLAN 10
!
line con 0
line vty 0 4
 login
!
end
```

RG-S3760-2#show running-config

```
Building configuration...
Current configuration : 1512 bytes
!
version  RGNOS  10.1.00(4),  Release(18437)(Tue  Jul  17  19:13:23  CST  2007
-ubu1server)
hostname RG-S3760-2
!
vlan 1
!
vlan 10
!
vlan 20
!
vlan 30
```

```
!
interface FastEthernet 0/1
!
interface FastEthernet 0/2
!
interface FastEthernet 0/3
!
interface FastEthernet 0/4
!
interface FastEthernet 0/5
!
interface FastEthernet 0/6
!
interface FastEthernet 0/7
!
interface FastEthernet 0/8
!
interface FastEthernet 0/9
 switchport access vlan 30
!
interface FastEthernet 0/10
 switchport mode trunk
!
interface FastEthernet 0/11
 no switchport
 ip address 172.16.1.1 255.255.255.252
!
interface FastEthernet 0/12
!
interface FastEthernet 0/13
!
interface FastEthernet 0/14
!
interface FastEthernet 0/15
!
interface FastEthernet 0/16
!
interface FastEthernet 0/17
!
interface FastEthernet 0/18
!
interface FastEthernet 0/19
!
interface FastEthernet 0/20
!
```

```
interface FastEthernet 0/21
!
interface FastEthernet 0/22
!
interface FastEthernet 0/23
!
interface FastEthernet 0/24
!
interface GigabitEthernet 0/25
!
interface GigabitEthernet 0/26
!
interface GigabitEthernet 0/27
!
interface GigabitEthernet 0/28
!
interface VLAN 10
 ip address 192.168.11.2 255.255.255.0
!
interface VLAN 20
 ip address 192.168.12.2 255.255.255.0
!
interface VLAN 30
 ip address 192.168.13.2 255.255.255.0
!
ip route  0.0.0.0 0.0.0.0  172.16.1.2
!
line con 0
line vty 0 4
 login
!
end
```

```
RG-RSR20-1#show running-config
Building configuration...
Current configuration : 1548 bytes
!
version  RGNOS  10.1.00(4),  Release(18443)(Tue  Jul  17  20:50:30  CST  2007
-ubu1server)
hostname RG-RSR20-1
!
ip access-list standard 10
 10 permit 192.168.11.0 0.0.0.255
 20 permit 192.168.12.0 0.0.0.255
 30 permit 192.168.13.0 0.0.0.255
!
```

```
!
ip access-list extended 115
 10 deny udp any any eq tftp
 20 deny tcp any any eq 135
 30 deny udp any any eq 135
 40 deny udp any any eq netbios-ns
 50 deny udp any any eq netbios-dgm
 60 deny tcp any any eq 139
 70 deny udp any any eq netbios-ss
 80 deny tcp any any eq 445
 90 deny tcp any any eq 593
 100 deny tcp any any eq 4444
 110 permit ip any any
!
interface FastEthernet 0/0
 ip nat inside
 ip address 172.16.1.2 255.255.255.252
 duplex auto
 speed auto
!
interface FastEthernet 0/1
 ip nat outside
 ip access-group 115 in
 ip access-group 115 out
 ip address 200.1.1.1 255.255.255.240
 duplex auto
 speed auto
!
ip nat pool internet 200.1.1.3 200.1.1.6 netmask 255.255.255.240
ip nat inside source static tcp 192.168.13.254 21 200.1.1.7 21 permit-inside
ip nat inside source static tcp 192.168.13.254 20 200.1.1.7 20 permit-inside
ip nat inside source list 10 pool internet overload
!
ip route  0.0.0.0 0.0.0.0  200.1.1.2
ip route  192.168.11.0 255.255.255.0  172.16.1.1
ip route  192.168.12.0 255.255.255.0  172.16.1.1
ip route  192.168.13.0 255.255.255.0  172.16.1.1
!
line con 0
line aux 0
line vty 0 4
 login
!
end
```

RG-RSR20-2#sh running-config

```
Building configuration...
Current configuration : 585 bytes
!
version  RGNOS  10.1.00(4),  Release(18443)(Tue  Jul  17  20:50:30  CST  2007
-ubu1server)
hostname RG-RSR20-2
!
enable password 7 08527d46
!
interface FastEthernet 0/0
 ip address 200.1.1.2 255.255.255.240
 duplex auto
 speed auto
!
interface FastEthernet 0/1
 duplex auto
 speed auto
!
interface Loopback 0
 ip address 210.1.1.1 255.255.255.0
!
ip route  0.0.0.0 0.0.0.0  200.1.1.1
!
line con 0
line aux 0
line vty 0 4
 login
 password 7 1443185e
!
end
```

综合实验2　混合篇

【实验名称】

综合实验2　混合篇。

【实验目的】

构建一个综合的、复杂的企业网络。应用已有的知识，构建和部署一个企业网络，同时必须考虑到网络的可用性、经济性、可靠性、安全性等因素，提高学员在构建企业网络方面的综合应用能力。

【背景描述】

你是某公司的 IT 管理人员，公司需要构建一个综合的企业网，公司有 4 个部门，从内网的安全考虑，使用 VLAN 技术将各部门划分到不同的 VLAN 中；为了提高公司的业务能力和增强企业的知名度，将公司的 Web 网站以及 FTP、Mail 服务发布到互联网上；公司在另外一个城市中也有一个分公司，公司与分公司需要进行一些业务的沟通，其业务数据很敏感，所以需要使用 IPSec VPN 技术保证数据的安全性；为了便于网络管理，公司内部的网络需要使用 OSPF 路由协议使用全网互通；公司需要能够访问互联网，并从 ISP 那里申请了一段公网 IP 地址 99.1.1.0/28；为了保证企业网内部安全，使用 ACL 技术保证内网安全。

【需求分析】

公司使用 RG-3760 为核心设备，并使用的双核心作为冗余，接入桌面的使用 RG-S2126G 交换机，RG-S2126-1 为 VLAN100 和 VLAN101 的接入设备，RG-S2126-2 为 VLAN102 和 VLAN103 的接入设备；采用 RG-RSR20 路由器接入 ISP 线路。

公司内部有 4 个部门，其分别划分到 VLAN100、VLAN101、VALN102、VLAN103，而服务群划分到 VLAN104，其网段分别为 192.168.100.0/24、192.168.101.0/24、192.168.102.0/14、192.168.103.0/24、192.168.104.0/24。

在两核心设备上配置链路聚合，增加带宽，配置 RSTP 协议，要求 RG-3760-1 为生成树的根，提高接口的收敛速度需要配置速端口；为了接入层安全，需要配置端口安全，VLAN100 和 VLAN102 限制最大连接数为 3，违规则 shutdown，VLAN101 和 VLAN103 限制最大连接数为 2，违规则 shutdown。

根据不同的部门使用不同公网 IP 地址去访问互联网，VLAN100 和 VLAN101 使用的公网地址段为 99.1.1.3 9～9.1.1.5，VLAN102 和 VLAN103 使用的公网地址段为 99.1.1.6～99.1.1.8。

FTP 服务器的内网地址为 192.168.104.252，其公网地址为 99.1.1.11；Web 服务器的内网地址为 192.168.104.254，其公网地址为 99.1.1.9，Mail 服务器的内网地址为 192.168.104.253，其公网地址为 99.1.1.10，将服务器发布互联网上并允许内网用户可以使用内部 IP 地址访问服务。

配置 IPSec VPN，公司总部传输业务数据的网段为 192.168.100.0/24，分公司传输业务数据的网段为 192.168.99.0/24。

在企业网内部运行 OSFP 路由协议，在总公司和分公司的两台路由器上配置 PPP 协议，并需要使用 CHAP 验证。

【实验拓扑】

实验的拓扑图，如图 28-1 所示。

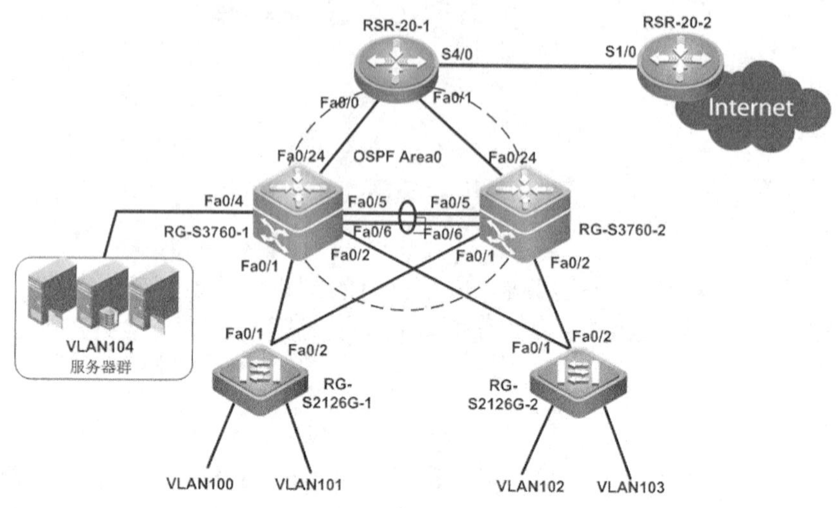

图 28-1

【实验设备】

二层交换机 RG-S2126G 2 台。

三层交换机 RG-S3760 2 台。

路由器 RG-RSR-20（带串口）2 台。

V.35 线缆（DTE/DCE）1 对。

计算机 1 台。

【IP 地址规划】

设备名称	端口名称	IP 地址
S3760-1	VLAN100	192.168.100.1/24
	VLAN101	192.168.101.1/24
	VLAN102	192.168.102.1/24
	VLAN103	192.168.103.1/24
	VLAN104	192.168.104.1/24
	FA0/24	10.1.1.2/30
S3760-2	VLAN100	192.168.100.2/24
	VLAN101	192.168.101.2/24
	VLAN102	192.168.102.2/24
	VLAN103	192.168.103.2/24
	VLAN104	192.168.104.2/24
	FA0/24	10.1.1.6/30
RSR20-1	Fa0/0	10.1.1.1/30
	Fa0/1	10.1.1.5/30
	S4/0	99.1.1.1/28
RSR20-2	S1/0	99.1.1.2/28
	Loopback 0	192.168.99.1/24

【VLAN 规划】

设备名称	端口	VLAN
RG-2126G-1	FA0/3-FA0/14	VLAN100
	FA0/15-FA0/24	VLAN101
RG-2126G-2	FA0/3-FA0/14	VLAN102
	FA0/15-FA0/24	VLAN103
S3760-1	FA0/4	VLAN104

【实验步骤】

步骤 1　配置接入层交换机。

```
Switch(config)#host RG-2126G-1
RG-S2126G-1 (config)#vlan 100
RG-S2126G-1 (config-vlan)#vlan 101
RG-S2126G-1 (config-vlan)#exit
RG-S2126G-1 (config)#spanning-tree
RG-S2126G-1 (config)#spanning-tree mode rstp
RG-S2126G-1 (config)#interface range FastEthernet0/1-2
RG-S2126G-1 (config-if-range)#switchport mode trunk
RG-S2126G-1 (config-if-range)#exit
RG-S2126G-1 (config)#interface range FastEthernet0/3-14
RG-S2126G-1 (config-if-range)#switchport access vlan 100
RG-S2126G-1 (config-if-range)# switchport port-security
RG-S2126G-1 (config-if-range)#switchport port-security maximum 2
RG-S2126G-1 (config-if-range)#switchport port-security violation shutdown
RG-S2126G-1 (config-if-range)#spanning-tree portfast
RG-S2126G-1 (config-if-range)#exit
RG-S2126G-1 (config)#interface range FastEthernet0/15-24
RG-S2126G-1 (config-if-range)#switchport access vlan 101
RG-S2126G-1 (config-if-range)# switchport port-security
RG-S2126G-1 (config-if-range)#switchport port-security maximum 3
RG-S2126G-1 (config-if-range)#switchport port-security violation shutdown
RG-S2126G-1 (config-if-range)#spanning-tree portfast
RG-S2126G-1 (config-if-range)#exit
Switch(config)#host RG-2126G-2
RG-S2126G-2 (config)#vlan 102
RG-S2126G-2 (config-vlan)#vlan 103
RG-S2126G-2 (config-vlan)#exit
RG-S2126G-2 (config)# spanning-tree
RG-S2126G-2 (config)#spanning-tree mode rstp
RG-S2126G-2 (config)#interface range FastEthernet0/1-2
RG-S2126G-2 (config-if-range)#switchport mode trunk
```

```
RG-S2126G-2 (config-if-range)#exit
RG-S2126G-2 (config)#interface range FastEthernet0/3-14
RG-S2126G-2 (config-if-range)#switchport access vlan 102
RG-S2126G-2 (config-if-range)# switchport port-security
RG-S2126G-2 (config-if-range)#switchport port-security maximum 2
RG-S2126G-2 (config-if-range)#switchport port-security violation shutdown
RG-S2126G-2 (config-if-range)#spanning-tree portfast
RG-S2126G-2 (config-if-range)#exit
RG-S2126G-2 (config)#interface range FastEthernet0/15-24
RG-S2126G-2 (config-if-range)#switchport access vlan 103
RG-S2126G-2 (config-if-range)# switchport port-security
RG-S2126G-2 (config-if-range)#switchport port-security maximum 3
RG-S2126G-2 (config-if-range)#switchport port-security violation shutdown
RG-S2126G-2 (config-if-range)#spanning-tree portfast
RG-S2126G-2 (config-if-range)#exit
```

步骤 2　配置核心交换机。

```
Switch(config)# hostname RG-S3760-1
RG-S3760-1(config)#vlan 100
RG-S3760-1(config-vlan)#vlan 101
RG-S3760-1(config-vlan)#vlan 102
RG-S3760-1(config-vlan)#vlan 103
RG-S3760-1(config-vlan)#vlan 104
RG-S3760-1(config-vlan)#exit
RG-S3760-1(config)# spanning-tree
RG-S3760-1(config)#spanning-tree mode rstp
RG-S3760-1(config)#spanning-tree priority 4096
RG-S3760-1(config)# interface range FastEthernet0/1-2
RG-S3760-1 (config-if-range)# switchport mode trunk
RG-S3760-1 (config-if-range)# exit
RG-S3760-1(config)# interface range FastEthernet0/5-6
RG-S3760-1 (config-if-range)# port-group 1
RG-S3760-1 (config-if-range)# exit
RG-S3760-1(config)# interface AggregatePort 1
RG-S3760-1(config-if)# switchport mode trunk
RG-S3760-1(config-if)#exit
RG-S3760-1(config)#interface VLAN 100
RG-S3760-1(config-if)# ip address 192.168.100.1 255.255.255.0
RG-S3760-1(config-if)#no shutdown
RG-S3760-1(config-if)#exit
RG-S3760-1(config)#interface VLAN 101
RG-S3760-1(config-if)#ip address 192.168.101.1 255.255.255.0
RG-S3760-1(config-if)#no shutdown
```

```
RG-S3760-1(config-if)#exit
RG-S3760-1(config)#interface VLAN 102
RG-S3760-1(config-if)#ip address 192.168.102.1 255.255.255.0
RG-S3760-1(config-if)#no shutdown
RG-S3760-1(config-if)#exit
RG-S3760-1(config)#interface VLAN 103
RG-S3760-1(config-if)#ip address 192.168.103.1 255.255.255.0
RG-S3760-1(config-if)#no shutdown
RG-S3760-1(config-if)#exit
RG-S3760-1(config)#interface VLAN 104
RG-S3760-1(config-if)#ip address 192.168.104.1 255.255.255.0
RG-S3760-1(config-if)#no shutdown
RG-S3760-1(config-if)#exit
RG-S3760-1(config)#interface FastEthernet 0/24
RG-S3760-1(config-if)#no switchport
RG-S3760-1(config-if)#ip address 10.1.1.2 255.255.255.252
RG-S3760-1(config-if)#no shutdown
RG-S3760-1(config-if)#exit
RG-S3760-1(config)#interface FastEthernet 0/4
RG-S3760-1(config-if)#switchport access vlan 104
RG-S3760-1(config-if)#exit
RG-S3760-1(config)#router ospf 10
RG-S3760-1(config-router)#network 10.1.1.0 0.0.0.3 area 0
RG-S3760-1(config-router)#network 192.168.100.0 0.0.0.255 area 0
RG-S3760-1(config-router)#network 192.168.101.0 0.0.0.255 area 0
RG-S3760-1(config-router)#network 192.168.102.0 0.0.0.255 area 0
RG-S3760-1(config-router)#network 192.168.103.0 0.0.0.255 area 0
RG-S3760-1(config-router)#network 192.168.104.0 0.0.0.255 area 0
RG-S3760-1(config-router)#exit
Switch(config)# hostname RG-S3760-2
RG-S3760-2(config)#vlan 100
RG-S3760-2(config-vlan)#vlan 101
RG-S3760-2(config-vlan)#vlan 102
RG-S3760-2(config-vlan)#vlan 103
RG-S3760-2(config-vlan)#vlan 104
RG-S3760-2(config-vlan)#exit
RG-S3760-2(config)# spanning-tree
RG-S3760-2(config)#spanning-tree mode rstp
RG-S3760-2(config)#spanning-tree priority 8192
RG-S3760-2(config)# interface range FastEthernet0/1-2
RG-S3760-2 (config-if-range)# switchport mode trunk
RG-S3760-2 (config-if-range)#exit
```

```
RG-S3760-2(config)#interface range FastEthernet0/5-6
RG-S3760-2 (config-if-range)#port-group 1
RG-S3760-2 (config-if-range)#exit
RG-S3760-2(config)#interface AggregatePort 1
RG-S3760-2(config-if)#switchport mode trunk
RG-S3760-2(config-if)#exit
RG-S3760-2(config)#interface VLAN 100
RG-S3760-2(config-if)#ip address 192.168.100.2 255.255.255.0
RG-S3760-2(config-if)#no shutdown
RG-S3760-2(config-if)#exit
RG-S3760-2(config)#interface VLAN 101
RG-S3760-2(config-if)#ip address 192.168.101.2 255.255.255.0
RG-S3760-2(config-if)#no shutdown
RG-S3760-2(config-if)#exit
RG-S3760-2(config)#interface VLAN 102
RG-S3760-2(config-if)#ip address 192.168.102.2 255.255.255.0
RG-S3760-2(config-if)#no shutdown
RG-S3760-2(config-if)#exit
RG-S3760-2(config)#interface VLAN 103
RG-S3760-2(config-if)#ip address 192.168.103.2 255.255.255.0
RG-S3760-2(config-if)#no shutdown
RG-S3760-2(config-if)#exit
RG-S3760-2(config)#interface VLAN 104
RG-S3760-2(config-if)#ip address 192.168.104.2 255.255.255.0
RG-S3760-2(config-if)#no shutdown
RG-S3760-2(config-if)#exit
RG-S3760-2(config)#interface FastEthernet 0/24
RG-S3760-2(config-if)#no switchport
RG-S3760-2(config-if)#ip address 10.1.1.6 255.255.255.252
RG-S3760-2(config-if)#no shutdown
RG-S3760-2(config-if)#exit
RG-S3760-2(config)#router ospf 10
RG-S3760-2(config-router)#network 10.1.1.4 0.0.0.3 area 0
RG-S3760-2(config-router)#network 192.168.100.0 0.0.0.255 area 0
RG-S3760-2(config-router)#network 192.168.101.0 0.0.0.255 area 0
RG-S3760-2(config-router)#network 192.168.102.0 0.0.0.255 area 0
RG-S3760-2(config-router)#network 192.168.103.0 0.0.0.255 area 0
RG-S3760-2(config-router)#network 192.168.104.0 0.0.0.255 area 0
RG-S3760-2(config-router)#exit
```

步骤 3　配置网络出口。

```
Router(config)#hostname RG-RSR20-1
RG-RSR20-1(config)#interface serial 4/0
```

```
RG-RSR20-1(config-if)#encapsulation ppp
RG-RSR20-1(config-if)#ip nat outside
RG-RSR20-1(config-if)# ip address 99.1.1.1 255.255.255.240
RG-RSR20-1(config-if)#exit
RG-RSR20-1(config)# interface FastEthernet 0/0
RG-RSR20-1(config-if)#ip nat inside
RG-RSR20-1(config-if)#ip address 10.1.1.1 255.255.255.252
RG-RSR20-1(config-if)#ip access-group 12 in
RG-RSR20-1(config-if)#exit
RG-RSR20-1(config)# interface FastEthernet 0/1
RG-RSR20-1(config-if)#ip nat inside
RG-RSR20-1(config-if)#ip address 10.1.1.5 255.255.255.252
RG-RSR20-1(config-if)#ip access-group 12 in
RG-RSR20-1(config-if)#exit
RG-RSR20-1(config)# username RG-RSR20-2 password 0 ruijie
RG-RSR20-1(config)#crypto isakmp policy 110
RG-RSR20-1(config -isakmp)#authentication pre-share
RG-RSR20-1(config -isakmp)#hash md5
RG-RSR20-1(config -isakmp)#group 2
RG-RSR20-1(config -isakmp)#exit
RG-RSR20-1(config)#crypto isakmp key 7 15581828182229 address 99.1.1.2
RG-RSR20-1(config)#crypto  ipsec  transform-set  vpn   ah-md5-hmac  esp-des
esp-md5-hmac
RG-RSR20-1 (cfg-crypto-trans)#mode tunel
RG-RSR20-1 (cfg-crypto-trans)#exit
RG-RSR20-1 (config)#crypto map site-to-site 10 ipsec-isakmp
RG-RSR20-1 (config-crypto-map)#set peer 99.1.1.2
RG-RSR20-1 (config-crypto-map)#set transform-set vpn
RG-RSR20-1 (config-crypto-map)#match address 110
RG-RSR20-1 (config-crypto-map)#exit
RG-RSR20-1  (config)#  ip  nat  pool  internet  99.1.1.3  99.1.1.5  netmask
255.255.255.240
RG-RSR20-1  (config)#ip  nat  pool  internet1  99.1.1.6  99.1.1.8  netmask
255.255.255.240
RG-RSR20-1 (config)#ip nat inside source static tcp 192.168.104.254 80 99.1.1.9
80 permit-inside
RG-RSR20-1 (config)#ip nat inside source static tcp 192.168.104.253 25 99.1.1.10
25 permit-inside
RG-RSR20-1 (config)#ip nat inside source static tcp 192.168.104.253 110 99.1.1.10
110 permit-inside
RG-RSR20-1 (config)#ip nat inside source static tcp 192.168.104.252 20 99.1.1.11
20 permit-inside
```

```
RG-RSR20-1 (config)#ip nat inside source static tcp 192.168.104.252 21 99.1.1.11
21 permit-inside
RG-RSR20-1 (config)#ip nat inside source list 11 pool internet1 overload
RG-RSR20-1 (config)#ip nat inside source list 100 pool internet overload
RG-RSR20-1 (config)# access-list 11 permit 192.168.102.0 0.0.0.255
RG-RSR20-1 (config)# access-list 11 permit 192.168.103.0 0.0.0.255
RG-RSR20-1 (config)# access-list 100 deny ip 192.168.100.0 0.0.0.255 192.168.99.0
0.0.0.255
RG-RSR20-1 (config)# access-list 100 permit ip 192.168.100.0 0.0.0.255 any
RG-RSR20-1 (config)#ip access-list 110 permit ip 192.168.100.0 0.0.0.255
192.168.99.0 0.0.0.255
RG-RSR20-1 (config)#router ospf 10
RG-RSR20-1 (config-router)#network 10.1.1.0 0.0.0.3 area 0
RG-RSR20-1 (config-router)#network 10.1.1.4 0.0.0.3 area 0
RG-RSR20-1 (config-router)#default-information originate
RG-RSR20-1 (config-router)#exit
RG-RSR20-1 (config)#ip route 0.0.0.0 0.0.0.0 99.1.1.2
RG-RSR20-1 (config)# access-list 12 permit 192.168.100.0 0.0.0.255
RG-RSR20-1 (config)# access-list 12 permit 192.168.101.0 0.0.0.255
RG-RSR20-1 (config)# access-list 12 permit 192.168.102.0 0.0.0.255
RG-RSR20-1 (config)# access-list 12 permit 192.168.103.0 0.0.0.255
RG-RSR20-1 (config)# access-list 12 deny any
```

步骤 4　配置分公司路由器。

```
Router(config)#hostname RG-RSR20-2
RG-RSR20-2(config)#interface serial 1/0
RG-RSR20-2(config-if)#ip address 99.1.1.2 255.255.255.240
RG-RSR20-2(config-if)#clock rate 64000
RG-RSR20-2(config-if)#encapsulation ppp
RG-RSR20-2(config-if)#ppp authentication chap
RG-RSR20-2(config-if)#crypto map site-to-site
RG-RSR20-2(config-if)#exit
RG-RSR20-2(config)#username RG-RSR20-1 password 0 ruijie
RG-RSR20-2(config)#interface Loopback 0
RG-RSR20-2(config-if)#ip address 192.168.99.1 255.255.255.0
RG-RSR20-2(config-if)#exit
RG-RSR20-2(config)#ip route  0.0.0.0 0.0.0.0  99.1.1.1
RG-RSR20-2(config)#crypto isakmp policy 110
RG-RSR20-2(config-isakmp)#authentication pre-share
RG-RSR20-2(config-isakmp)#hash md5
RG-RSR20-2(config-isakmp)#group 2
RG-RSR20-2(config-isakmp)#exit
RG-RSR20-2(config)#crypto isakmp key 7 1007072e0f1b4f address 99.1.1.1
```

```
RG-RSR20-2(config)#crypto ipsec transform-set vpn ah-md5-hmac esp-des esp-md5-hmac
RG-RSR20-2(config)#crypto map site-to-site 10 ipsec-isakmp
RG-RSR20-2(cfg-crypto-map)#set peer 99.1.1.1
RG-RSR20-2(cfg-crypto-map)#set transform-set vpn
RG-RSR20-2(cfg-crypto-map)#match address 110
RG-RSR20-2(cfg-crypto-map)#exit
RG-RSR20-2(config)# access-list 110 permit ip 192.168.99.0 0.0.0.255 192.168.100.0 0.0.0.255
```

步骤 5　配置服务器。

配置内网的 Web、FTP、Mail 服务器，使外网可以访问。

（1）安装服务器。

在 Windows server 2003 服务器上，选择“开始”→“控制面板”→“添加删除程序”命令，如图 28-2 所示。

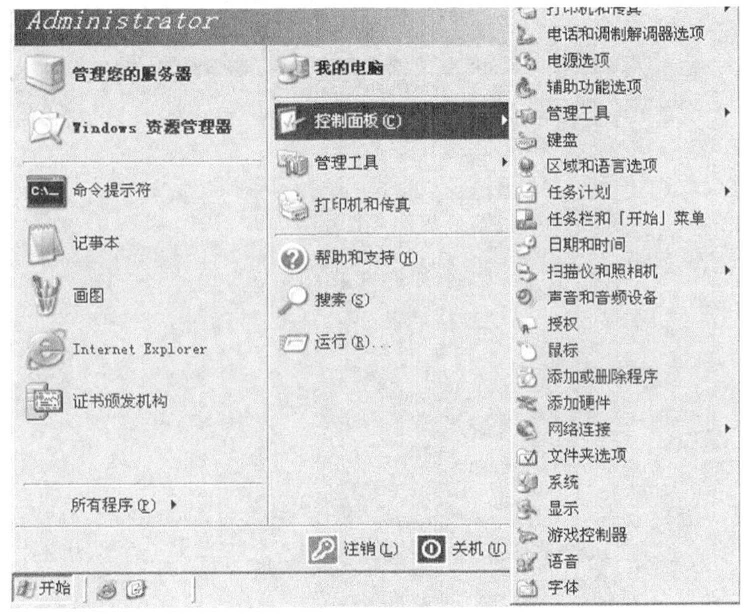

图 28-2

单击“添加删除组件”按钮，如图 28-3 所示。

图 28-3

选择“应用程序服务器”→选择“Internet 信息服务”命令，单击“详细信息”按钮，选择“SMTP Service”和“文件传输协议（FTP）服务”，单击“确定”，如图 28-4 所示。

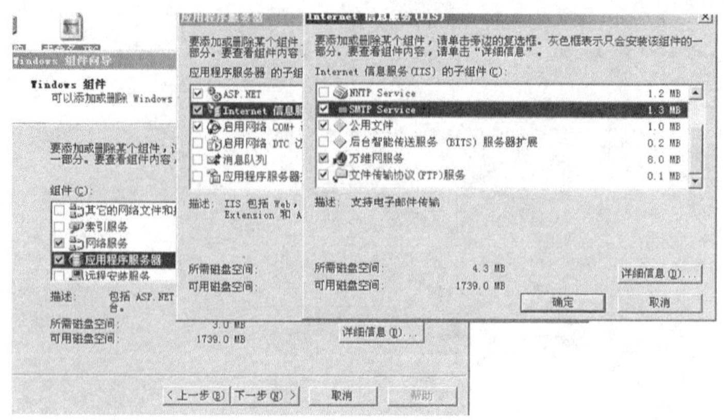

图 28-4

在 Windows 组件向导中，选择“电子邮件服务”，单击“详细信息”按钮，在弹出的“电子邮件服务”中选择“POP3 服务”和“POP3 服务 Web 管理”，如图 28-5 所示。

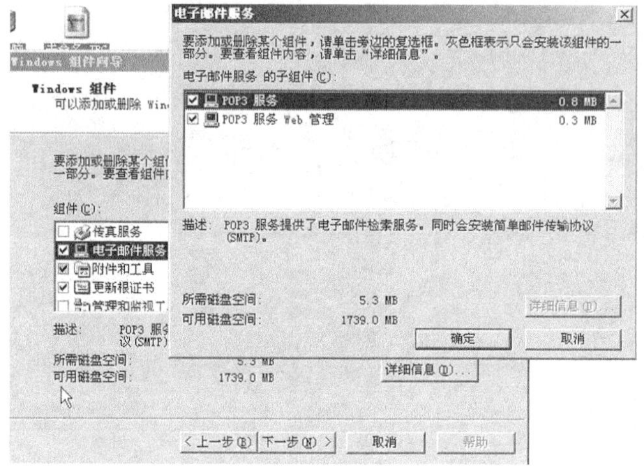

图 28-5

然后单击确定，进行安装，如图 28-6 所示。安装完成后，如图 28-7 所示。

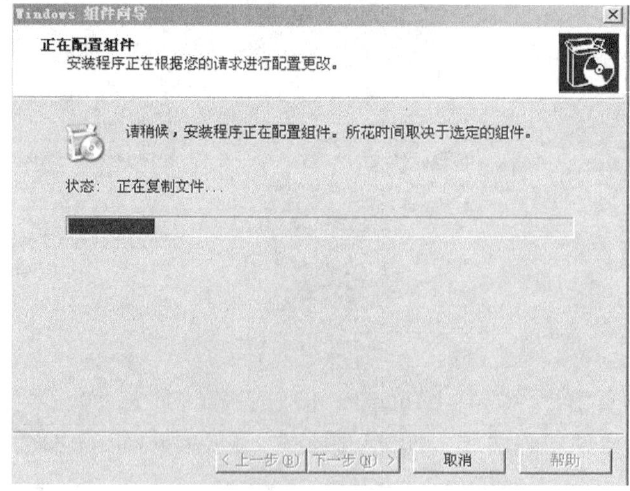

图 28-6

图 28-7

（2）配置服务器。

安装完服务之后，单击“开始”→“管理工具”→“Internet 信息服务（IIS）管理器”命令，如图 28-8 所示。

创建 Web 站点，右击在弹出的快捷菜单中选择“属性”，在路径下创建 Web 站点主页，如图 28-9 所示。

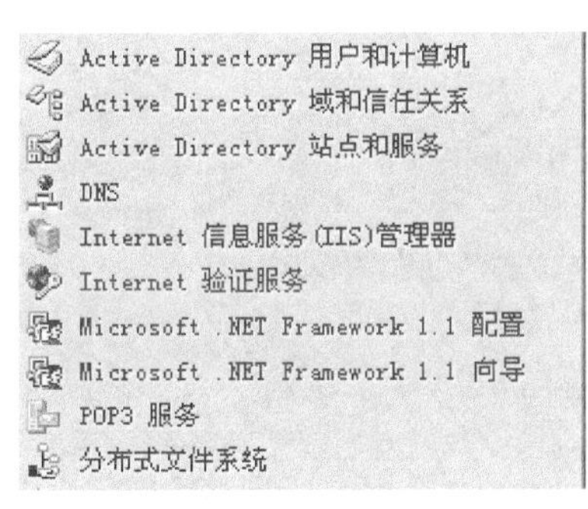

图 28-8

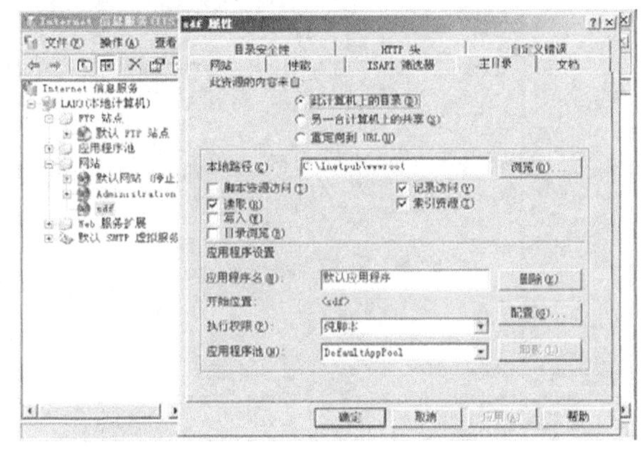

图 28-9

创建 FTP 站点，右击在弹出的快捷键菜单中选择“属性”，配置主目录路径，在路径下创建文件目录，如图 28-10 所示。

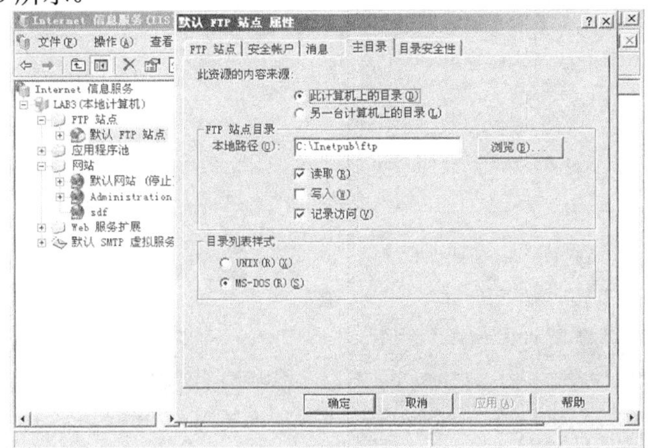

图 28-10

配置 SMTP 邮件服务器，如图 28-11 所示。

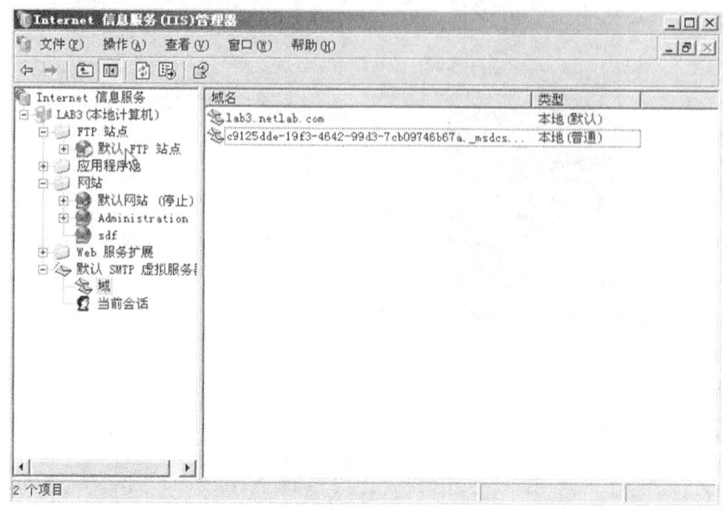

图 28-11

配置 POP3 服务器，如图 28-12 所示。

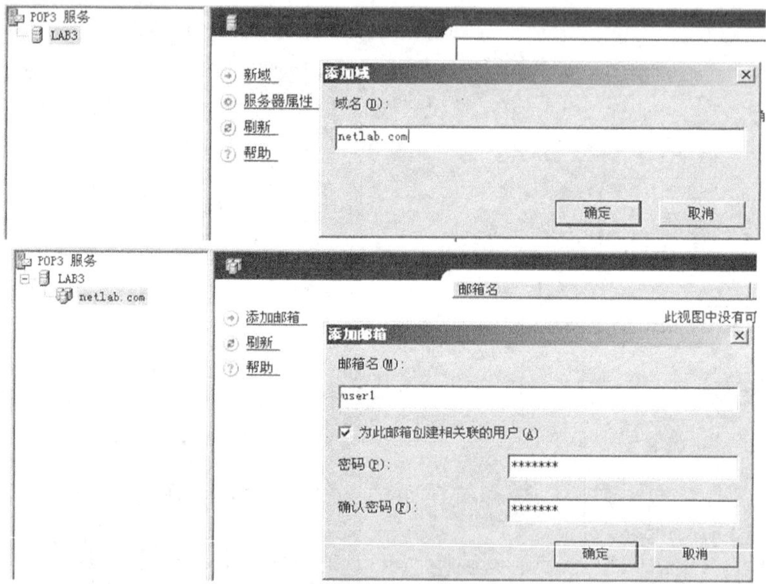

图 28-12

步骤 6　验证测试。

路由的测试：查看路由表。

```
RG-RSR20-1#show ip route
Codes:  C - connected, S - static,  R - RIP B - BGP
        O - OSPF, IA - OSPF inter area
        N1 - OSPF NSSA external type 1, N2 - OSPF NSSA external type 2
        E1 - OSPF external type 1, E2 - OSPF external type 2
        i - IS-IS, L1 - IS-IS level-1, L2 - IS-IS level-2, ia - IS-IS inter area
        * - candidate default
Gateway of last resort is 99.1.1.2 to network 0.0.0.0
```

```
S*   0.0.0.0/0 [1/0] via 99.1.1.2
C    10.1.1.0/30 is directly connected, GigabitEthernet 0/0
C    10.1.1.1/32 is local host.
C    99.1.1.0/28 is directly connected, serial 4/0
C    99.1.1.1/32 is local host.
O    192.168.100.0/24 [110/2] via 10.1.1.2, 00:03:37, GigabitEthernet 0/0
O    192.168.101.0/24 [110/2] via 10.1.1.2, 00:03:37, GigabitEthernet 0/0
O    192.168.102.0/24 [110/2] via 10.1.1.2, 00:03:37, GigabitEthernet 0/0
O    192.168.103.0/24 [110/2] via 10.1.1.2, 00:03:37, GigabitEthernet 0/0
O    192.168.104.0/24 [110/2] via 10.1.1.2, 00:03:37, GigabitEthernet 0/0
```

RG-S3760-1#show ip route

```
Codes:  C - connected, S - static,  R - RIP B - BGP
       O - OSPF, IA - OSPF inter area
       N1 - OSPF NSSA external type 1, N2 - OSPF NSSA external type 2
       E1 - OSPF external type 1, E2 - OSPF external type 2
       i - IS-IS, L1 - IS-IS level-1, L2 - IS-IS level-2, ia - IS-IS inter area
       * - candidate default
Gateway of last resort is 10.1.1.1 to network 0.0.0.0
O*E2 0.0.0.0/0 [110/10] via 10.1.1.1, 00:07:16, FastEthernet 0/24
C    10.1.1.0/30 is directly connected, FastEthernet 0/24
C    10.1.1.2/32 is local host.
C    192.168.100.0/24 is directly connected, VLAN 100
C    192.168.100.1/32 is local host.
C    192.168.101.0/24 is directly connected, VLAN 101
C    192.168.101.1/32 is local host.
C    192.168.102.0/24 is directly connected, VLAN 102
C    192.168.102.1/32 is local host.
C    192.168.103.0/24 is directly connected, VLAN 103
C    192.168.103.1/32 is local host.
C    192.168.104.0/24 is directly connected, VLAN 104
C    192.168.104.1/32 is local host.
```

VPN 的验证测试

RG-RSR20-2#ping

Protocol [ip]:

Target IP address: 192.168.100.1

Repeat count [5]:

Datagram size [100]:

Timeout in seconds [2]:

Extended commands [n]: y

Source address:192.168.99.1

Time to Live [1, 64]:

Type of service [0, 31]:

```
Data Pattern [0xABCD]:0xabcd
Sending 5, 100-byte ICMP Echoes to 192.168.100.1, timeout is 2 seconds:
  < press Ctrl+C to break >
!!!!!
Success rate is 100 percent (5/5), round-trip min/avg/max = 120/168/230 ms
RG-RSR20-2#show crypto ipsec sa
Interface: serial 4/0
        Crypto map tag:site-to-site, local addr 99.1.1.2
        media mtu 1500
        ==================================
        item type:static, seqno:10, id=32
        local  ident (addr/mask/prot/port): (192.168.99.0/0.0.0.255/0/0))
        remote  ident (addr/mask/prot/port): (192.168.100.0/0.0.0.255/0/0))
        PERMIT
        #pkts encaps: 5, #pkts encrypt: 5, #pkts digest 10
        #pkts decaps: 5, #pkts decrypt: 5, #pkts verify 10
        #send errors 0, #recv errors 0
        Inbound esp sas:
            spi:0x4f0db14a (1326297418)
             transform: esp-des esp-md5-hmac
             in use settings={Tunnel,}
             crypto map site-to-site 10
             sa timing: remaining key lifetime (k/sec): (4606998/960)
             IV size: 8 bytes
             Replay detection support:Y

        Inbound ah sas:
            spi:0x27d22a8d (668084877)
             transform: ah-null ah-md5-hmac
             in use settings={Tunnel,}
             crypto map site-to-site 10
             sa timing: remaining key lifetime (k/sec): (4606998/960)
             IV size: 0 bytes
             Replay detection support:Y
        Outbound esp sas:
            spi:0x5807d305 (1476907781)
             transform: esp-des esp-md5-hmac
             in use settings={Tunnel,}
             crypto map site-to-site 10
             sa timing: remaining key lifetime (k/sec): (4606998/960)
             IV size: 8 bytes
             Replay detection support:Y
```

```
        Outbound ah sas:
            spi:0x30ce9b27 (818846503)
             transform: ah-null ah-md5-hmac
             in use settings={Tunnel,}
             crypto map site-to-site 10
             sa timing: remaining key lifetime (k/sec): (4606998/960)
             IV size: 0 bytes
             Replay detection support:Y
RG-RSR20-2#sh crypto isakmp sa
 destination     source          state       conn-id        lifetime(second)
 99.1.1.1       99.1.1.2       QM IDLE         33              76641
  96591ce0a3672aed  e767e768e869e969
```

验证 VLAN 配置。

```
RG-S3760-2#show vlan
VLAN Name                        Status    Ports
---- ------------------------- --------- ----------------------------------
   1 VLAN0001                 STATIC    Fa0/1, Fa0/2, Fa0/3, Fa0/4
                                       Fa0/7, Fa0/8, Fa0/9, Fa0/10
                                       Fa0/11, Fa0/12, Fa0/13, Fa0/14
                                       Fa0/15, Fa0/16, Fa0/17, Fa0/18
                                         Fa0/19, Fa0/20, Fa0/21, Fa0/22
                                         Fa0/23, Gi0/25, Gi0/26, Gi0/27
                                         Gi0/28, Ag1
 100 VLAN0100                            STATIC    Fa0/1, Fa0/2, Ag1
 101 VLAN0101                            STATIC    Fa0/1, Fa0/2, Ag1
 102 VLAN0102                            STATIC    Fa0/1, Fa0/2, Ag1
 103 VLAN0103                            STATIC    Fa0/1, Fa0/2, Ag1
 104 VLAN0104                            STATIC    Fa0/1, Fa0/2, Ag1
```

验证生成树的配置。

```
RG-S3760-1#show spanning-tree
StpVersion : RSTP
SysStpStatus : ENABLED
MaxAge : 20
HelloTime : 2
ForwardDelay : 15
BridgeMaxAge : 20
BridgeHelloTime : 2
BridgeForwardDelay : 15
MaxHops: 20
TxHoldCount : 3
PathCostMethod : Long
BPDUGuard : Disabled
```

```
BPDUFilter : Disabled
BridgeAddr : 00d0.f806.1c91
Priority: 4096
TimeSinceTopologyChange : 0d:2h:41m:46s
TopologyChanges : 4
DesignatedRoot : 1000.00d0.f806.1c91
RootCost : 0
RootPort : 0
RG-S3760-2#show spanning-tree
StpVersion : RSTP
SysStpStatus : ENABLED
MaxAge : 20
HelloTime : 2
ForwardDelay : 15
BridgeMaxAge : 20
BridgeHelloTime : 2
BridgeForwardDelay : 15
MaxHops: 20
TxHoldCount : 3
PathCostMethod : Long
BPDUGuard : Disabled
BPDUFilter : Disabled
BridgeAddr : 00d0.f8b4.e54b
Priority: 8192
TimeSinceTopologyChange : 0d:2h:41m:32s
TopologyChanges : 4
DesignatedRoot : 1000.00d0.f806.1c91
RootCost : 190000
RootPort : 29
```

验证链路聚合。

```
RG-S3760-1#sh aggregatePort 1 summary
AggregatePort MaxPorts SwitchPort Mode   Ports
------------- -------- ---------- ------ -----------------------------------
Ag1           8        Enabled    TRUNK  Fa0/5   ,Fa0/6
RG-S3760-1#show interface aggregateport 1
Index(dec):29 (hex):1d
AggregatePort 1 is UP  , line protocol is UP
Hardware is Aggregate Link AggregatePort
Interface address is: no ip address
  MTU 1500 bytes, BW 1000000 Kbit
  Encapsulation protocol is Bridge, loopback not set
  Keepalive interval is 10 sec , set
```

```
  Carrier delay is 2 sec
  RXload is 1 ,Txload is 1
  Queueing strategy: WFQ
  Switchport attributes:
    interface's description:""
    medium-type is copper
    lastchange time:345 Day:17 Hour:15 Minute:23 Second
    Priority is 0
    admin duplex mode is AUTO, oper duplex is Full
    admin speed is AUTO, oper speed is 100Mbps
    flow control admin status is AUTO,flow control oper status is ON
    broadcast Strom Control is OFF,multicast Strom Control is OFF,unicast Strom
Control is OFF
Aggregate Port Informations:
        Aggregate Number: 1
        Name: "AggregatePort 1"
        Refs: 2
        Members: (count=2)
        FastEthernet 0/5 Link Status: Up
        FastEthernet 0/6 Link Status: Up
```

验证访问控制。

```
RG-RSR20-1#show access-lists
ip access-list standard 11
 10 permit 192.168.102.0 0.0.0.255
 20 permit 192.168.103.0 0.0.0.255
ip access-list standard 12
 10 permit 192.168.100.0 0.0.0.255
 20 permit 192.168.101.0 0.0.0.255
30 permit 192.168.102.0 0.0.0.255
 40 permit 192.168.103.0 0.0.0.255
ip access-list extended 100
 10 deny ip 192.168.100.0 0.0.0.255 192.168.99.0 0.0.0.255
 20 permit ip 192.168.100.0 0.0.0.255 any
ip access-list extended 110
 10 permit ip 192.168.100.0 0.0.0.255 192.168.99.0 0.0.0.255
RG-RSR20-1#sh ip access-group
ip access-group 12 in
Applied On interface FastEthernet 0/0.
Applied On interface FastEthernet 0/1.
```

验证网络地址转换。

在外部网络中使用 PC 访问 FTP 和 WWW 网站，在 IE 浏览器中输入 http://99.1.1.9，可以访问内网的网站，如图 28-13 所示。

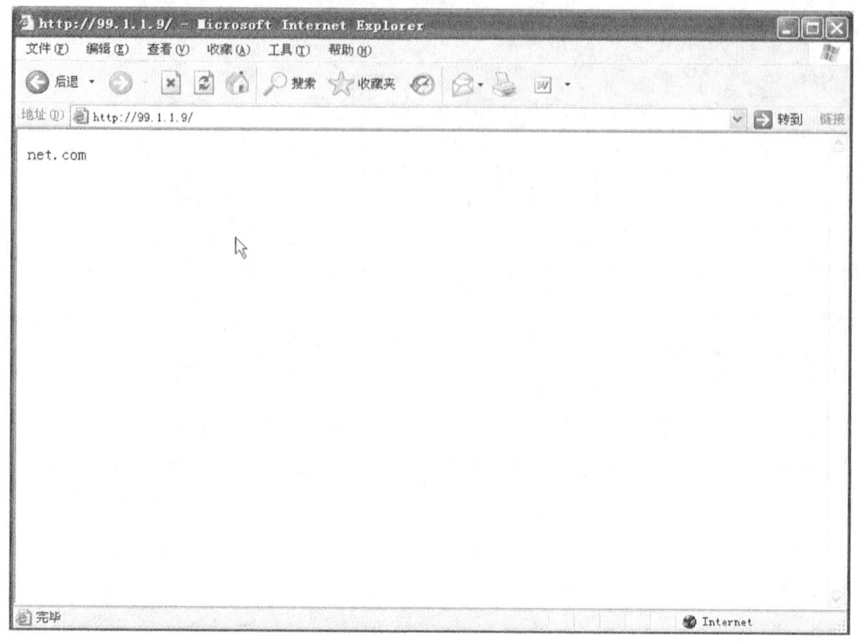

图 28-13

在 IE 浏览器中输入 ftp://99.1.1.11，可以访问内网的 FTP 服务，如图 28-14 所示。

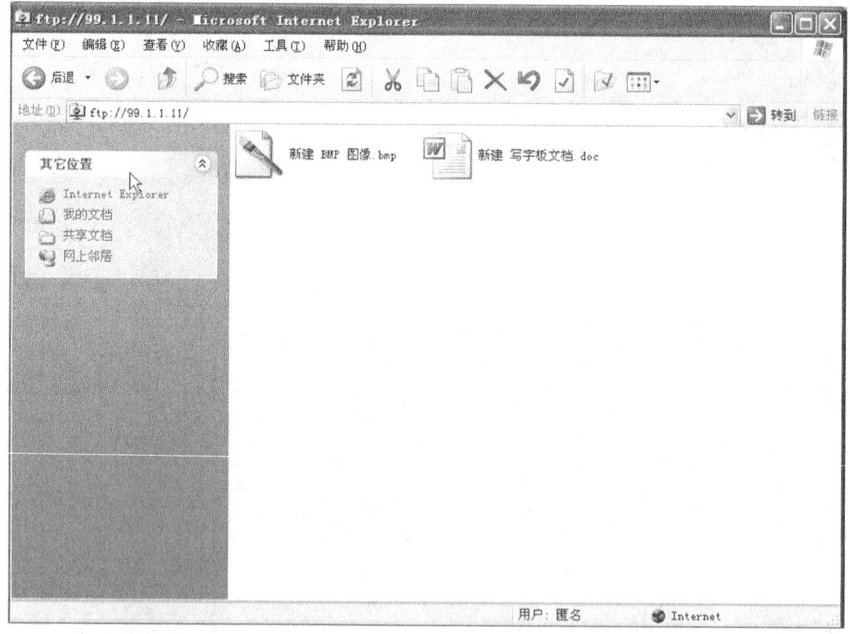

图 28-14

【参考配置】

```
RG-RSR20-1#show running-config
Building configuration...
Current configuration : 2300 bytes
!
version RGNOS 10.1.00(4), Release(18443)(Tue Jul 17 21:16:17 CST 2007
```

```
-ubu1server)
hostname RG-RSR20-1
!
ip access-list standard 11
 10 permit 192.168.102.0 0.0.0.255
 20 permit 192.168.103.0 0.0.0.255
!
ip access-list standard 12
 10 permit 192.168.100.0 0.0.0.255
 20 permit 192.168.101.0 0.0.0.255
30 permit 192.168.102.0 0.0.0.255
 40 permit 192.168.103.0 0.0.0.255
!
ip access-list extended 100
 10 deny ip 192.168.100.0 0.0.0.255 192.168.99.0 0.0.0.255
 20 permit ip 192.168.100.0 0.0.0.255 any
!
ip access-list extended 110
 10 permit ip 192.168.100.0 0.0.0.255 192.168.99.0 0.0.0.255
!
crypto isakmp policy 110
 authentication pre-share
 hash md5
 group 2
!
crypto isakmp key 7 15581828182229 address 99.1.1.2
crypto ipsec transform-set vpn  ah-md5-hmac esp-des esp-md5-hmac
crypto map site-to-site 10 ipsec-isakmp
 set peer 99.1.1.2
 set transform-set vpn
 match address 110
!
interface serial 4/0
 ip nat outside
 ip address 99.1.1.1 255.255.255.240
 crypto map site-to-site
!
interface serial 4/1
 clock rate 64000
!
interface FastEthernet 0/0
```

```
 ip nat inside
ip access-group 12 in
 ip address 10.1.1.1 255.255.255.252
 duplex auto
 speed auto
!
interface FastEthernet 0/1
 ip nat inside
ip access-group 12 in
 ip address 10.1.1.5 255.255.255.252
 duplex auto
 speed auto
!
ip nat pool internet 99.1.1.3 99.1.1.5 netmask 255.255.255.240
ip nat pool internet1 99.1.1.6 99.1.1.8 netmask 255.255.255.240
ip nat inside source static tcp 192.168.104.254 80 99.1.1.9 80 permit-inside
ip nat inside source static tcp 192.168.104.253 25 99.1.1.10 25 permit-inside
ip nat inside source static tcp 192.168.104.253 110 99.1.1.10 110 permit-inside
ip nat inside source static tcp 192.168.104.252 20 99.1.1.11 20 permit-inside
ip nat inside source static tcp 192.168.104.252 21 99.1.1.11 21 permit-inside
ip nat inside source list 11 pool internet1 overload
ip nat inside source list 100 pool internet overload
!
router ospf 10
 network 10.1.1.0 0.0.0.3 area 0
 network 10.1.1.4 0.0.0.3 area 0
 default-information originate
!
ip route  0.0.0.0 0.0.0.0  99.1.1.2
!
line con 0
line aux 0
line vty 0 4
 login
!
end
```
RG-RSR20-2#show running-config
```
Building configuration...
Current configuration : 1208 bytes
!
version  RGNOS  10.1.00(4),  Release(18443)(Tue  Jul  17  21:16:17  CST  2007
```

```
-ubu1server)
hostname RG-RSR20-2
!
ip access-list extended 110
 10 permit ip 192.168.99.0 0.0.0.255 192.168.100.0 0.0.0.255
!
enable password 7 13544019
!
crypto isakmp policy 110
 authentication pre-share
 hash md5
 group 2
!
crypto isakmp key 7 1007072e0f1b4f address 99.1.1.1
crypto ipsec transform-set vpn  ah-md5-hmac esp-des esp-md5-hmac
crypto map site-to-site 10 ipsec-isakmp
 set peer 99.1.1.1
 set transform-set vpn
 match address 110
!
interface serial 1/0
 ip address 99.1.1.2 255.255.255.240
 crypto map site-to-site
 clock rate 64000
!
interface serial 1/1
 clock rate 64000
!
interface FastEthernet 0/0
 duplex auto
 speed auto
!
interface FastEthernet 0/1
 duplex auto
 speed auto
!
interface Loopback 0
 ip address 192.168.99.1 255.255.255.0
!
ip route  0.0.0.0 0.0.0.0  99.1.1.1
!
```

```
line con 0
line aux 0
line vty 0 4
 login
 password 7 11437556
!
end
```

RG-S3760-1#show running-config

```
Building configuration...
Current configuration : 2100 bytes
!
version  RGNOS  10.1.00(4),  Release(18437)(Tue  Jul  17  19:13:23  CST  2007
-ubu1server)
hostname RG-S3760-1
!
vlan 1
!
vlan 100
!
vlan 101
!
vlan 102
!
vlan 103
!
vlan 104
!
spanning-tree
spanning-tree mode rstp
spanning-tree mst 0 priority 4096
interface FastEthernet 0/1
 switchport mode trunk
!
interface FastEthernet 0/2
 switchport mode trunk
!
interface FastEthernet 0/3
!
interface FastEthernet 0/4
 switchport access vlan 104
!
```

```
interface FastEthernet 0/5
 port-group 1
!
interface FastEthernet 0/6
 port-group 1
!
interface FastEthernet 0/7
!
interface FastEthernet 0/8
!
interface FastEthernet 0/9
!
interface FastEthernet 0/10
!
interface FastEthernet 0/11
!
interface FastEthernet 0/12
!
interface FastEthernet 0/13
!
interface FastEthernet 0/14
!
interface FastEthernet 0/15
!
interface FastEthernet 0/16
!
interface FastEthernet 0/17
!
interface FastEthernet 0/18
!
interface FastEthernet 0/19
!
interface FastEthernet 0/20
!
interface FastEthernet 0/21
!
interface FastEthernet 0/22
!
interface FastEthernet 0/23
!
interface FastEthernet 0/24
```

```
 no switchport
 ip address 10.1.1.2 255.255.255.252
!
interface GigabitEthernet 0/25
!
interface GigabitEthernet 0/26
!
interface GigabitEthernet 0/27
!
interface GigabitEthernet 0/28
!
interface AggregatePort 1
 switchport mode trunk

 medium-type copper
!
interface VLAN 100
 ip address 192.168.100.1 255.255.255.0
!
interface VLAN 101
 ip address 192.168.101.1 255.255.255.0
!
interface VLAN 102
 ip address 192.168.102.1 255.255.255.0
!
interface VLAN 103
 ip address 192.168.103.1 255.255.255.0
!
interface VLAN 104
 ip address 192.168.104.1 255.255.255.0
!
router ospf 10
 network 10.1.1.0 0.0.0.3 area 0
 network 192.168.100.0 0.0.0.255 area 0
 network 192.168.101.0 0.0.0.255 area 0
 network 192.168.102.0 0.0.0.255 area 0
 network 192.168.103.0 0.0.0.255 area 0
 network 192.168.104.0 0.0.0.255 area 0
!
line con 0
line vty 0 4
```

```
 login
!
end
```

RG-S3760-2#show running-config

```
Building configuration...
Current configuration : 2071 bytes
!
version  RGNOS  10.1.00(4),  Release(18437)(Tue  Jul  17  19:13:23  CST  2007
-ubu1server)
hostname RG-S3760-2
!
vlan 1
!
vlan 100
!
vlan 101
!
vlan 102
!
vlan 103
!
vlan 104
!
spanning-tree
spanning-tree mode rstp
spanning-tree mst 0 priority 8192
interface FastEthernet 0/1
 switchport mode trunk
!
interface FastEthernet 0/2
 switchport mode trunk
!
interface FastEthernet 0/3
!
interface FastEthernet 0/4
!
interface FastEthernet 0/5
 port-group 1
!
interface FastEthernet 0/6
 port-group 1
```

```
!
interface FastEthernet 0/7
!
interface FastEthernet 0/8
!
interface FastEthernet 0/9
!
interface FastEthernet 0/10
!
interface FastEthernet 0/11
!
interface FastEthernet 0/12
!
interface FastEthernet 0/13
!
interface FastEthernet 0/14
!
interface FastEthernet 0/15
!
interface FastEthernet 0/16
!
interface FastEthernet 0/17
!
interface FastEthernet 0/18
!
interface FastEthernet 0/19
!
interface FastEthernet 0/20
!
interface FastEthernet 0/21
!
interface FastEthernet 0/22
!
interface FastEthernet 0/23
!
interface FastEthernet 0/24
 no switchport
 ip address 10.1.1.6 255.255.255.252
!
interface GigabitEthernet 0/25
!
```

```
interface GigabitEthernet 0/26
!
interface GigabitEthernet 0/27
!
interface GigabitEthernet 0/28
!
interface AggregatePort 1
 switchport mode trunk

 medium-type copper
!
interface VLAN 100
 ip address 192.168.100.2 255.255.255.0
!
interface VLAN 101
 ip address 192.168.101.2 255.255.255.0
!
interface VLAN 102
 ip address 192.168.102.2 255.255.255.0
!
interface VLAN 103
 ip address 192.168.103.2 255.255.255.0
!
interface VLAN 104
 ip address 192.168.104.2 255.255.255.0
!
router ospf 10
 network 10.1.1.4 0.0.0.3 area 0
 network 192.168.100.0 0.0.0.255 area 0
 network 192.168.101.0 0.0.0.255 area 0
 network 192.168.102.0 0.0.0.255 area 0
 network 192.168.103.0 0.0.0.255 area 0
 network 192.168.104.0 0.0.0.255 area 0
!
line con 0
line vty 0 4
 login
!
end
```

RG-S2126G-1#show running-config

```
Building configuration...
```

```
Current configuration : 4855 bytes
!
version  RGNOS  10.1.00(4),  Release(18443)(Tue  Jul  17  19:51:54  CST  2007
-ubu6server)
hostname RG-S2126G-1
!
vlan 1
!
vlan 100
!
vlan 101
!
spanning-tree
spanning-tree mode rstp
interface FastEthernet 0/1
 switchport mode trunk
!
interface FastEthernet 0/2
 switchport mode trunk
!
interface FastEthernet 0/3
 switchport access vlan 100
 switchport port-security maximum 2
 switchport port-security violation shutdown
 switchport port-security
 spanning-tree portfast
!
interface FastEthernet 0/4
 switchport access vlan 100
 switchport port-security maximum 2
 switchport port-security violation shutdown
 switchport port-security
 spanning-tree portfast
!
interface FastEthernet 0/5
 switchport access vlan 100
 switchport port-security maximum 2
 switchport port-security violation shutdown
 switchport port-security
 spanning-tree portfast
!
```

```
interface FastEthernet 0/6
 switchport access vlan 100
 switchport port-security maximum 2
 switchport port-security violation shutdown
 switchport port-security
 spanning-tree portfast
!
interface FastEthernet 0/7
 switchport access vlan 100
 switchport port-security maximum 2
 switchport port-security violation shutdown
 switchport port-security
 spanning-tree portfast
!
interface FastEthernet 0/8
 switchport access vlan 100
 switchport port-security maximum 2
 switchport port-security violation shutdown
 switchport port-security
 spanning-tree portfast
!
interface FastEthernet 0/9
 switchport access vlan 100
 switchport port-security maximum 2
 switchport port-security violation shutdown
 switchport port-security
 spanning-tree portfast
!
interface FastEthernet 0/10
 switchport access vlan 100
 switchport port-security maximum 2
 switchport port-security violation shutdown
 switchport port-security
 spanning-tree portfast
!
interface FastEthernet 0/11
 switchport access vlan 100
 switchport port-security maximum 2
 switchport port-security violation shutdown
 switchport port-security
 spanning-tree portfast
```

```
!
interface FastEthernet 0/12
 switchport access vlan 100
 switchport port-security maximum 2
 switchport port-security violation shutdown
 switchport port-security
 spanning-tree portfast
!
interface FastEthernet 0/13
 switchport access vlan 100
 switchport port-security maximum 2
 switchport port-security violation shutdown
 switchport port-security
 spanning-tree portfast
!
interface FastEthernet 0/14
 switchport access vlan 100
 switchport port-security maximum 2
 switchport port-security violation shutdown
 switchport port-security
 spanning-tree portfast
!
interface FastEthernet 0/15
 switchport access vlan 101
 switchport port-security maximum 3
 switchport port-security violation shutdown
 switchport port-security
 spanning-tree portfast
!
interface FastEthernet 0/16
 switchport access vlan 101
 switchport port-security maximum 3
 switchport port-security violation shutdown
 switchport port-security
 spanning-tree portfast
!
interface FastEthernet 0/17
 switchport access vlan 101
 switchport port-security maximum 3
 switchport port-security violation shutdown
 switchport port-security
```

```
 spanning-tree portfast
!
interface FastEthernet 0/18
 switchport access vlan 101
 switchport port-security maximum 3
 switchport port-security violation shutdown
 switchport port-security
 spanning-tree portfast
!
interface FastEthernet 0/19
 switchport access vlan 101
 switchport port-security maximum 3
 switchport port-security violation shutdown
 switchport port-security
 spanning-tree portfast
!
interface FastEthernet 0/20
 switchport access vlan 101
 switchport port-security maximum 3
 switchport port-security violation shutdown
 switchport port-security
 spanning-tree portfast
!
interface FastEthernet 0/21
 switchport access vlan 101
 switchport port-security maximum 3
 switchport port-security violation shutdown
 switchport port-security
 spanning-tree portfast
!
interface FastEthernet 0/22
 switchport access vlan 101
 switchport port-security maximum 3
 switchport port-security violation shutdown
 switchport port-security
 spanning-tree portfast
!
interface FastEthernet 0/23
 switchport access vlan 101
 switchport port-security maximum 3
 switchport port-security violation shutdown
```

```
 switchport port-security
 spanning-tree portfast
!
interface FastEthernet 0/24
 switchport access vlan 101
 switchport port-security maximum 3
 switchport port-security violation shutdown
 switchport port-security
 spanning-tree portfast
!
line con 0
line vty 0 4
 login
!
end
RG-S21226G-2#show running-config
Building configuration...
Current configuration : 4586 bytes
!
version RGNOS 10.1.00(4), Release(18443)(Tue Jul 17 19:51:54 CST 2007
-ubu6server)
hostname RG-S21226G-2
!
vlan 1
!
vlan 102
!
vlan 103
!
spanning-tree
spanning-tree mode rstp
interface FastEthernet 0/1
 switchport mode trunk
!
interface FastEthernet 0/2
 switchport mode trunk
!
interface FastEthernet 0/3
 switchport access vlan 102
 switchport port-security maximum 2
 switchport port-security violation shutdown
```

```
 switchport port-security
 spanning-tree portfast
!
interface FastEthernet 0/4
 switchport access vlan 102
 switchport port-security maximum 2
 switchport port-security violation shutdown
 switchport port-security
 spanning-tree portfast
!
interface FastEthernet 0/5
 switchport access vlan 102
 switchport port-security maximum 2
 switchport port-security violation shutdown
 switchport port-security
 spanning-tree portfast
!
interface FastEthernet 0/6
 switchport access vlan 102
 switchport port-security maximum 2
 switchport port-security violation shutdown
 switchport port-security
 spanning-tree portfast
!
interface FastEthernet 0/7
 switchport access vlan 102
 switchport port-security maximum 2
 switchport port-security violation shutdown
 switchport port-security
 spanning-tree portfast
!
interface FastEthernet 0/8
 switchport access vlan 102
 switchport port-security maximum 2
 switchport port-security violation shutdown
 switchport port-security
 spanning-tree portfast
!
interface FastEthernet 0/9
 switchport access vlan 102
 switchport port-security maximum 2
```

```
 switchport port-security violation shutdown
 switchport port-security
 spanning-tree portfast
!
interface FastEthernet 0/10
 switchport access vlan 102
 switchport port-security maximum 2
 switchport port-security violation shutdown
 switchport port-security
 spanning-tree portfast
!
interface FastEthernet 0/11
 switchport access vlan 102
 switchport port-security maximum 2
 switchport port-security violation shutdown
 switchport port-security
 spanning-tree portfast
!
interface FastEthernet 0/12
 switchport access vlan 102
 switchport port-security maximum 2
 switchport port-security violation shutdown
 switchport port-security
 spanning-tree portfast
!
interface FastEthernet 0/13
 switchport access vlan 102
 switchport port-security maximum 2
 switchport port-security violation shutdown
 switchport port-security
 spanning-tree portfast
!
interface FastEthernet 0/14
 switchport access vlan 102
 switchport port-security maximum 2
 switchport port-security violation shutdown
 switchport port-security
 spanning-tree portfast
!
interface FastEthernet 0/15
 switchport access vlan 103
```

```
 switchport port-security maximum 3
 switchport port-security violation shutdown
 spanning-tree portfast
!
interface FastEthernet 0/16
 switchport access vlan 103
 switchport port-security maximum 3
 switchport port-security violation shutdown
 spanning-tree portfast
!
interface FastEthernet 0/17
 switchport access vlan 103
 switchport port-security maximum 3
 switchport port-security violation shutdown
 spanning-tree portfast
!
interface FastEthernet 0/18
 switchport access vlan 103
 switchport port-security maximum 3
 switchport port-security violation shutdown
 spanning-tree portfast
!
interface FastEthernet 0/19
 switchport access vlan 103
 switchport port-security maximum 3
 switchport port-security violation shutdown
 spanning-tree portfast
!
interface FastEthernet 0/20
 switchport access vlan 103
 switchport port-security maximum 3
 switchport port-security violation shutdown
 spanning-tree portfast
!
interface FastEthernet 0/21
 switchport access vlan 103
 switchport port-security maximum 3
 switchport port-security violation shutdown
 spanning-tree portfast
!
interface FastEthernet 0/22
```

```
 switchport access vlan 103
 switchport port-security maximum 3
 switchport port-security violation shutdown
 spanning-tree portfast
!
interface FastEthernet 0/23
 switchport access vlan 103
 switchport port-security maximum 3
 switchport port-security violation shutdown
 spanning-tree portfast
!
interface FastEthernet 0/24
 switchport access vlan 103
 switchport port-security maximum 3
 switchport port-security violation shutdown
 spanning-tree portfast
!
line con 0
line vty 0 4
 login
!
end
```

扩展实验 1 利用 TFTP 升级现有交换机操作系统

【实验名称】

利用 TFTP 升级现有交换机操作系统。

【实验目的】

能够利用 TFTP 升级现有交换机操作系统。

【背景描述】

你是锐捷网络产品的用户，现在你拥有锐捷交换机 S2126G 或 S3550-24，你的交换机的操作系统比较老，已经不能支持网络中新的功能的需要，为了满足网络需求，要升级交换机操作系统。

【需求分析】

计算机的 Com 口通过 Console 线缆与交换机的 Console 口连接在一起，同时计算机的网卡也通过直连线连接到交换机的以太网端口上，在计算机上安装 TFTP 服务器，通过 TFTP 升级交换机的操作系统。

【实验拓扑】

实验的拓扑图，如图 29-1 所示。

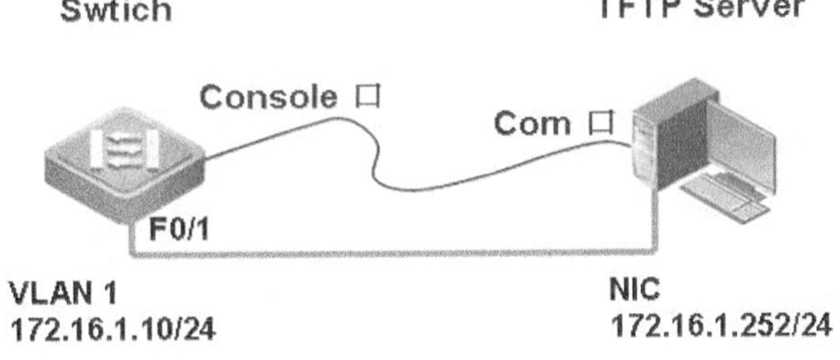

图 29-1

【实验设备】

交换机 1 台。

计算机 1 台。

【预备知识】

交换机的组成结构、TFTP 的操作方法。

【实验步骤】

步骤 1 检查交换机现有的操作系统版本。

```
Switch#show version
System description       : Red-Giant Gigabit Intelligent Switch(S2126G) By
                           Ruijie Network
System uptime            : 0d:0h:1m:42s
```

```
System hardware version : 3.3
System software version : 1.66 Build Jun 29 2006 Release
System BOOT version       : RG-S2126G-BOOT   03-02-02
System CTRL version       : RG-S2126G-CTRL   03-11-02
Running Switching Image : Layer2
```

从中可以看到，现在的交换机操作系统版本为 1.66。

步骤 2　运行 TFTP 服务器并将升级文件存放在 TFTP 服务器目录下。

安装 TFTP Server 的计算机 IP 地址为 172.16.1.252/24，将升级文件（s2126g-v1.68.bin）放置在 TFTP Server 的安装目录下，如图 29-2 所示。

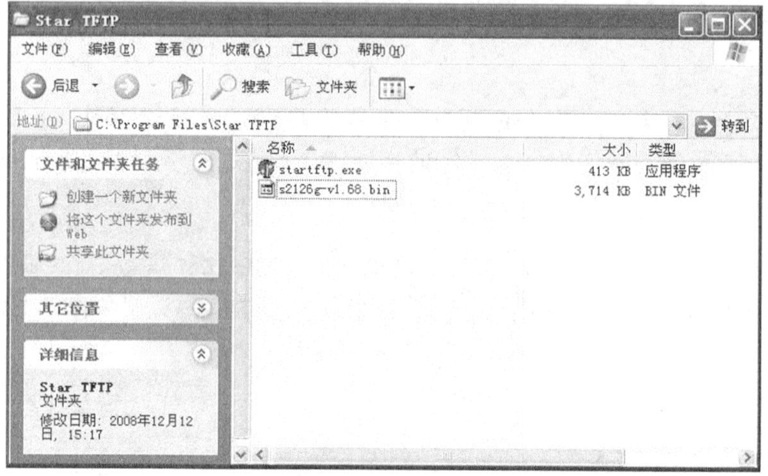

图 29-2

运行 TFTP Server，Star TFTP Server 的界面如图 29-3 所示。

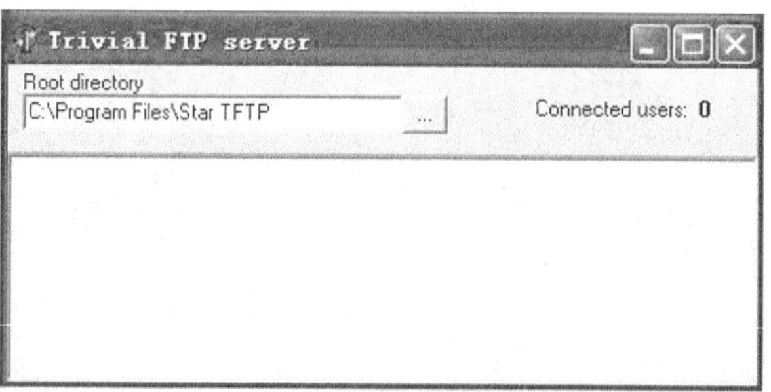

图 29-3

步骤 3　配置交换机管理 IP 地址。

```
Switch#configure  terminal
Switch(config)#interface vlan 1
Switch(config-if)#ip address 172.16.1.10 255.255.255.0
Switch(config-if)#no shutdown
Switch(config-if)#end
```

验证交换机的配置和 TFTP Server 的连通性。

```
Switch#ping 172.16.1.252
Sending 5, 100-byte ICMP Echos to 172.16.1.252,
```

```
timeout is 2000 milliseconds.
!!!!!
Success rate is 100 percent (5/5)
Minimum = 1ms Maximum = 5ms, Average = 2ms
Switch#show running-config
System software version : 1.66 Build Jun 29 2006 Release
Building configuration...
Current configuration : 131 bytes
!
version 1.0
!
hostname Switch
vlan 1
!
interface vlan 1
 no shutdown
 ip address 172.16.1.10 255.255.255.0
!
end
```

步骤 4　升级交换机操作系统。

使用 copy 命令从 TFTP Server 复制操作系统文件到 Flash，可以使用两种方式，第一种将 TFTP Server 的地址和升级文件名直接写入命令中。

```
Switch#copy tftp://172.16.1.252/s2126g-v1.68.bin flash:s2126g.bin
!!!!!!!!!!!!!!!!!!!!!!!!!!!!!!!!!!!!!!!!!!!!!!!!!!!!!!!!!!!!!!!!!!!!!!!!!!!!!!!!!!!!
!!!!!!!!!!!!!!!!!!!!!!!!!!!!!!!!!!!!!!!!!!!!!!!!!!!!!!!!!!!!!!!!!!!!!!!!!!!!!!!!!!!!
!!!!!!!!!!!!!!!!!!!!!!!!!!!!!!!!!!!!!!!!!!!!!!!!!!!!!!!!!!!!!!!!!!!!!!!!!!!!!!!!!!!!
!!!!!!!!!!!!!!!!!!!!!!!!!!!!!!!!!!!!!!!!!!!!!!!!!!!!!!!!!!!!!!!!!!!!!!!!!!!!!!!!!!!!
!!!!!!!!!!!!!!!!!!!!!!!!!!!!!!!!!!!!!!!!!!!!!!!!!!!!!!!!!!!!!!!!!!!!!!!!!!!!!!!!!!!!
!!!!!!!!!!!!!!!!!!!!!!!!!
%Success : Transmission success,file length 3802676
```

第二种可以不在命令中写入，而是等待提示输入。

```
Switch#copy tftp flash:s2126g.bin
Source filename []?s2126g-v1.68.bin
Address of remote host []172.16.1.252
!!!!!!!!!!!!!!!!!!!!!!!!!!!!!!!!!!!!!!!!!!!!!!!!!!!!!!!!!!!!!!!!!!!!!!!!!!!!!!!!!!!!
!!!!!!!!!!!!!!!!!!!!!!!!!!!!!!!!!!!!!!!!!!!!!!!!!!!!!!!!!!!!!!!!!!!!!!!!!!!!!!!!!!!!
!!!!!!!!!!!!!!!!!!!!!!!!!!!!!!!!!!!!!!!!!!!!!!!!!!!!!!!!!!!!!!!!!!!!!!!!!!!!!!!!!!!!
!!!!!!!!!!!!!!!!!!!!!!!!!!!!!!!!!!!!!!!!!!!!!!!!!!!!!!!!!!!!!!!!!!!!!!!!!!!!!!!!!!!!
!!!!!!!!!!!!!!!!!!!!!!!!!!!!!!!!!!!!!!!!!!!!!!!!!!!!!!!!!!!!!!!!!!
%Success : Transmission success,file length 3802676
```

步骤 5　重新启动交换机，验证结果。

```
Switch#reload
```

```
System configuration has been modified. Save? [yes/no]:n
Proceed with reload? [confirm]
```

使用 reload 命令重新启动交换机，待启动完毕可以验证操作系统版本，可以看到已经升级到了 1.68 版本。

```
Switch#show version
System description : Red-Giant Gigabit Intelligent Switch(S2126G) By
                     Ruijie Network
System uptime : 0d:0h:0m:33s
System hardware version : 3.3
System software version : 1.68 Build Apr 25 2007 Release
System BOOT version : RG-S2126G-BOOT  03-02-02
System CTRL version : RG-S2126G-CTRL  03-11-02
Running Switching Image : Layer2
```

【注意事项】

（1）保证交换机的管理 IP 地址和 TFTP 服务器的 IP 地址在同一个网段。

（2）确保在升级过程中不断电，否则可能造成操作系统被破坏。

扩展实验 2　利用 TFTP 升级现有路由器操作系统

【实验名称】

利用 TFTP 升级现有路由器操作系统。

【实验目的】

能够利用 TFTP 升级现有路由器操作系统。

【背景描述】

你是锐捷网络产品的用户，现在你拥有锐捷路由器 RSR20，你的路由器的操作系统比较老，已经不能支持网络中新的功能的需要，为了满足网络需求，要升级路由器操作系统。

【需求分析】

计算机的 Com 口通过 Console 线缆与路由器的 Console 口连接在一起，同时计算机的网卡也通过交叉线连接到路由器的以太网端口上，在计算机上安装 TFTP 服务器，通过 TFTP 升级路由器的操作系统。

【实验拓扑】

实验的拓扑图，如图 30-1 所示。

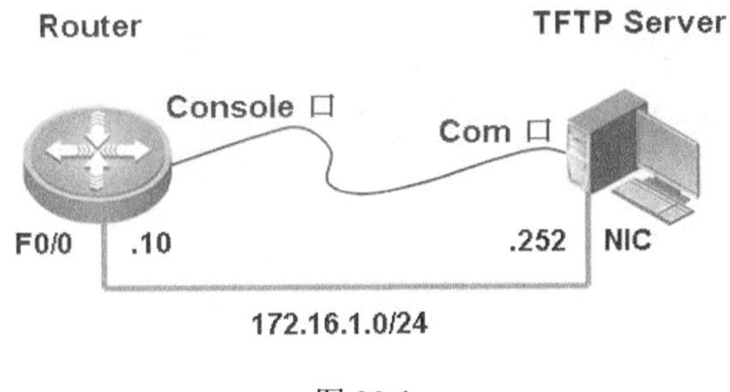

图 30-1

【实验设备】

路由器 1 台。

计算机 1 台。

【预备知识】

路由器的组成结构、TFTP 的操作方法。

【实验步骤】

步骤 1　检查现有路由器的操作系统版本。

```
RSR20#show version
System description      : Ruijie Router(RSR20-04) by Ruijie Network
System start time       : 2009-8-18 1:29:17
System hardware version : 1.0
```

```
System software version : RGNOS 10.1.00(4), Release(18443)
System boot version     : 10.1.17980
System serial number    : 1234942570002
```

从中可以看到，现在路由器使用的操作系统 RGNOS 的版本为 10.1。

步骤 2　运行 TFTP 服务器并将升级文件存放在 TFTP 服务器目录下。

安装 TFTP Server 的计算机 IP 地址为 172.16.1.252/24，将升级文件（RSR10_20_rgnos10.2(27335).bin）放置在 TFTP Server 的安装目录下，如图 30-2 所示。

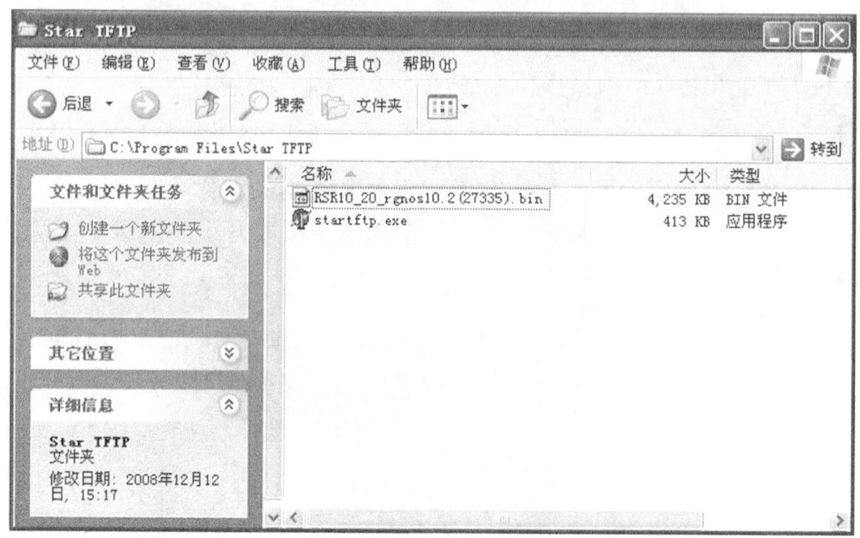

图 30-2

运行 TFTP Server，Star TFTP Server 的运行界面图 30-3 所示。

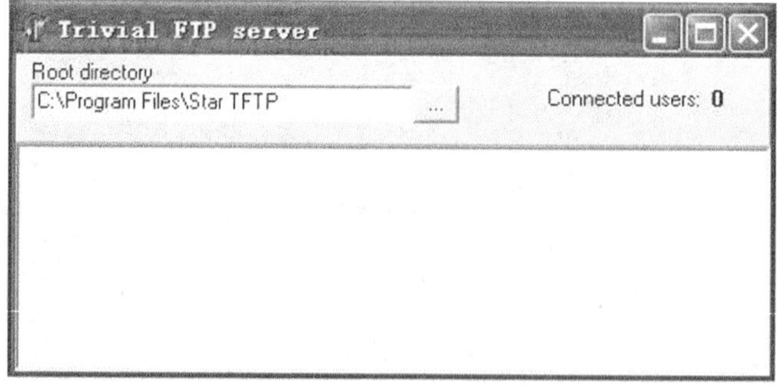

图 30-3

步骤 3　配置路由器接口 IP 地址。

```
RSR20#configure terminal
Enter configuration commands, one per line.  End with CNTL/Z.
RSR20(config)#hostname Router
Router(config)#interface fastEthernet 0/0
Router(config-if)#ip address 172.16.1.10 255.255.255.0
Router(config-if)#no shutdown
Router(config-if)#end
```

验证路由器的配置和 TFTP Server 的连通性。

```
Router#ping 172.16.1.252
Sending 5, 100-byte ICMP Echoes to 172.16.1.252, timeout is 2 seconds:
  < press Ctrl+C to break >
!!!!!
Success rate is 100 percent (5/5), round-trip min/avg/max = 1/2/10 ms
Router#show running-config
Building configuration...
Current configuration : 427 bytes
!
version RGNOS 10.1.00(4), Release(18443)(Tue Jul 17 20:50:30 CST 2007
-ubu1server)
hostname Router
!
!
interface FastEthernet 0/0
 ip address 172.16.1.10 255.255.255.0
 duplex auto
 speed auto
!
interface FastEthernet 0/1
 duplex auto
 speed auto
!
line con 0
line aux 0
line vty 0 4
 login
!
end
```

步骤 4　升级路由器操作系统。

```
Router#copy tftp flash
Address of remote host []?172.16.1.252
Source filename []?RSR10_20_rgnos10.2(27335).bin
Extended commands [n]:
Destination filename [RSR10_20_rgnos10.2(27335).bin]?rgnos.bin
Accessing tftp://172.16.1.252/RSR10 20 rgnos10.2(27335).bin...
!!!!!!!!!!!!!!!!!!!!!!!!!!!!!!!!!!!!!!!!!!!!!!!!!!!!!!!!!!!!!!!!!!!!!!!!!!!!
!!!!!!!!!!!!!!!!!!!!!!!!!!!!!!!!!!!!!!!!!!!!!!!!!!!!!!!!!!!!!!!!!!!!!!!!!!!!
!!!!!!!!!!!!!!!!!!!!!!!!!!!!!!!!!!!!!!!!!!!!!!!!!!!!!!!!!!!!!!!!!!!!!!!!!!!!
!!!!!!!!!!!!!!!!!!!!!!!!!!!!!!!!!!!!!!!!!!!!!!!!!!!!!!!!!!!!!!!
Success : Transmission success,file length 4335680
```

按照提示逐步输入 TFTP 服务器的 IP 地址、升级文件名和保存在路由器 Flash 中的文件名之后，即开始从 TFTP 服务器中复制升级文件到路由器中。

步骤 5　重新启动路由器，验证结果。

```
Router#reload
Processed with reload? [no]y
```

使用 reload 命令重新启动路由器，待启动完毕可以验证操作系统版本，还可以看到已经升级到了 RGNOS 10.2 版本。

```
Router#show version
System description      : Ruijie Router(RSR20-04) by Ruijie Network
System start time       : 2009-8-18 2:16:35
System hardware version : 1.0
System software version : RGNOS 10.2.00(2), Release(27335)
System boot version     : 10.2.24515
System serial number    : 1234942570002
```

【注意事项】

（1）要保证路由器和 TFTP 服务器的连通性。

（2）路由器如果和计算机直连要使用交叉线，除非路由器端口支持智能反转，即可以实现对直连线、交叉线的自适应。

（3）确保在升级过程中不断电，否则可能造成操作系统被破坏。

扩展实验 3　利用 ROM 方式重写交换机操作系统

【实验名称】

利用 ROM 方式重写交换机操作系统。

【实验目的】

当交换机操作系统丢失后，能够利用 ROM 方式重写交换机操作系统。

【背景描述】

某锐捷网络用户，有一台锐捷交换机 S2126G 或 S3550-24，现在交换机的操作系统因某种原因丢失，交换机不能正常工作，必须为交换机重新写入新的系统。

【需求分析】

计算机的 Com 口通过 Console 线缆与交换机的 Console 口连接在一起，同时计算机的网卡也通过直连线连接到交换机的以太网端口上，在计算机上安装 TFTP 服务器，通过 XModem 方式重写交换机的操作系统。

【实验拓扑】

实验的拓扑图，如图 31-1 所示。

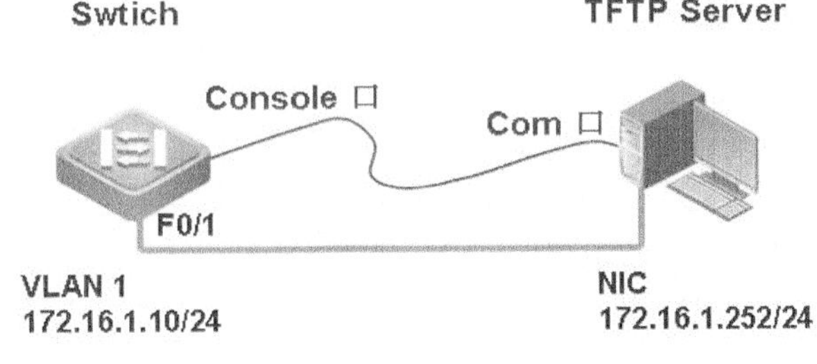

图 31-1

【实验设备】

交换机 1 台。

计算机 1 台。

【预备知识】

交换机的组成结构、TFTP 的操作方法。

【实验步骤】

步骤 1　启动 TFTP 服务器，保持 TFTP 服务器和交换机连接正常。

步骤 2　设置超级终端的每秒位数为 57600。

启动 TFTP 服务器上的超级终端，设置 Com 口属性的每秒位数为“57600”，数据位为“8”，奇偶校验为“无”，停止位为“1”，数据流控制为“无”，如图 31-2 所示。

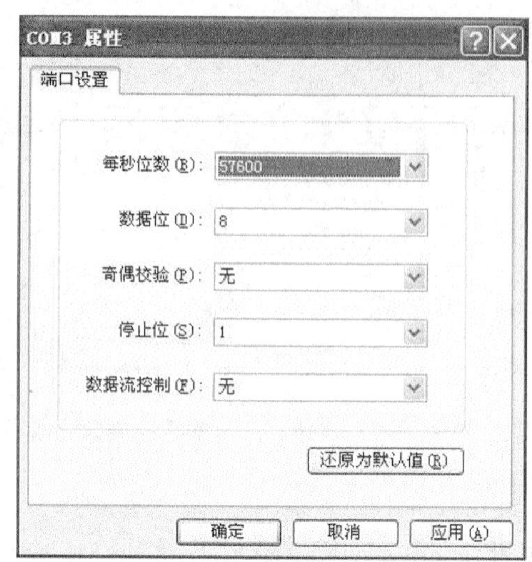

图 31-2

步骤 3　交换机加电。

交换机加电后立刻有节奏的按 Esc 键，会出现如图 31-3 所示的提示界面，此时输入“y”。

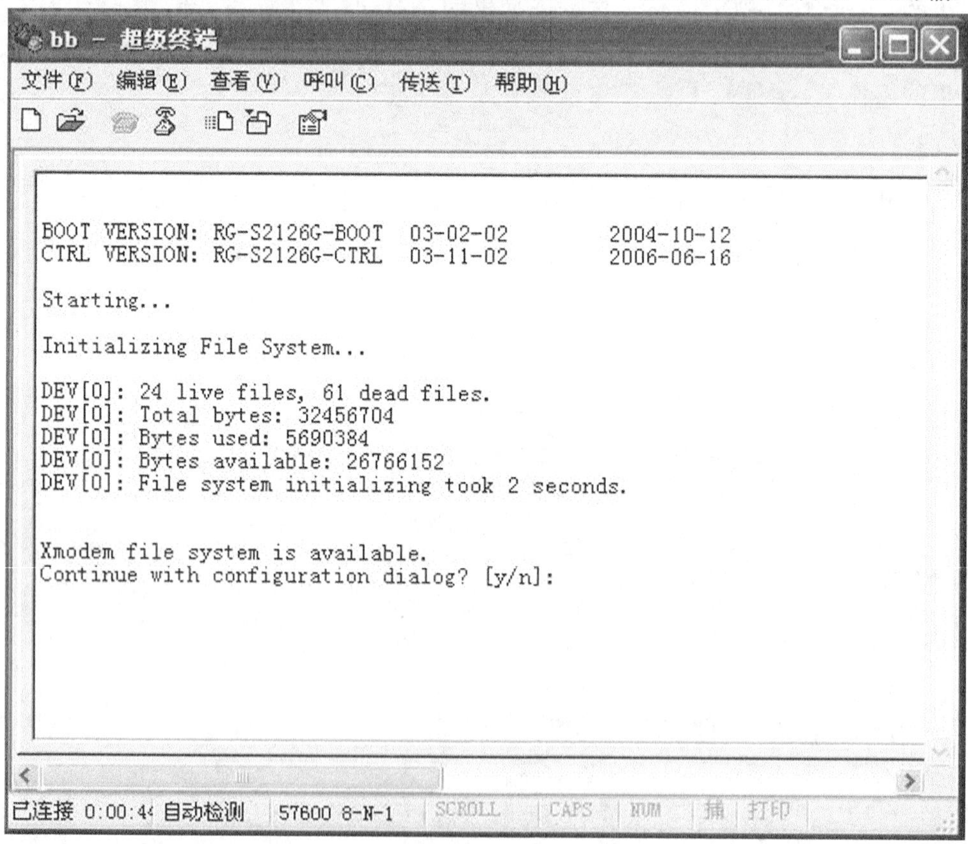

图 31-3

出现如图 31-4 所示的选项菜单，选择“1--Download”，并按照提示输入交换机操作系统文件的文件名。

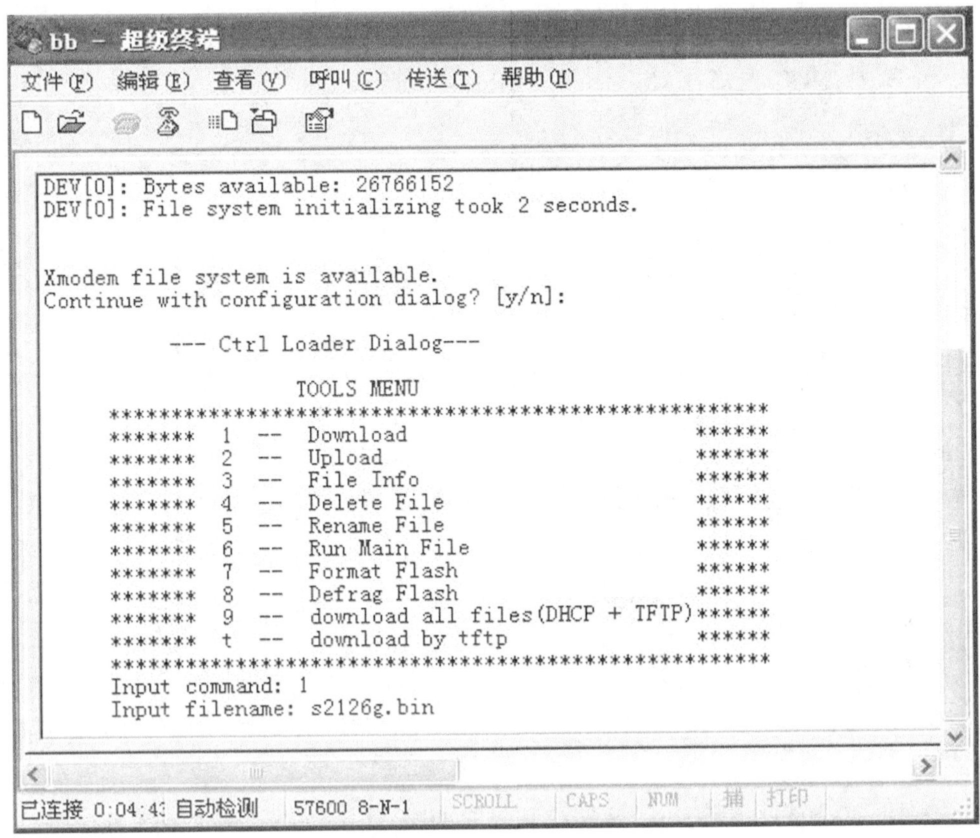

图 31-4

输入文件名后按 Enter 键，超级终端开始传送文件，在出现一连串“⊥”符号时，立刻在超级终端窗口的菜单中选择“传送”→“发送文件”命令，如图 31-5 所示。

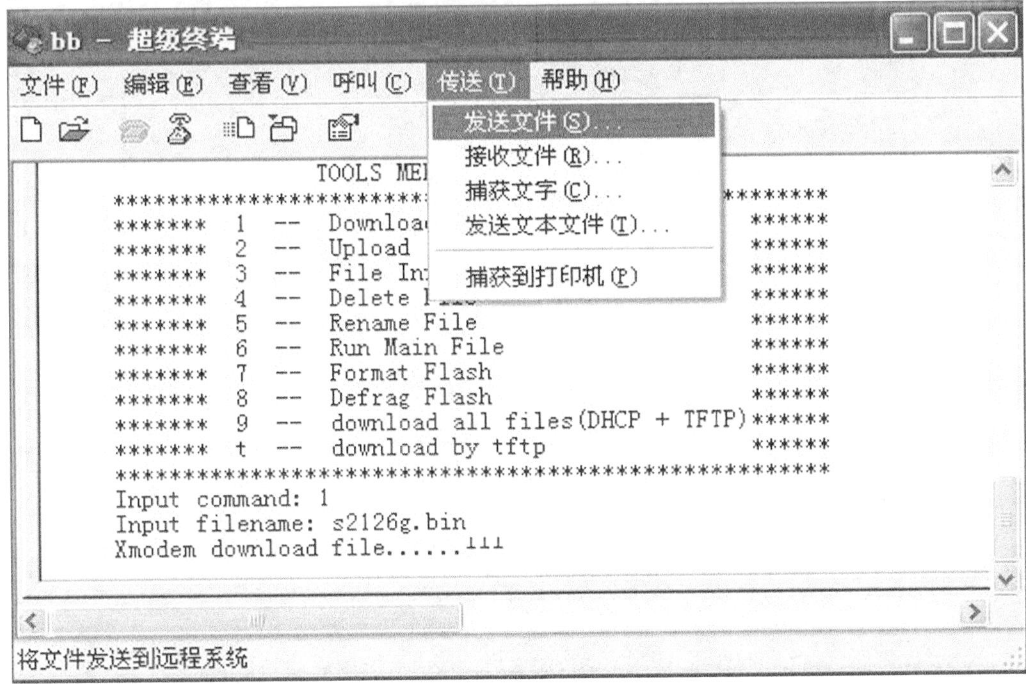

图 31-5

这时会弹出一个对话框，在“文件名”框中输入文件名和完整的路径（或通过“浏览”找到），在“协议”下拉列表框中选择“XModem”，如图 31-6 所示。

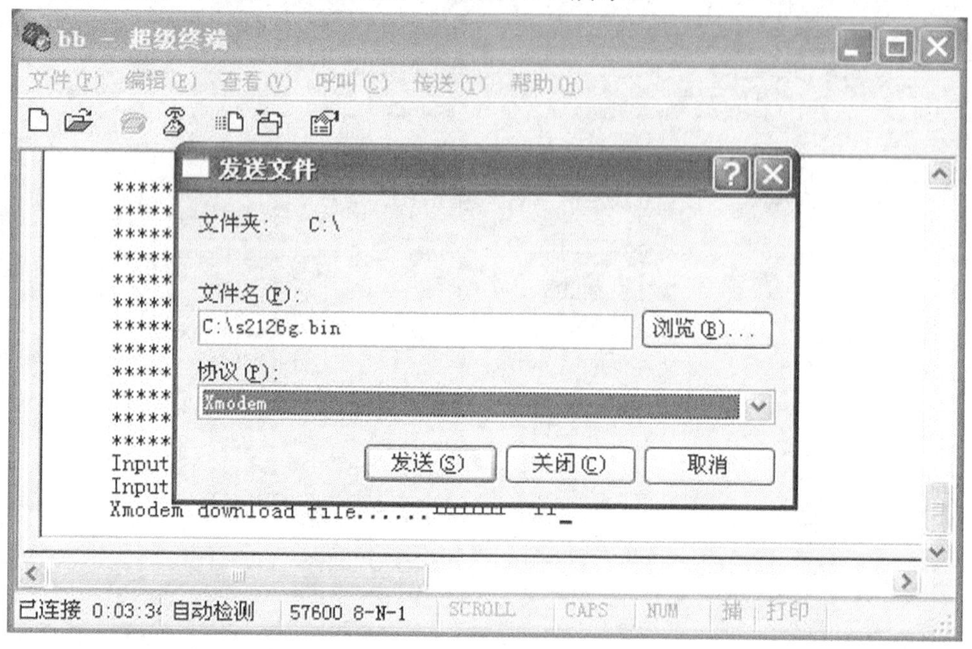

图 31-6

单击“发送”按钮后，开始传送文件，其界面如图 31-7 所示。整个传送过程用时较长，可以看到提示需要 36 分钟左右。

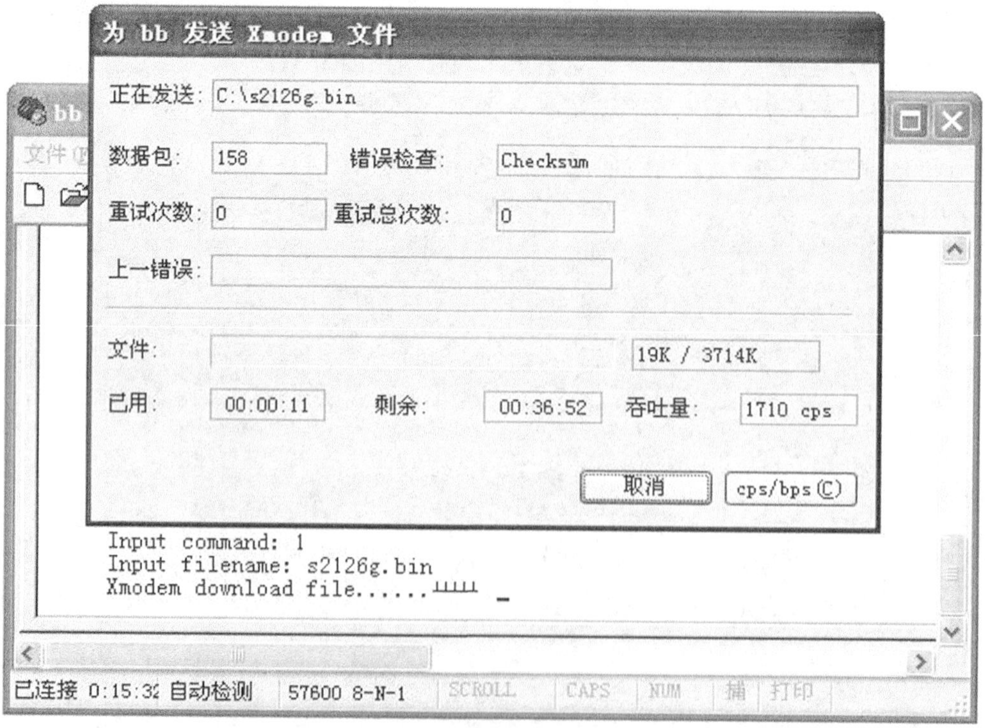

图 31-7

传输完成后，会显示“Download OK”的提示，如图 31-8 所示。

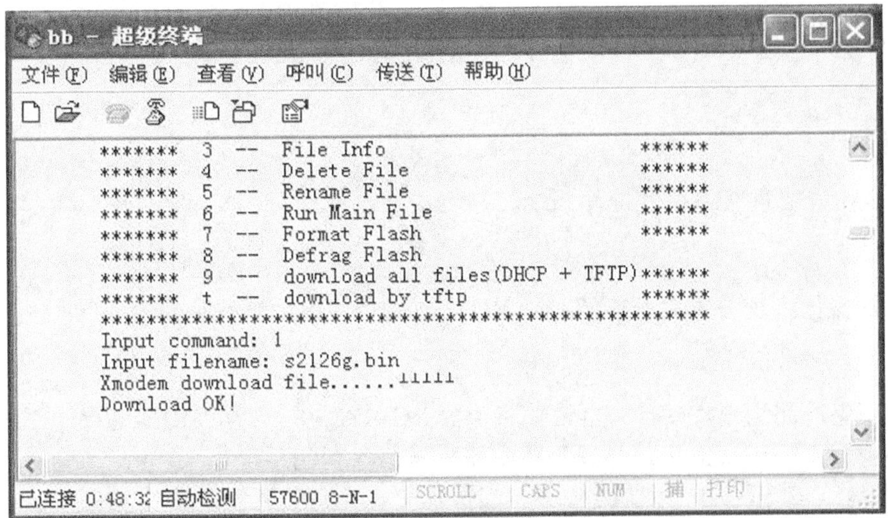

图 31-8

步骤 4　重新启动交换机，能够看到操作系统已经重新写入 Flash，即可正常启动和引导交换机。

【注意事项】

（1）这种方法速度较慢。

（2）建议将连接 TFTP 服务器的网线接在交换机的 0/1 端口。

（3）操作系统文件传输过程中，注意不能断电。

（4）操作系统传输完毕后，重新启动交换机。

扩展实验 4　利用 ROM 方式重写路由器操作系统

【实验名称】

利用 ROM 方式重写路由器操作系统。

【实验目的】

当路由器操作系统丢失后，能够利用 ROM 方式重写路由器操作系统。

【背景描述】

某锐捷网络产品用户，拥有一台锐捷路由器 RG-RSR20，现在路由器的操作系统因为某种原因丢失，路由器不能正常工作，必须为路由器重新写入新的系统。

【需求分析】

计算机的 Com 口通过 Console 线缆与路由器的 Console 口连接在一起，同时计算机的网卡也通过交叉线连接到路由器的以太网端口上，在计算机上安装 TFTP 服务器，通过 XModem 方式重写路由器的操作系统。

【实验拓扑】

实验的拓扑图，如图 32-1 所示。

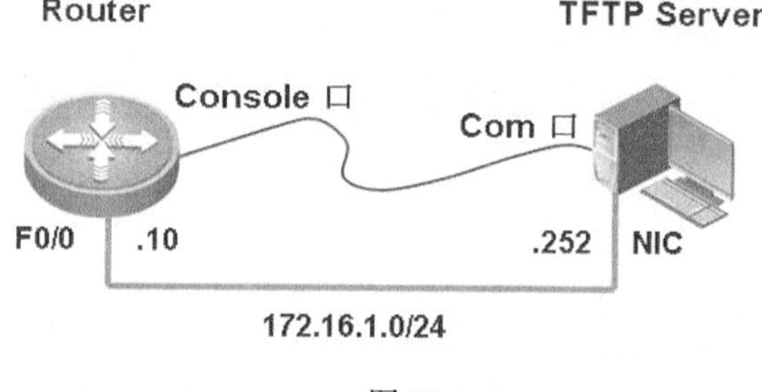

图 32-1

【实验设备】

路由器 1 台。
计算机 1 台。

【预备知识】

路由器的组成结构、TFTP 的操作方法。

【实验步骤】

步骤 1　启动 TFTP 服务器。

启动 TFTP 服务器，保持 TFTP 服务器和路由器正常连接。

步骤 2　设置超级终端中 Com 口属性。

启动 TFTP 服务器上的超级终端，设置 Com 口属性的每秒位数为“9600”，数据位为“8”，奇偶校验为“无”，停止位为“1”，数据流控制为“无”（或者单击“还原为默认值”按钮即可），如图 32-2 所示。

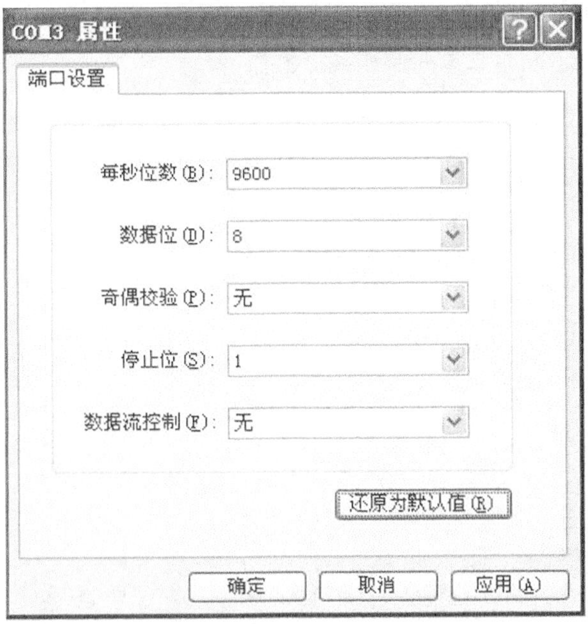

COM3 属性

端口设置

每秒位数(B): 9600

数据位(D): 8

奇偶校验(P): 无

停止位(S): 1

数据流控制(F): 无

还原为默认值(R)

确定　取消　应用(A)

图 32-2

步骤 3　路由器加电后立刻按 Ctrl+C 键，即可进入路由器的监控模式，如图 32-3 所示。

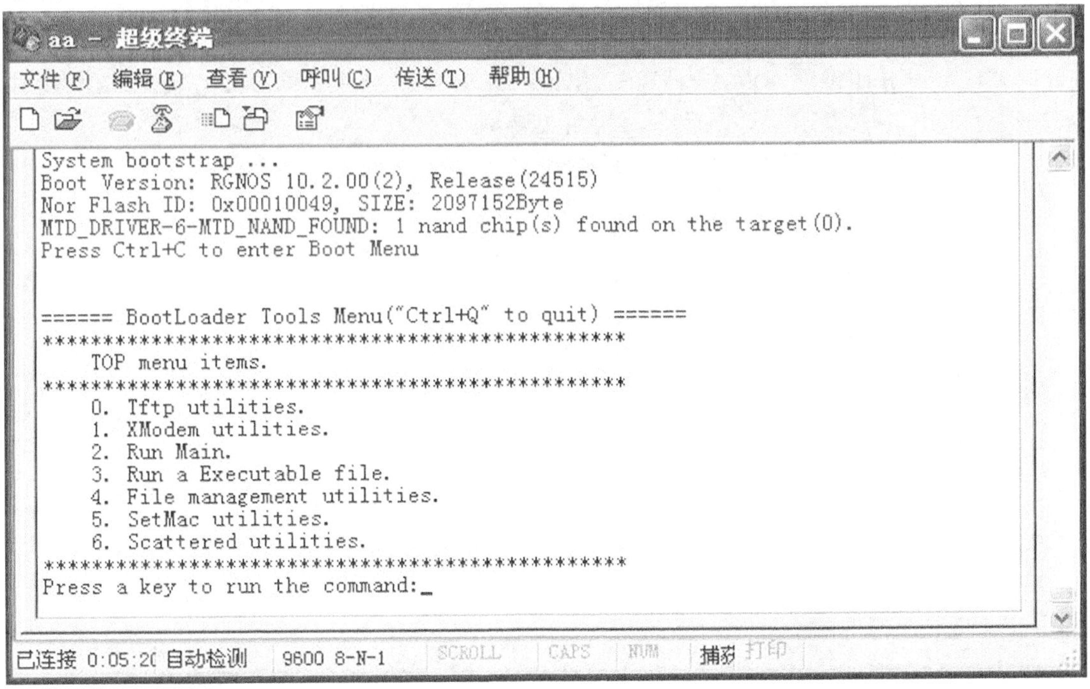

aa - 超级终端

文件(F)　编辑(E)　查看(V)　呼叫(C)　传送(T)　帮助(H)

```
System bootstrap ...
Boot Version: RGNOS 10.2.00(2), Release(24515)
Nor Flash ID: 0x00010049, SIZE: 2097152Byte
MTD_DRIVER-6-MTD_NAND_FOUND: 1 nand chip(s) found on the target(0).
Press Ctrl+C to enter Boot Menu


====== BootLoader Tools Menu("Ctrl+Q" to quit) ======
*************************************************
    TOP menu items.
*************************************************
    0. Tftp utilities.
    1. XModem utilities.
    2. Run Main.
    3. Run a Executable file.
    4. File management utilities.
    5. SetMac utilities.
    6. Scattered utilities.
*************************************************
Press a key to run the command:_
```

已连接 0:05:20 自动检测　9600 8-N-1　SCROLL　CAPS　NUM　捕获　打印

图 32-3

在监控模式的菜单中会给出 0，1，2，3，4…等命令选择键，在菜单中根据提示输入相应的命令选择键可执行相应的命令或者进入下一级菜单，使用 CTRL+Z 命令可从子菜单退到上一级菜单。

在这里选择“1”，使用 XModem 方式重写路由器的操作系统，在出现的子菜单中选择“1”，升级路由器的主程序，也就是操作系统，如图 32-4 所示。

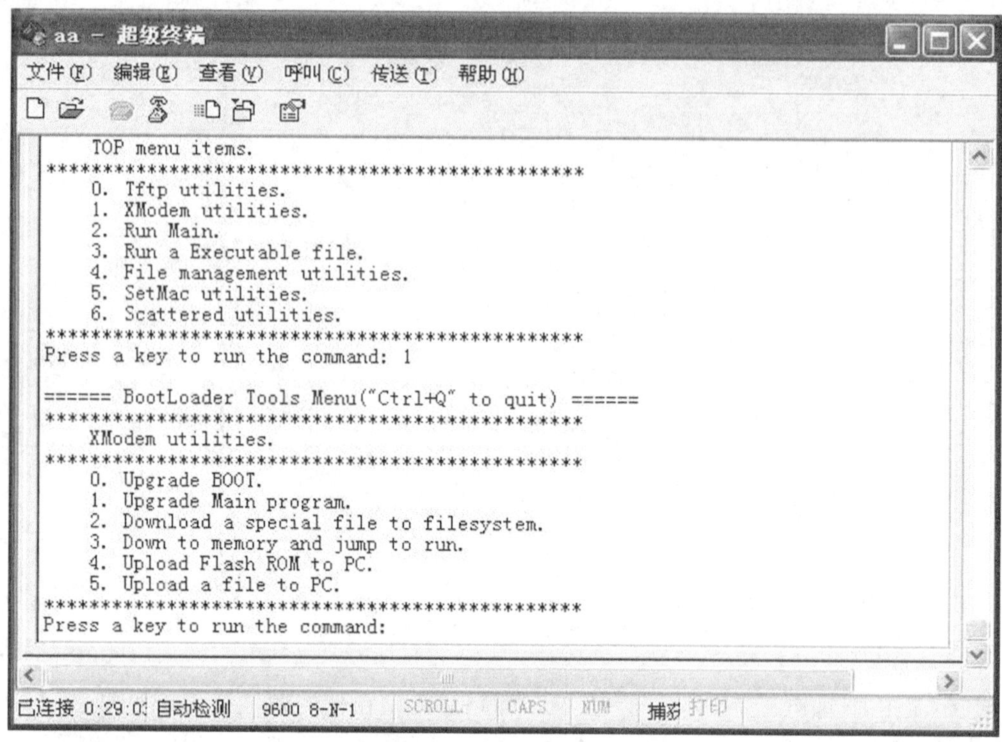

图 32-4

此时超级终端开始发送文件，当界面中出现一连串的“C”提示符时，立刻在超级终端的菜单中选择“传送”→“发送文件”命令，如图 32-5 所示。

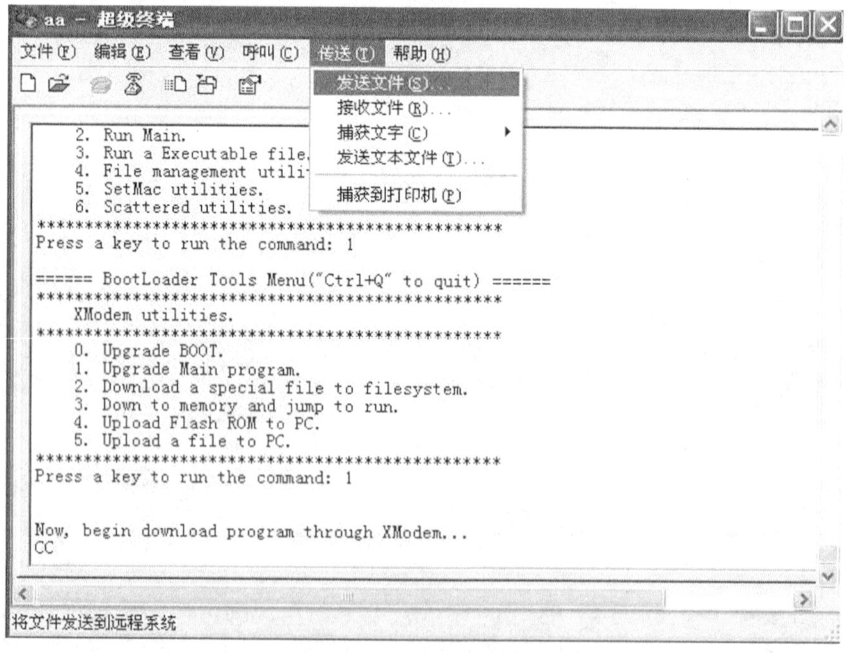

图 32-5

这时会弹出一个对话框，在“文件名”框中输入文件名和完整的路径（或通过“浏览”找到），在“协议”下拉列表框中选择“Xmodem”，如图 32-6 所示。

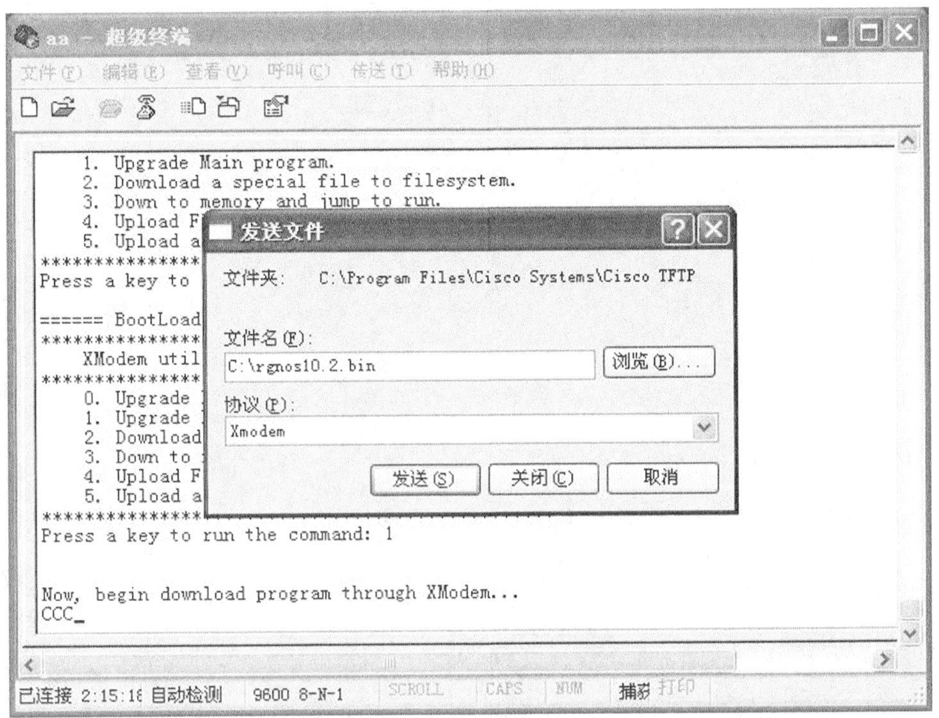

图 32-6

单击“发送”按钮后，开始传送文件，其界面如图 32-7 所示。整个传送过程用时较长，可以看到提示需要 1 小时 40 分钟左右。

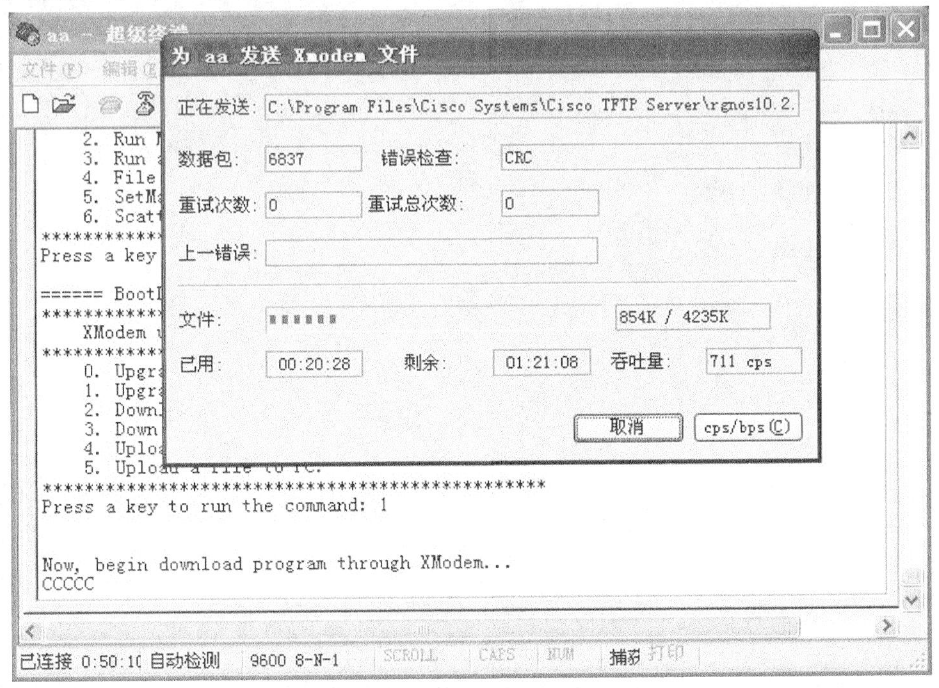

图 32-7

文件传输完毕后，会给出图 32-8 所显示的信息，包括操作系统文件的版本、大小等，然后询问是否要升级操作系统（路由器中旧的操作系统为 10.1 版本，新传进去的为 10.2 版本），默认是“Y”。

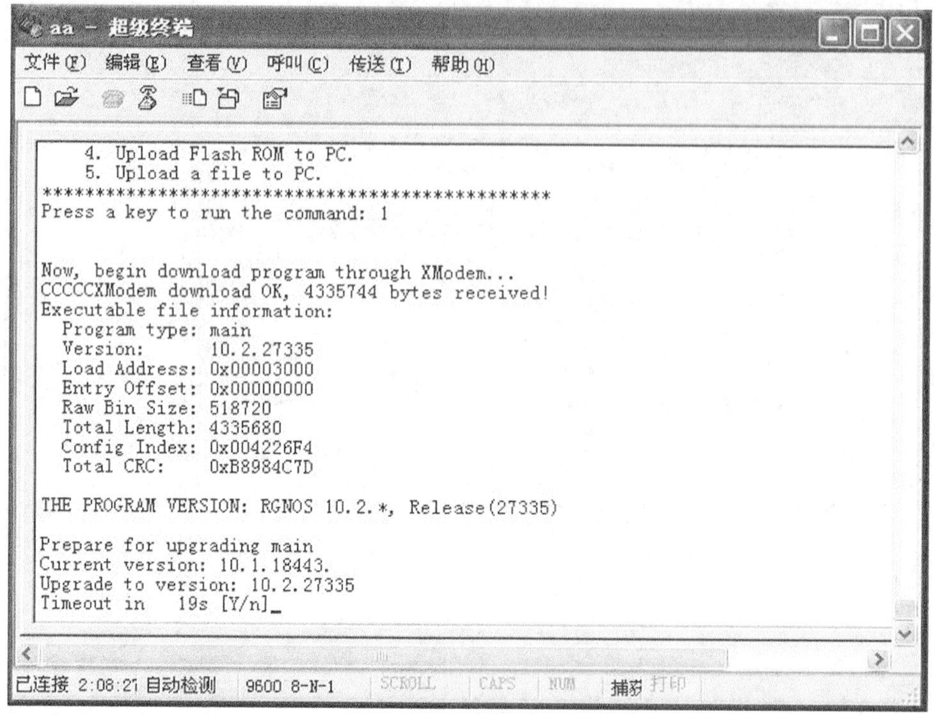

图 32-8

随后开始一个写入新操作系统的过程，结束后返回到菜单界面，如图 32-9 所示。

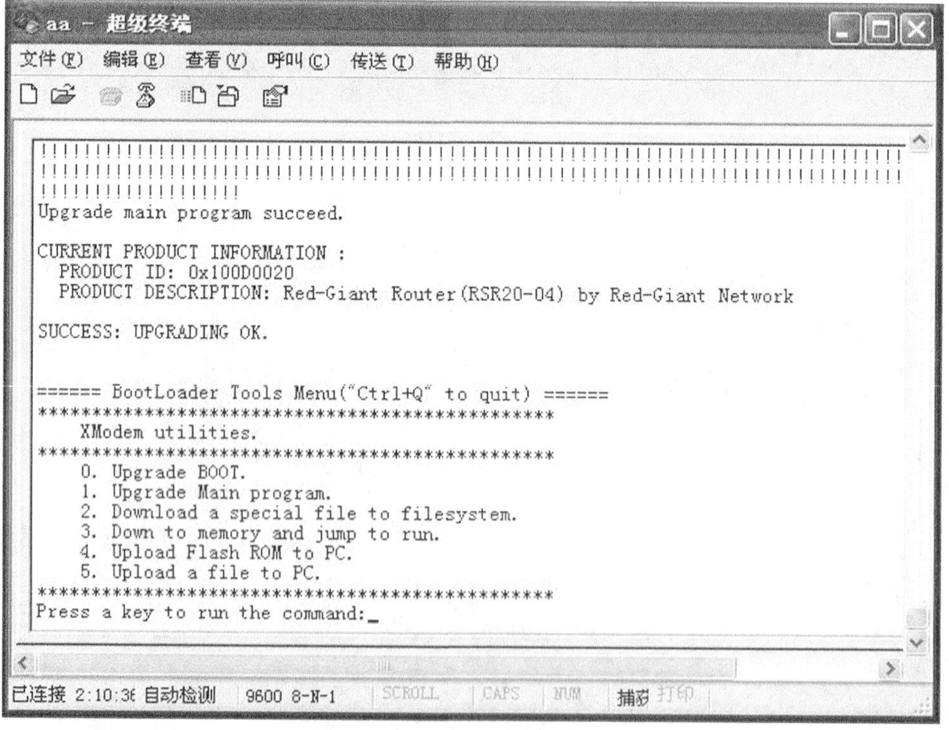

图 32-9

步骤 4　重新启动路由器，能够发现操作系统已经重新写入 Flash，即可正常启动和引导交换机。

【注意事项】

（1）这种方法速度较慢。

（2）建议将连接 TFTP 服务器的网线接在路由器集成的第一个以太网口。

（3）升级完毕后，重新启动路由器。

（4）不同的路由器型号，采用 ROM 方式升级的方法也不同，详见操作系统的升级说明。

扩展实验 5　利用 TFTP 备份还原交换机配置文件

【实验名称】

利用 TFTP 备份还原交换机配置文件。

【实验目的】

能够从 TFTP 服务器备份、还原设备配置。

【背景描述】

假设某台交换机的配置文件由于误操作或其他某种原因被破坏了，现在需要从 TFTP 服务器上的备份配置文件中恢复。

本实验以 1 台 S2126G 交换机为例，交换机名为 SW-A。1 台 PC 机通过串口（Com）连接到交换机的控制（Console）端口，通过网卡（NIC）连接到交换机的 fastethernet 0/1 端口。假设 PC 机的 IP 地址为 172.16.1.252/24，交换机的 fastethernet0/1 端口的 IP 地址为 172.16.1.10/24，PC 机上已安装和打开了 TFTP Server 程序。

【需求分析】

计算机的 Com 口通过 Console 线缆与交换机的 Console 口连接在一起，同时计算机的网卡也通过直连线连接到交换机的以太网端口上，在计算机上安装 TFTP 服务器，通过 TFTP 升级交换机的操作系统。

【实验拓扑】

实验的拓扑图，如图 33-1 所示。

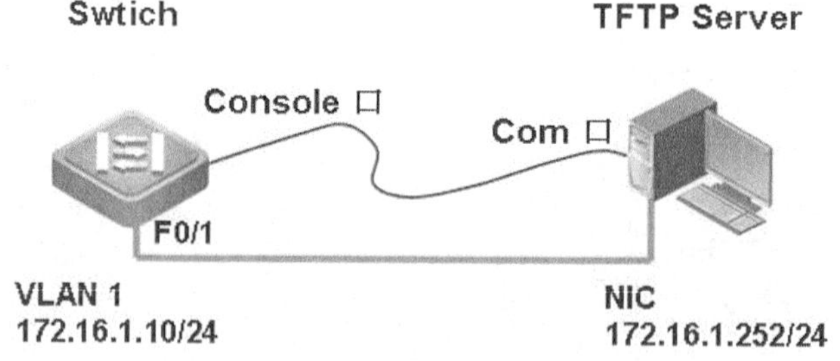

图 33-1

【实验设备】

交换机 1 台。

计算机 1 台。

【预备知识】

交换机的组成结构、TFTP 的操作方法。

【实验步骤】

步骤 1　配置交换机的主机名和管理 IP 地址，验证配置并保存。

```
Switch#configure terminal
Enter configuration commands, one per line.  End with CNTL/Z.
Switch(config)#hostname SW-A
SW-A (config)#interface vlan 1
SW-A (config-if)#ip address 172.16.1.10 255.255.255.0
SW-A (config-if)#no shutdown
SW-A (config-if)#end
SW-A #show ip interface
Interface              : VL1
Description            : Vlan 1
OperStatus             : up
ManagementStatus       : Enabled
Primary Internet address: 172.16.1.10/24
Broadcast address      : 255.255.255.255
PhysAddress            : 00d0.f88b.ca33
SW-A# copy running-config startup-config
Building configuration...
[OK]
SW-A#
```

步骤 2　打开 TFTP 服务器，验证和 TFTP 服务器的连通性。

安装 TFTP Server 的计算机 IP 地址为 172.16.1.252/24，打开 TFTP Server，验证交换机和 TFTP Server 的连通性。

```
SW-A#ping 172.16.1.252
Sending 5, 100-byte ICMP Echos to 172.16.1.252,
timeout is 2000 milliseconds.
!!!!!
Success rate is 100 percent (5/5)
Minimum = 1ms Maximum = 5ms, Average = 2ms
```

步骤 3　备份交换机配置文件并验证。

```
SW-A#copy startup-config tftp
Address of remote host []172.16.1.252
Destination filename [config.text]?s2126g-config.text
!
%Success : Transmission success,file length 129
```

此时在 TFTP Server 的安装目录下可以看到文件 s2126g-config.text，如图 33-2 所示。

文件 s2126g-config.text 的内容如图 33-3 所示。

步骤 4　删除交换机配置文件并重新启动。

```
SW-A#delete flash:config.text
SW-A#reload
System configuration has been modified. Save? [yes/no]:n
```

```
Proceed with reload? [confirm]
```

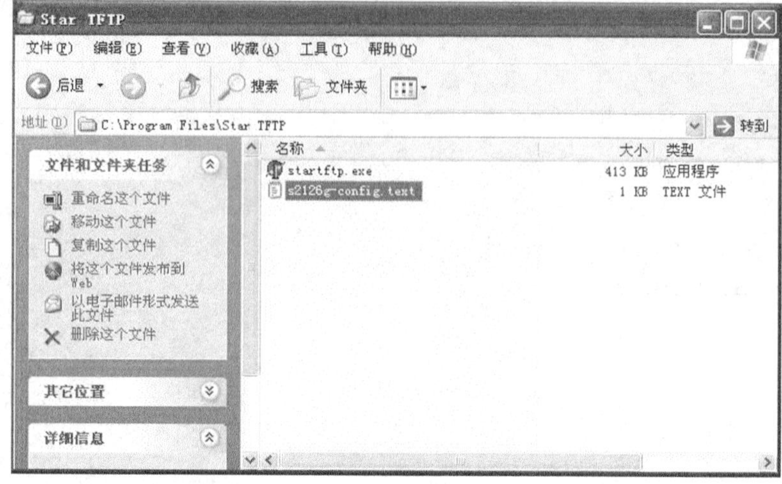

图 33-2

s2126g-config.text - 记事本

文件(F) 编辑(E) 格式(O) 查看(V) 帮助(H)

```
!
version 1.0
!
hostname SW-A
vlan 1
!
interface vlan 1
 no shutdown
 ip address 172.16.1.10 255.255.255.0
!
end
```

图 33-3

交换机重新启动后失去配置文件，会提示是否进入对话模式，选择“n”进入命令行模式配置交换机的管理 IP 地址，必须保证和 TFTP 服务器的连通性。

```
At any point you may enter a question mark '?' for help.
Use ctrl-c to abort configuration dialog at any prompt.
Default settings are in square brackets '[]'.
Continue with configuration dialog? [yes/no]:n
Switch>enable
Switch# configure terminal
Switch(config)#interface vlan 1
Switch(config-if)#ip address 172.16.1.1 255.255.255.0
Switch(config-if)#no shutdown
Switch(config-if)#end
```

```
Switch#
Switch#ping 172.16.1.252
Sending 5, 100-byte ICMP Echos to 172.16.1.252,
timeout is 2000 milliseconds.
!!!!!
Success rate is 100 percent (5/5)
Minimum = 1ms Maximum = 12ms, Average = 3ms
Switch#
```

步骤 5　从 TFTP 服务器恢复交换机配置。

```
Switch#copy tftp://172.16.1.252/s2126g-config.text startup-config
!
%Success : Transmission success,file length 129
```

将配置文件的备份复制到 startup-config 之后，用 show configure 查看结果，可以看到已经成功恢复配置文件。

```
Switch#show configure
Using 129 out of 6291456 bytes

!
version 1.0
!
hostname SW-A
vlan 1
!
interface vlan 1
 no shutdown
 ip address 172.16.1.10 255.255.255.0
!
end
```

不过用 show running-config 可以看到此时内存中生效的配置却是当前的配置。

```
Switch#show running-config
System software version : 1.68 Build Apr 25 2007 Release
Building configuration...
Current configuration : 130 bytes
!
version 1.0
!
hostname Switch
vlan 1
!
interface vlan 1
 no shutdown
 ip address 172.16.1.1 255.255.255.0
```

```
!
end
```

步骤 6　重新启动交换机使新的配置生效。

```
Switch#reload
System configuration has been modified. Save? [yes/no]:n
Proceed with reload? [confirm]
RG21 Ctrl Loader Version 03-11-02
Base ethernet MAC Address: 00:D0:F8:8B:CA:33
Initializing File System...

DEV[0]: 25 live files, 59 dead files.


DEV[0]: Total bytes: 32456704


DEV[0]: Bytes used: 5690681


DEV[0]: Bytes available: 26765855


DEV[0]: File system initializing took 7 seconds.


Executing file: flash:s2126g.bin CRC ok
Loading "flash:s2126g.bin"..................................................OK
Entry point: 0x00014000
executing...
RuiJie Internetwork Operating System Software
S2126G(50G26S) Software (RGiant-21-CODE) Version 1.68
Copyright (c) 2001-2005 by RuiJie Network Inc.
Compiled Apr 25 2007, 14:51:50.
Entry point: 0x00014000
Initializing File System...
DEV[1]: 25 live files, 59 dead files.


DEV[1]: Total bytes: 32456704


DEV[1]: Bytes used: 5690681


DEV[1]: Bytes available: 26765855



Initializing...
Done
2008-12-12 10:25:04  @5-WARMSTART:System warmstart
```

```
2008-12-12 10:25:05  @5-LINKUPDOWN:Fa0/1 changed state to up
2008-12-12 10:25:05  @5-LINKUPDOWN:VL1 changed state to up
SW-A>enable
SW-A#show running-config
System software version : 1.68 Build Apr 25 2007 Release
Building configuration...
Current configuration : 129 bytes
!
version 1.0
!
hostname SW-A
vlan 1
!
interface vlan 1
 no shutdown
 ip address 172.16.1.10 255.255.255.0
!
end
SW-A#
```

【注意事项】

保证交换机的管理 IP 地址和 TFTP 服务器的 IP 地址在同一个网段。

扩展实验 6　利用 TFTP 备份还原路由器配置文件

【实验名称】

利用 TFTP 备份还原路由器配置文件。

【实验目的】

能够从 TFTP 服务器备份，并恢复路由器配置文件。

【背景描述】

假设某台路由器的配置文件由于误操作或其他某种原因被破坏了，现在需要从 TFTP 服务器上的备份配置文件中恢复。

本实验以 1 台 RSR20 路由器为例，路由器命名为 Router。1 台 PC 机通过串口（Com）连接到路由器的控制（Console）端口，通过网卡（NIC）连接到路由器的 fastethernet0/0 端口。假设 PC 机的 IP 地址为 172.16.1.252/24，路由器的 fastethernet0/0 端口的 IP 地址为 172.16.1.10/24，PC 机上已安装和打开了 TFTP Server 程序。

【需求分析】

计算机的 Com 口通过 Console 线缆与路由器的 Console 口连接在一起，同时计算机的网卡也通过交叉线连接到路由器的以太网端口上，在计算机上安装 TFTP 服务器，通过 TFTP 升级路由器的操作系统。

【实验拓扑】

实验的拓扑图，如图 34-1 所示。

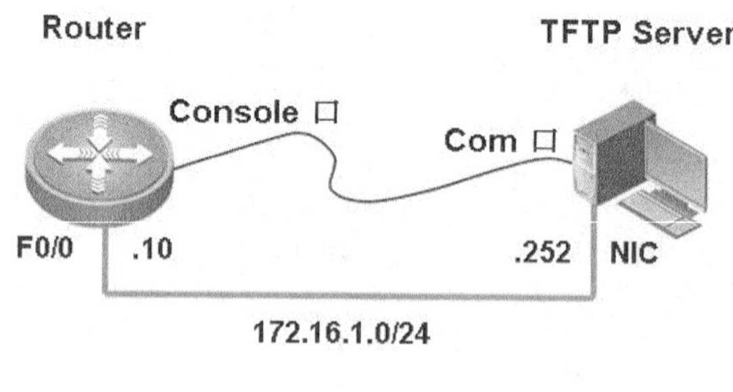

图 34-1

【实验设备】

路由器 1 台。

计算机 1 台。

【预备知识】

路由器的组成结构、TFTP 的操作方法。

【实验步骤】

步骤 1　配置路由器的主机名和接口 IP 地址，验证配置并保存。

```
RSR20#configure terminal
Enter configuration commands, one per line.  End with CNTL/Z.
RSR20(config)#hostname Router
Router(config)#interface fastEthernet 0/0
Router(config-if)#ip address 172.16.1.10 255.255.255.0
Router(config-if)#no shutdown
Router(config-if)#end
Router#show ip interface fastEthernet 0/0
FastEthernet 0/0
  IP interface state is: UP
  IP interface type is: BROADCAST
  IP interface MTU is: 1500
  IP address is:
    172.16.1.10/24 (primary)
  IP address negotiate is: OFF
  Forward direct-boardcast is: OFF
  ICMP mask reply is: ON
  Send ICMP redirect is: ON
  Send ICMP unreachabled is: ON
  DHCP relay is: OFF
  Fast switch is: ON
  Help address is:
  Proxy ARP is: ON
Router#copy running-config startup-config
Building configuration...
[OK]
```

步骤 2　打开 TFTP 服务器，验证和 TFTP 服务器的连通性。

安装 TFTP Server 的计算机 IP 地址为 172.16.1.252/24，打开 TFTP Server，验证路由器和 TFTP Server 的连通性。

```
Router#ping 172.16.1.252
Sending 5, 100-byte ICMP Echoes to 172.16.1.252, timeout is 2 seconds:
  < press Ctrl+C to break >
!!!!!
Success rate is 100 percent (5/5), round-trip min/avg/max = 1/1/1 ms
```

步骤 3　备份路由器配置文件并验证。

```
Router#copy running-config tftp
Address of remote host []?172.16.1.252
Destination filename []?running-config.text
Building configuration...
Accessing running-config...
```

```
Success : Transmission success,file length 427
```

或者如下。

```
Router#copy startup-config tftp
Address of remote host []?172.16.1.252
Destination filename [config.text]?startup-config.text
Accessing startup-config...
Success : Transmission success,file length 427
```

此时在 TFTP Server 的安装目录下可以看到文件 running-config.text 或者文件 startup-config.text，如图 34-2 所示。

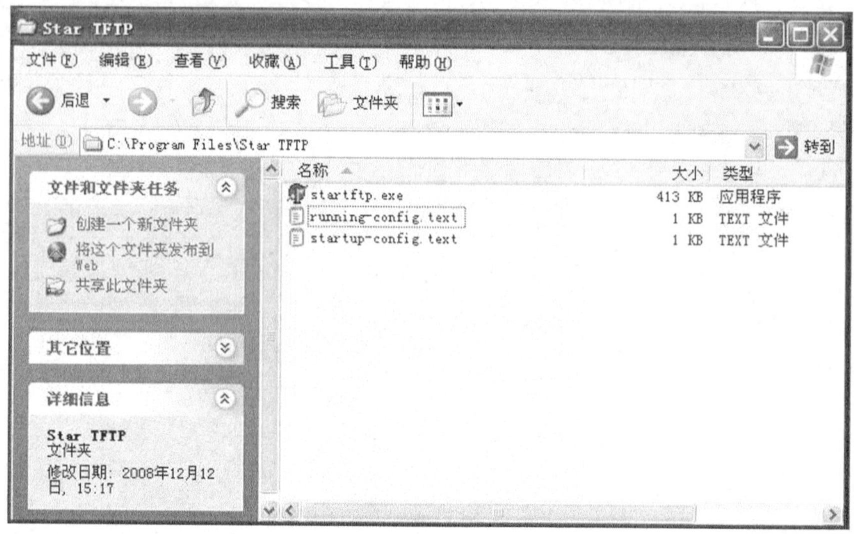

图 34-2

文件 startup-config.text 的内容如图 34-3 所示。

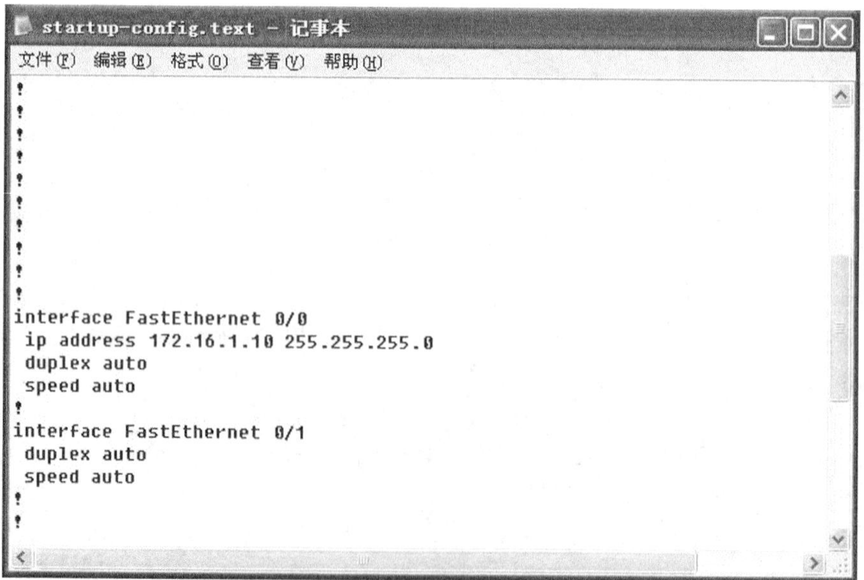

图 34-3

步骤 4　删除路由器配置文件并重新启动。

```
Router#del config.text
Router#
Router#reload
Processed with reload? [no]y
```

路由器重新启动后丢失配置文件，需要重新配置接口IP地址以实现和TFTP Server的连通性。

```
RSR20#configure terminal
Enter configuration commands, one per line.  End with CNTL/Z.
RSR20(config)#interface fastEthernet 0/0
RSR20(config-if)#ip address 172.16.1.1 255.255.255.0
RSR20(config-if)#no shutdown
RSR20(config-if)#end
RSR20#
RSR20#ping 172.16.1.252
Sending 5, 100-byte ICMP Echoes to 172.16.1.252, timeout is 2 seconds:
  < press Ctrl+C to break >
!!!!!
Success rate is 100 percent (5/5), round-trip min/avg/max = 1/2/10 ms
```

步骤 5　从 TFTP 服务器恢复路由器配置。

```
RSR20#copy tftp startup-config
Address of remote host []?172.16.1.252
Source filename []?startup-config.text
Extended commands [n]:
Accessing tftp://172.16.1.252/startup-config.text...
Success : Transmission success,file length 427
```

利用 TFTP 服务器将备份的 startup-config.text 文件恢复到路由器中后，用 show startup-config.text 查看结果，可以看到已经成功恢复配置文件。

```
RSR20#show startup-config
!
version RGNOS 10.1.00(4), Release(18443)(Tue Jul 17 20:50:30 CST 2007
-ubu1server)
hostname Router
!
!
interface FastEthernet 0/0
 ip address 172.16.1.10 255.255.255.0
 duplex auto
 speed auto
!
interface FastEthernet 0/1
 duplex auto
```

```
 speed auto
!
line con 0
line aux 0
login
!
end
```

不过用 show running-config 可以看到此时内存中生效的配置却是当前的配置：

```
RSR20#show running-config
Building configuration...
Current configuration : 409 bytes
!
version RGNOS 10.1.00(4), Release(18443)(Tue Jul 17 20:50:30 CST 2007
-ubu1server)
!
!
interface FastEthernet 0/0
 ip address 172.16.1.1 255.255.255.0
 duplex auto
 speed auto
!
interface FastEthernet 0/1
 duplex auto
 speed auto
!
line con 0
line aux 0
line vty 0 4
 login
!
end
```

步骤 6　重新启动路由器使新的配置生效。

```
RSR20#reload
Processed with reload? [no]y
System Reload Now......
Reload Reason:
System bootstrap ...
Boot Version: RGNOS 10.2.00(2), Release(24515)
Nor Flash ID: 0x00010049, SIZE: 2097152Byte
MTD_DRIVER-6-MTD_NAND_FOUND: 1 nand chip(s) found on the target(0).
```

```
Press Ctrl+C to enter Boot Menu ......
Main Program File Name rgnos.bin, Load Main Program ...
Executing program, launch at: 0x00010000
Ruijie Network Operating System Software
Release Software (tm), RGNOS 10.1.00(4), Release(18443), Compiled Tue Jul 17
20:50:30 CST 2007 by ubu1server

Copyright (c) 1998-2007 by Ruijie Networks.

All Rights Reserved.

Neither Decompiling Nor Reverse Engineering Shall Be Allowed.
Aug 18 02:31:16 RSR20 %7:1 nand chip(s) found on the target(0).
 Aug 18 02:31:42 RSR20 %7:%LINK CHANGED: Interface FastEthernet 0/0, changed state
to up
 Aug 18 02:31:42 RSR20 %7:%LINE PROTOCOL CHANGE: Interface FastEthernet 0/0,
changed state to UP
 Router>enable
Router#show running-config
Building configuration...
Current configuration : 427 bytes
!
version RGNOS 10.1.00(4), Release(18443)(Tue Jul 17 20:50:30 CST 2007
-ubu1server)
hostname Router
!
!
interface FastEthernet 0/0
 ip address 172.16.1.10 255.255.255.0
 duplex auto
 speed auto
!
interface FastEthernet 0/1
 duplex auto
 speed auto
!
line con 0
line aux 0
line vty 0 4
 login
```

```
!
end
Router#
```

【注意事项】

（1）要保证路由器和 TFTP 服务器的连通性。

（2）路由器如果和计算机直连要使用交叉线，除非路由器端口支持智能反转，即可以实现对直连线、交叉线的自适应。

附录A 术语表

术语	解释
ACK	TCP首部中的确认（ACKnowledgment）标志
ANSI	American National Standards Institute，美国国家标准协会
API	Application Programming Interface，应用编程接口
ARP	Address Resolution Protocol，地址解析协议
ARPANET	Advanced Research Projects Agency NETwork，远景研究规划局（美国国防部）网
AS	Autonomous System，自治系统
ASCII	American Standard Code for Information Interchange，美国信息交换标准代码
BGP	Border Gateway Protocol，边界网关协议
BIND	Berkeley Internet Name Domain，伯克利Internet域名
BOOTP	BOOTstrap Protocol，引导程序协议
BSD	Berkeley Software Distribution，伯克利软件发布
CIDR	Classless InterDomain Routing，无类型域间选路
CIX	Commercial Internet Exchange，商业互联网交换
CLNP	Connection Less Network Protocol，无连接网络协议
CRC	Cyclic Redundancy Check，循环冗余检验
CSLIP	Compressed SLIP，压缩的SLIP
CSMA	Carrier Sense Multiple Access，载波侦听多路访问
DCE	Data Circuit-terminating Equipment，数据电路端接设备
DDN	Defense Data Network，国防数据网
DF	TCP首部中的不分段（Don't Fragment）标志
DHCP	Dynamic Host Configuration Protocol，动态主机配置协议
DLPI	Data Link Provider Interface，数据链路提供者接口
DNS	Domain Name System，域名系统
DSAP	Destination Service Access Point，目的服务访问点
DSLAM	DSL Access Multiplexer，数字用户线接入复用器
DSSS	Direct Sequence Spread Spectrum，直接序列扩频
DTS	Distributed Time Service，分布式时间服务
DVMRP	Distance Vector Multicast Routing Protocol，距离向量多播选路协议
EOL	End of Option List，选项表结束
EGP	External Gateway Protocol，外部网关协议
EIA	Electronic Industries Association，美国电子工业协会
FAQ	Frequently Asked Question，经常提出的问题
FCS	Frame Check Sequence，帧检验序列
FDDI	Fiber Distributed Data Interface，光纤分布式数据接口

（续表）

术语	解释
FIFO	First In, First Out，先进先出
FIN	TCP首部中的终止（FINish）标志
FQDN	Full Qualified Domain Name，完全合格的域名
FTP	File Transfer Protocol，文件传送协议
GIF	Graphics Interchange Format，图形交换格式
HDLC	High-level Data Link Control，高级数据链路控制
HELLO	选路协议
HTML	Hyper Text Markup Language，超文本置标语言
HTTP	Hyper Text Transfer Protocol，超文本传送协议
IAB	Internet Architecture Board，Internet体系结构委员会
IANA	Internet Assigned Numbers Authority，Internet号分配机构
ICMP	Internet Control Message Protocol，因特网控制报文协议
IDRP	Inter Domain Routing Protocol，域间选路协议
IEEE	Institute of Electrical and Electronics Engineers，电气和电子工程师学会（美国）
IESG	Internet Engineering Steering Group，Internet工程指导小组
IETF	Internet Engineering Task Force，Internet工程任务小组
IGMP	Internet Group Management Protocol，Internet组管理协议
IGP	Interior Gateway Protocol，内部网关协议
IMAP	Internet Message Access Protocol，Internet报文存取协议
INN	InterNet News，因特网新闻
IP	Internet Protocol，网际协议
IPC	Inter Process Communication，进程间通信
IRTF	Internet Research Task Force，Internet研究专门小组
IRTP	Internet Reliable Transaction Protocol，因特网可靠事务协议
IS-IS	Intermediate System to Intermediate System Protocol，中间系统到中间系统协议
ISN	Initial Sequence Number，初始序号
ISO	International Organization for Standardization，国际标准化组织
ISOC	Internet SOCiety，Internet协会
ISS	Initial Send Sequence number，初始发送序号
JPEG	Joint Photographic Experts Group，联合图片专家组
LAN	Local Area Network，局域网
LCP	Link Control Protocol，链路控制协议
LF	Line Feed，换行
LIFO	Last In, First Out，后进先出
LLC	Logical Link Control，逻辑链路控制
LSRR	Loose Source and Record Route，宽松的源站及记录路由
MAN	Metropolitan Area Network，城域网

（续表）

术语	解释
MBONE	Multicast Backbone On the InterNEt，Internet上的多播主干网
MIB	Management Information Base，管理信息库
MILNET	MILitary NETwork，军用网
MIME	Multipurpose Internet Mail Extensions，通用因特网邮件扩充
MSL	Maximum Segment Lifetime，报文段最大生存时间
MSS	Maximum Segment Size，报文段最大长度
MTA	Message Transfer Agent，报文传送代理
MTU	Maximum Transmission Unit，最大传输单元
NCP	Network Control Protocol，网络控制协议
NCSA	National Center for Supercomputing Applications，国家超级计算中心（美国）
NFS	Network File System，网络文件系统
NIC	Network Information Center，网络信息中心
NIT	Network Interface Tap，网络接口栓（Sun公司的一个程序）
NNRP	Network News Reading Protocol，网络新闻读取协议
NNTP	Network News Transfer Protocol，网络新闻传送协议
NOP	No Operation，无操作
NSFNET	National Science Foundation NETwork，国家科学基金网络
NSI	NASA Science Internet，（美国）国家宇航局Internet
NTP	Network Time Protocol，网络时间协议
NVT	Network Virtual Terminal，网络虚拟终端
OSF	Open Software Foundation，开放软件基金
OSI	Open Systems Interconnection，开放系统互连
OSPF	Open Shortest Path First，开放最短通路优先
PCB	Protocol Control Block，协议控制块
PDU	Protocol Data Unit，协议数据单元
PNG	Portable Network Graphics，可携式网络图像
POSIX	Portable Operation System Interface，可移植操作系统接口
PPP	Point-to-Point Protocol，点对点协议
PSH	TCP首部中的急迫（PuSH ）标志
RARP	Reverse Address Resolution Protocol，逆地址解析协议
RDP	Reliable Datagram Protocol，可靠数据报
RFC	Request For Comment，是Internet的文档，意思是“请提意见”
RIP	Routing Information Protocol，路由信息协议
RPC	Remote Procedure Call，远程过程调用
RST	TCP首部中的重建（ReSeT）标志
RTO	Retransmission Time Out，重传超时
RTT	Round-Trip Time，往返时间

（续表）

术语	解释
SACK	Selective ACKnowledgment，有选择的确认
SLIP	Serial Line Internet Protocol，串行线路因特网协议
SMI	Structure of Management Information，管理信息结构
SMTP	Simple Mail Transfer Protocol，简单邮件传送协议
SNMP	Simple Network Management Protocol，简单网络管理协议
SPT	Server Processing Time，服务器处理时间
SSAP	Source Service Access Point，源服务访问点
SSRR	Strict Source and Record Route，严格的源站及记录路由
SVR4	System V Release 4，系统版本4
SYN	TCP首部中的序号同步（SYNchronous sequence number ）标志
TAO	TCP Accelerated Open，TCP加速打开
TCP	Transmission Control Protocol，传输控制协议
TFTP	Trivial File Transfer Protocol，简单文件传送协议
TLI	Transport Layer Interface，运输层接口
TTL	Time-To-Live，寿命或生存时间
Telnet	远程登录协议
UA	User Agent，用户代理
UDP	User Datagram Protocol，用户数据报协议
URG	TCP首部中的紧急(URGent)指针
URI	Universal Resource Identifier，通用资源标识符
URL	Uniform Resource Locator，统一资源定位符
URN	Uniform Resource Name，统一资源名字
UTC	Coordinated Universal Time，协调的统一时间
VMTP	Versatile Message Transaction Protocol，通用报文事务协议
WAN	Wide Area Network，广域网
WWW	World Wide Web，万维网
XDR	eXternal Data Representation，外部数据表示
XID	transaction ID，事务标识符

附录 B　锐捷职业认证体系

锐捷职业认证是 IT 领域的一项网络专业技能认证，拥有锐捷职业认证资格的专业人士将具有专业的网络知识和网络技能，并且能为雇佣他们的管理者、组织、企业带来巨大的价值和报酬。

锐捷认证体系包括通用技术认证和专项技术认证。通用技术认证包括网络工程方向及网络安全方向，专项技术认证包括 IPv6、存储、无线和 IP 通信方向。其中通用技术认证是目前国内外需求量最大，考取人数最多的认证。

一、锐捷职业认证体系概述

1. 通用技术认证

锐捷在通用技术认证中的网络工程方向提供了 5 个认证等级，它们所代表的专业水平逐级提升：网络管理员、网络工程师、调试工程师、资深网络工程师和互联网专家。

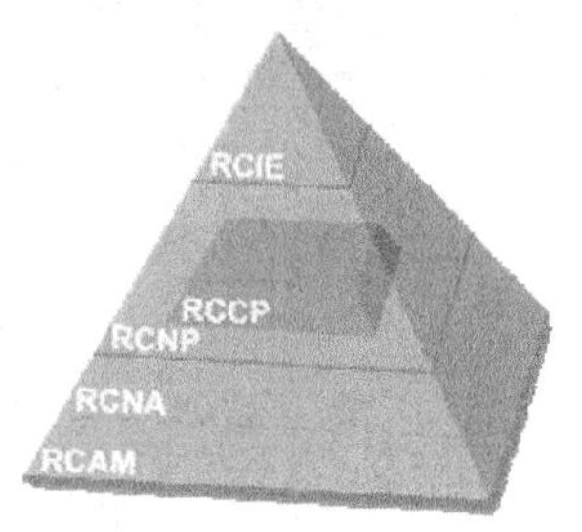

- **网络管理员（RCAM）**：锐捷职业认证的第一步首先从网络管理员级别开始，其代表网络技术的入门等级，适用于网络技术的初学者。
- **网络工程师（RCNA）**：网络工程领域的初级资格认证，获得 RCNA 资格的人员可以搭建和维护 100 个以下节点的中小型网络。
- **调试工程师（RCCP）**：网络工程领域的中级资格认证。获得 RCCP 资格的人员具备丰富的网络知识和实践操作技能，能够熟练的配置和调试多种网络设备。具有 RCCP 认证的工程师能够设计和构建超过 100 个节点的大中型园区网络。
- **资深网络工程师（RCNP）**：网络工程领域的高级资格认证。获得 RCNP 认证的人员能够驾驭路由器、交换机、WLAN 等产品，熟练的对其各种功能和特性进行配置和调试，并在网络中部署高级的路由选择协议和各种安全特性、冗余机制、优化技术等。具有 RCNP 认证的工程师能够设计和构建超过 500 个节点的大中型园区网络。
- **互联网专家（RCIE）**：网络工程领域的顶级认证。获得 RCIE 认证的人员作为网络技术领域的专家，不仅具有丰富的网络理论知识和实践操作技能，并能够对网络中出现的故障和疑难问题进行分析及排错。获得 RCIE 认证的人员将具备实施大型网络中所需要的各种技能。

2. 专项技术认证

锐捷职业认证还提供了多个专项技术认证，以考察相关人员在特定的技术领域方面具备的知识和技能。锐捷专项技术认证包括 IPv6、存储、无线和 IP 通信。通过 4 门专项认证课程的学习，学习者能够在 IPv6、存储、无线及 IP 通信技术领域具有专家级的知识和技能，并拥有驾驭相关产品的能力。

二、锐捷职业认证和途径

锐捷认证面向的是锐捷合作伙伴、经销商、网络技术专业人士以及对网络技术感兴趣的人群。要获得职业认证体系中的不同等级的认证，都需要通过一些必须的笔试、Lab 考试以及具备必要的必备资格。

1．通用技术认证

认证	认证课程	必需的考试	必备资格
RCAM	网络基础 Fundamental	Fundamental Written	—
RCNA	网络设备互连（IND） Interconnecting Networking Devices	IND Written IND Lab	—
RCCP	设备调试与网络优化（DOND） Debugging and Optimizing Networking Devices	DOND Written DOND Lab	—
RCNP	构建高级的路由互联网络（BARI） Building Advanced Routing Internetworks	BARI Written BARI Lab	具有生效的 RCNA 或 RCCP 证书
	构建高级的交换网络（BASN） Building Advanced Switched Networks	BASN Written BASN Lab	
	构建优化的互联网络（BOI） Building Optimized Internetworks	BOI Written BOI Lab	
	网络服务架构的设计与实施（DINSA） Designing and Implementing Network Service Architectures	DINSA Written DINSA Lab	
RCIE	—	RCIE Written RCIE Lab	—

2．专项技术认证

认证	认证课程	必需的考试
IPv6 Specialist	部署 IPv6 网络 Deploying IPv6 Networks	IPv6 Written IPv6 Lab
Storage Specialist	构建存储网络 Building Storage Networks	Storage Written Storage Lab
WLAN Specialist	无线局域网的设计与实施 Designing and Implementing Wireless LAN	WLAN Written WLAN Lab
IP Communication Specialist	IP 通信技术 IP Communication Technology	IP Communication Written IP Communication Lab

注：对于锐捷专项技术认证，不需要考生预先具有任何认证证书，只需要通过相应的专项技术考试即可。

锐捷网络大学

参 考 文 献

1 RFC 网站 http://www.ietf.org/

2 IEEE 网站 http://www.ietf.org/

3 锐捷网络网站 http://www.ruijie.com.cn/

4 史蒂文斯著. TCP/IP 详解. 卷 1：协议. 范建华等译. 北京：机械工业出版社，2000

5 杨靖，刘亮主编. 实用网络技术配置指南初级篇. 北京：北京希望电子出版社，2006